THE EINSTEIN PARADOX

The famously controversial 1935 paper by Einstein, Podolsky, and Rosen (EPR) took aim at the heart of the flourishing field of quantum mechanics. The paper provoked responses from the leading theoretical physicists of the day, and brought entanglement and nonlocality to the forefront of discussion. This book looks back at the seminal year in which the EPR paper was published and explores the intense debate it unleashed. These conversations in print and in private correspondence offer significant insight into the minds of pioneering quantum physicists including Niels Bohr, Erwin Schrödinger and Albert Einstein himself. Offering the most complete collection of sources to date – many published or translated here for the first time – this text brings a rich new context to this pivotal moment in physics history. Both researchers and students in the history and philosophy of science, and enthusiasts alike, will find this book illuminating.

GUIDO BACCIAGALUPPI is Associate Professor of Foundations of Physics at Utrecht University. His research interests lie in the philosophy of quantum mechanics, the philosophy of science, and the history of quantum mechanics. He co-authored a monograph on the 1927 Solvay conference with Antony Valentini, *Quantum Theory at the Crossroads* (Cambridge University Press, 2009), and previously collaborated with Elise Crull on *Grete Hermann: Between Physics and Philosophy* (Springer, 2017).

ELISE CRULL is Associate Professor of Philosophy at The City College of New York. Her research focuses on the philosophy of physics, the intersection between physics and metaphysics, and on general topics within history and philosophy of science. She co-authored a volume on the work of Grete Hermann with Guido Bacciagaluppi, *Grete Hermann: Between Physics and Philosophy* (Springer, 2017), and continues to promote scholarly work at the intersection of science, history, and philosophy.

THE EINSTEIN PARADOX

The Debate on Nonlocality and Incompleteness in 1935

GUIDO BACCIAGALUPPI
Utrecht University

ELISE CRULL
The City College of New York

CAMBRIDGE
UNIVERSITY PRESS

CAMBRIDGE
UNIVERSITY PRESS

Shaftesbury Road, Cambridge CB2 8EA, United Kingdom

One Liberty Plaza, 20th Floor, New York, NY 10006, USA

477 Williamstown Road, Port Melbourne, VIC 3207, Australia

314–321, 3rd Floor, Plot 3, Splendor Forum, Jasola District Centre,
New Delhi – 110025, India

103 Penang Road, #05–06/07, Visioncrest Commercial, Singapore 238467

Cambridge University Press is part of Cambridge University Press & Assessment,
a department of the University of Cambridge.

We share the University's mission to contribute to society through the pursuit of
education, learning and research at the highest international levels of excellence.

www.cambridge.org
Information on this title: www.cambridge.org/9781107014459

DOI: 10.1017/9781139028493

First published 2024

A catalogue record for this publication is available from the British Library

A Cataloging-in-Publication data record for this book is available from the Library of Congress

ISBN 978-1-107-01445-9 Hardback

To Don Howard, who brought us together on this project.

Contents

Part V Correspondence

Preface

> What amused Einstein [about EPR] was that, while all the scientists were quite positive that the argument was wrong, they all gave different reasons for their belief!
>
> *Conversation with Banesh Hoffmann in 1937/38, from Jammer (1974, p. 187)*

In our current understanding of the philosophy and foundations of quantum mechanics, the 1935 paper by Einstein, Podolsky and Rosen, or EPR as they are universally known, embodies more than any other the focus on entanglement and (non)locality that has been and still is a defining feature of the field. It is now commonplace – but not out of place – to emphasise how, by elaborating on EPR's concerns, the work of J. S. Bell in the 1960s revived the foundations of quantum mechanics, and how original fields of research have now emerged, roughly since the 1990s, in which entanglement has turned from a source of embarrassment to a new physical resource.

This book looks back at the seminal year 1935, when the EPR paper was published, and at the debate it unleashed then, in both published and unpublished form. While some of the sources of this debate are well known and some of its aspects are well studied, the book contains the most complete collection of sources to date – many published or translated here for the first time, including minor responses to the EPR paper – and offers both a fresh analysis of known material, and at least a preliminary analysis of lesser-known aspects of the debate.

The material made available for the first time (at least in full and in English) includes Heisenberg's draft reply to EPR, the correspondence between Einstein and Schrödinger from the summer of 1935, short notes by Einstein and Schrödinger in the run-up to EPR, and a wealth of other correspondence between Schrödinger, Bohr, Heisenberg and others, providing a fascinating glimpse into the thoughts and exchanges prompted by the publication of the EPR paper. We note among other things that although the Einstein Project has made excellent progress in recent years, it is still several years away from publishing Einstein's correspondence with Schrödinger translated in this volume.

Our main focus is the panorama of responses and reactions elicited by the EPR paper and their background. The material is quite vast and rather disparate, with some responses spelled out in full, as was Bohr's reply, others consisting of scattered remarks in correspondence, as was Pauli's reaction. As we indicate in the subtitle to this volume, two perspectives from which to analyse this material readily suggest themselves: those of (in)completeness and (non)locality. Although it is certainly true that the topics of completeness, completability and nonlocality continue to be foci of much investigation in the foundations literature, we do not discuss later developments except in as much as they illuminate the early period of concern to us here: those few years between 1927 and 1935 in which these topics so central to the interpretation of quantum mechanics transitioned from being amorphous worries to fully articulated, hotly debated matter within the physics community.

What EPR were questioning was the completeness of quantum mechanics. This question was not in itself new, and one can analyse responses in the context of other discussions of completeness. So, for instance, in his letter to Schrödinger of 9 July 1935 (translated in this volume), Pauli points out how EPR's inference to values for both position and momentum leads to contradictions with statistical predictions of quantum mechanics, if we assume these values to obey the functional relations between observables. This argument (further refined by Schrödinger himself) turns out to be a variant of von Neumann's no-go theorem against hidden variables. A useful contrast is provided by the rather neglected work of Grete Hermann. In an essay published in *Naturwissenschaften* a few weeks before Schrödinger's famous 'cat' paper (both also included in translation in this volume), she argues for the completeness of quantum mechanics in the context of Weizsäcker's analysis of the Heisenberg γ-ray microscope, which (as pointed out already by Jammer) contains the delayed-choice elements of the EPR thought experiment. A longer version of the same essay was published two months before the EPR paper, containing an overview and criticism of the then-current arguments for the completeness of quantum mechanics (including a trenchant criticism of von Neumann's proof). We shall use Hermann's analysis to shine further light on both Heisenberg's and Bohr's replies to EPR.

From the point of view of later developments in foundations and of the historical investigation of Einstein's concerns, the crucial point of the EPR argument, however, is the focus on locality and on entanglement. As for Einstein himself, we suggest that the way he rules out the 'Schrödingerian' view of the wavefunction in his correspondence with Schrödinger (and less explicitly in the EPR paper) is an early form of the 'insolubility theorem': no quantum-mechanical description of the measurement process can explain the *distant* measurement results. But a crucial role in the story of entanglement is played by Schrödinger, and we analyse in considerable detail the various phases of his own struggles with entanglement, at least since 1927, up to and including 1935. Another significant but lesser-known contribution to the study of entanglement is provided by Wendell Furry's reply to EPR, which we also reprint and analyse below. Among the 'orthodox' physicists we discuss, Pauli is perhaps the most sensitive and certainly the most explicit about the issue of systems and subsystems, as expressed, for example, in his treatment of measurement theory in his 1933 handbook article. Finally, Bohr's famous reply (which we analyse afresh at least to our own satisfaction[1]) also takes entanglement very seriously, if treating it less explicitly than Pauli – or so we argue. We find, however, that even Bohr (like many others) fails to fully take on board the implications of EPR for nonlocality.

★

Part I of the book contains our treatment of the background leading up to the publication of EPR, and of the EPR paper itself and the debate it unleashed (roughly 1926–36). In these chapters we combine extant literature with novel primary source material in order to re-examine the thinking of key players in this history.

We begin, naturally, with a chapter on Einstein. The development of Einstein's thinking in relation to EPR has been already described in detail, in particular the origins of his worries about nonlocality, the development of EPR out of the photon-box thought experiment of 1930, and the relation between the version of the EPR argument published in 1935 and his own preferred formulations. Chapter 1 details not only the prehistory of EPR, but also examines the structure and logic of the EPR paper – including Einstein's own preferred version of the argument for incompleteness. We here attempt a seamless interweaving of the excellent extant literature with additional details that have emerged from our work and the recent work of others. Some examples of new aspects in this pre-history of EPR include evidence of a 'proto' photon-box thought experiment Einstein had developed in connection with his ill-starred collaboration with Emil Rupp in 1926. We also describe the potential importance to this pre-history of Einstein's paper with Tolman and Podolsky and of Einstein's seminar and discussions with Schrödinger in Berlin in the early 1930s. The paper by Einstein, Tolman and Podolsky is reprinted in Part II, and the report of the Berlin seminar is translated here for the first time (also in Part II). We also

[1] While extremely influential, Bohr's reply also counts as notoriously obscure. Part of this famed obscurity may be due to the fact that in the important collection of sources by Wheeler and Zurek (1983), pp. 699 and 700 of Bohr's reply are printed in reverse order as pp. 149 and 148 (of course, the editorial mishap may have been due to obscurity of the original in the first place).

include in full the correspondence between Einstein and Schrödinger analysed by Fine and by Howard.

Chapter 2 provides a complete list and brief analyses of published and unpublished responses to EPR written in 1935. Virtually all of the responses themselves are included in our Parts III and IV, and our discussion in Chapter 2 invites a renewed consideration of certain contributors not much discussed elsewhere in the literature. These include going beyond Kemble's short criticism of EPR to his ensuing disagreement with Margenau about the viability of an ensemble interpretation of the wavefunction, and also a response to Kemble's note on EPR by Podolsky himself. We also examine the correspondence between Margenau and Einstein in the wake of EPR, discussing the role of the collapse postulate, and finally we discuss the little-known papers by Furry, which although not entirely satisfactory *qua* a response to EPR's arguments, are nevertheless of great potential interest for the foundations literature more generally.

The one author who clearly appreciated the role of (non)locality in the EPR argument was Schrödinger, and it is to him we turn in Chapter 3. That Schrödinger understood EPR's true worries is plainly evident in such prominent instances as his first paper on entanglement and the 'Schrödinger cat' paper (both included in Part III), as well as in his correspondence, and has been analysed perceptively by Fine. Still, Schrödinger's reaction to the EPR paper appears to be less widely known than, say, Bohr's, and this despite the fact that it fits rather nicely with later concerns in foundations. Taking the lead both from the EPR paper and from Pauli's remarks in the correspondence, Schrödinger shows that EPR's locality considerations lead to the assignment of values to *all* quantum mechanical observables, but that under what would appear to be mild assumptions this then leads to contradictions of the von Neumann type. This dilemma (as he explicitly calls it) is thus similar to those familiar from the more recent literature on foundations of quantum mechanics between nonlocality, on the one hand, and no-go results, on the other (whether through violation of the Bell inequalities, the Kochen–Specker theorem, or what you will). We shall first look at Schrödinger's fundamental worries in the years leading up to 1935. The chapter then discusses in detail the direct reaction by Schrödinger to EPR. It will, however, not exhaust our discussion of Schrödinger, who will be a recurring character in the book, having poked and prodded his peers on EPR during the whole summer and autumn of 1935, leaving behind a folder in his *Nachlaß* explicitly labelled 'Korrespondenz betreffend das Einstein-Paradoxon' ['Correspondence Concerning the Einstein Paradox'].

Chapter 4 introduces in larger fashion than in the extant literature the interesting role of Heisenberg in this debate. Although we have already published an analysis of Heisenberg's posthumously published draft response to EPR, only now are we able to situate this excellent primary source in its fullest context, by contributing a chapter describing, for example, Heisenberg's thinking prior to EPR about interacting systems and hidden variables, the crucial role of Grete Hermann for Heisenberg's thinking about separability, completeness and observational context, and describing the correspondence between Heisenberg and Bohr discussing (read: debating) Heisenberg's manuscript.

The final chapter of Part I dives deep into Bohr's views on (in)completeness and (non)locality. Perhaps the most outspoken and famous respondent to EPR, in this chapter we add subtle shading to the canonical analysis of Bohr's philosophical approach to quantum mechanics in part by re-evaluating the role of disturbance in his published response to EPR and comparing it with the measurement theory of Pauli.

The rest of the book contains primary source material, much of it new and/or translated newly into English. Part II includes selected pre-EPR papers, including the Einstein–Tolman–Podolsky paper, an anonymous report of Einstein's seminar in Berlin in 1931, and unpublished notes by Schrödinger titled 'Bohr–Einstein Example', translated and published here for the first time.

Parts III and IV consist of the EPR paper and responses to it. Part III contains in particular some of the most famous papers in the foundations of physics from the 1930s. Besides the EPR paper itself, it includes also Schrödinger's entanglement paper, Schrödinger's cat paper, Bohr's reply to EPR, and also Heisenberg's manuscript reply. We also include two further papers we believe to be of special interest: Hermann's in *Naturwissenschaften* and Furry's extensive reply to EPR (and short sequel). Part IV then contains the further responses to EPR discussed in Chapter 2 (including the riposte to Kemble penned by Podolsky and some interesting jottings from Schrödinger on the topic of canonical conjugates).

Part V is our translation of the heap of correspondence by Schrödinger and other physicists on EPR, including those letters collected in Schrödinger's folder. These letters, though appearing last in the present work, were its very inspiration.

★

A project like the one undertaken here is, in many ways, a marathon. A slow, painstakingly detailed marathon, perhaps, but a richly rewarding one nonetheless. As we have covered this terrain – some of it wonderful and new, some of it comfortingly familiar – we have accrued many debts of gratitude. Our list of those to whom we owe thanks, and in how many ways, is surely as vast as was the journey itself. The brevity of these sentences fails to capture the magnitude of your support.[2]

[2] Some more detailed thanks will appear at appropriate places in the text or footnotes.

We owe special thanks to some fellow researchers for sharing their insights, drafts, notes and transcriptions with us: in particular to Martin Jähnert, Christian Joas, Dennis Lehmkuhl, Christoph Lehner, and Jos Uffink (who was of great help also with some of the archival material in Vienna and by reading a full draft of Part I), and also in various ways to Harvey Brown, Jeroen van Dongen, Arthur Fine, Chris Fuchs, Nancy Greenspan, Helge Kragh, Terri Seager, and Paul Teller.

Big thanks go also to various institutions and the people there who provided assistance (financial and not), hospitality and more. In no very systematic order: The Leverhulme Trust (grant F/00 152/AN), and June Middleton – our splendid research officer at the University of Aberdeen. The Max Planck Institute for the History of Science in Berlin, where we spent three extremely productive summers in 2011, 2012 and 2013, and there especially Massimiliano Badino, Alex Blum, Carmen Hammer, Jürgen Renn, and again Martin Jähnert, Christian Joas and Christoph Lehner. The Edelstein Centre at Hebrew University for support for EC through an Edelstein Centre Fellowship for the year 2013–14, along with members of the philosophy of physics group there (Yemima Ben-Menahem, Meir Hemmo, Guy Hetzroni and Orly Shenker). The Centre for Time, University of Sydney for hospitality to EC in 2008; the Reilly Center for Science, Technology, and Values, University of Notre Dame for hospitality to GB in 2011; the Visiting Scholars Programme of the College of Arts and Sciences, University of Aberdeen for support for EC also in 2011. The lovely folks at the Albert Einstein Archives (Chaya Becker, Roni Grosz and Barbara Wolff); Christian Joas, Felicity Pors and Rob Sunderland at the Niels Bohr Archive in Copenhagen; Diana Kormos-Buchwald and Dennis Lehmkuhl at the Einstein Papers Project; Lindsay Claro and very specially Lisa Black at Princeton University Press; Paul and Wendy Teller; Georgina and Sebastian Born; Dorothea Barfknecht and Gabriele Stefanski at the Staatsbibliothek zu Berlin – Preußischer Kulturbesitz; Graeme D. Eddie at the Centre for Research Collections – University of Edinburgh Library; Allen Packwood and Gemma Cook at the Churchill Archives Centre – Churchill College Cambridge; Bob Podolsky; Stephen Kemble and siblings; Mary McDonald at the American Philosophical Society; Bettina Loycke at Wiley-VCH; Johannes Blümer at the Heisenberg-Gesellschaft; Anna Krutsch at the Archiv Deutsches Museum; staff at Springer, very specially Ties Nijssen; Thomas Notthoff at the Max-Planck-Gesellschaft; Anita Hollier at the CERN Archive; and staff at the Zentralbibliothek für Physik Wien, very specially Christoph Capellaro. EC also acknowledges both research and travel support in the summers of 2015 and 2018 from the Professional Staff Congress at the City University of New York ('Trad-A' grant cycles 46 and 49, respectively). And of course and very specially our much forbearing and ever-encouraging editors at Cambridge University Press, first Simon Capelin and Róisín

Munnelly, then Nicholas Gibbons, Rowan Groat and Sarah Armstrong, and to our great copy-editor, Gary Smith.

On a more personal note, we thank our families – not only for being the water-bearers in the final stretch, but for cheering us on through many, many miles. Thanks to Katie, who, because she is nearly perfectly co-extant with this project, has thereby continually reminded us that great things do come from small beginnings; to Jen, for innumerable things (but among them, much-needed research breaks spent discussing – or participating in – classical music); to Elise's family – for loving me even though I've spent the better part of a decade rambling on about 'The Book with Guido'.

Lastly, we have dedicated this book to Don Howard as but a small token of gratitude in exchange for his '*furchtbar nicht* trivial' role in this project. Sie sind Doktorvater, Verschränker und Freund.

Permissions and Copyright Notices

(a) By kind permission of the American Philosophical Society: J. Trimmer's translation of Schrödinger (1935f), including the figure, originally published as: Trimmer, J. D. (1980). The Present Situation in Quantum Mechanics: A Translation of Schrödinger's 'Cat Paradox'. *Proceedings of the American Philosophical Society* **124** (5), 323–338 (see also (k)).

(b) Courtesy of the Niels Bohr Archive, Copenhagen: Quotation and two figures from Bohr's talk of 5 October 1931 (AHQP-BMSS 12, section 5); quotation from the German version of Bohr's reply to EPR (AHQP BMSS-14, section 1); translations from the letter by Ehrenfest to Bohr of 9 July 1931 (BSC-EHR-310709t); and translations of the letters from Bohr to Heisenberg of 10 and 15 September 1935 (AHQP-BSC 20, section 2), from Bohr to Einstein of 25 October 1935 (BSCSupp-EIN-351025f), from Heisenberg to Bohr of 19 September 1935 (AHQP-BSC 20, section 2), and from Heisenberg to Margrethe Bohr of 13 September 1935 (BSCSupp-HEI-350913t) (see also (e)).

(c) By kind permission of Princeton University Press: From COLLECTED PAPERS OF ALBERT EINSTEIN (various, published and unpublished). Albert Einstein Archives – archival call numbers 2-100, 5-207, 10-162, 10-231, 10-232 TTRL, 10-581, 10-583, 10-588, 16-112, 17-066, 22-047, 22-047 TTRL, 22-049, 22-049 TLC, 22-052, 22-052 TLC, 22-174, 71-836, 71-836 ALSX and 91-989. Copyright © Hebrew University of Jerusalem. Published by Princeton University Press.

(d) By kind permission of Wiley-VCH: Translation of the report of Einstein's Berlin seminar (Anonymous). Physikalisches Colloquium. Berlin, 4. November 1931. *Angewandte Chemie*. 1932. **45**. 23. Copyright Wiley-VCH GmbH. Reproduced with permission.

(e) By kind permission of the Heisenberg-Gesellschaft: Translations of the letters from Heisenberg to Pauli of 2 July 1935 (CERN Archive, Pauli Letter Collection heisenberg_0017-092r), from Heisenberg to Bohr of 19 September 1935 (AHQP-BSC 20, section 2), and from Heisenberg to Margrethe Bohr of 13 September 1935 (BSCSupp-HEI-350913t) (see also (b)); and translations from the letters from Heisenberg to Einstein of 10 June 1927 (Albert Einstein Archives 12-174), and from Heisenberg to Hermann of 9 July 1936 (Archiv der sozialen Demokratie, Bad Godesberg, Nachlaß Grete Henry-Hermann 1/GHAJ).

(f) By kind permission of the Kemble family: Reproduction of Kemble's typescript 'Physical Reality and Quantum Mechanics' (AHQP 54, section 3-019); and quotations from the letters from Kemble to Einstein of 15 October 1935 (AHQP 53, section 7-034), and from Kemble to Margenau of 9 November 1935 (AHQP 54, section 3-017) and 20 November 1935 (AHQP 54, section 3-019).

(g) By kind permission of the Max-Planck-Gesellschaft: Translation of the manuscript 'Ist eine deterministische Ergänzung der Quantenmechanik möglich?' by Werner Heisenberg (AMPG, III. Abt., Rep. 93, Nr. 668).

(h) By kind permission of *The New York Times*: Quotations from 'Conservative Einstein' (*The New York Times*, 4 May 1935).

(i) By kind permission of the Pauli Committee at CERN: Translations of the letters from Pauli to Heisenberg of 15 June 1935 (CERN Archive, Pauli Letter Collection heisenberg_0215), and from Pauli to Schrödinger of 9 July 1935, including two figures (see also (j2)).

(j) Printed with kind permission of the Braunizer family:

(j1) Source – Austrian Central Library for Physics: Translation of Erwin Schrödinger's manuscript 'Bohr-Einstein Beispiel' (ZBP-Wien W33-781/4); translations from his notebooks 'Wellenstatistik I' and 'Wellenstatistik II' (ZBP-Wien W33-737 and ZBP-Wien W33-738), and from the letter by Sommerfeld to Schrödinger of 15 December 1931 (ZBP-Wien B33-377).

(j2) Source – Austrian Central Library for Physics (copy): Translations of the letters from Schrödinger to Teller of 14 June 1935, from Teller to Schrödinger of 11 June 1935 (see also (l)), from Schrödinger to Born of 29 June 1935, from Schrödinger to Pauli of early July 1935, from Pauli to Schrödinger of 9 July 1935, including the two figures (see also (i)), from Schrödinger to Einstein of 13 July 1935, from Schrödinger to Berliner of 25 July 1935, from Schrödinger to Berliner of 11 August 1935, from Schrödinger to Einstein of 19 August 1935 (including the figure), from Schrödinger to Einstein of 4 October 1935, and from Schrödinger to Bohr of 13 October 1935.

(j3) Erwin Schrödinger's manuscript 'The Observation of Canonically Conjugates'; quotation from the letter from Schrödinger to the Editor of *Nature* of 12 June 1935; translations of the letters from Schrödinger to Sommerfeld of 11 December 1931 (including the figure), from Schrödinger to Laue of 25 July 1935, from Berliner to Schrödinger of 1 July, 29 July and 14 August 1935, from Bohr to Schrödinger of 26 October 1935, from Born to Schrödinger of 28 June 1935 (see also (p)), and from Einstein to Schrödinger of 17 June, 19 June, 8 August and 4 September 1935.

(k) Reprinted or translated by permission from Springer Nature Customer Service Centre GmbH: Springer Nature, *Nature*, Quantum Mechanics and Physical Reality, N. Bohr, 1935; Springer Nature, *Nature*, Quantum Mechanics as a Physical Theory, H. T. F., 1935; Springer Nature, *Naturwissenschaften*, Fig. 1 from Atomkern und periodisches System der Elemente, Lise Meitner, 1934; Springer Nature, *Naturwissenschaften*, Die gegenwärtige Situation in der Quantenmechanik, Erwin Schrödinger, 1935 (see also (a)). These permissions will be displayed also in the text.

(l) By kind permission of Paul and Wendy Teller: Translation of the letter from Edward Teller to Schrödinger of 11 June 1935 (see also (j2)).

(m) Reprinted with permission from: A. Einstein, R. C. Tolman and B. Podolsky, *Physical Review* **37**, 780–781 (1931); A. Einstein, B. Podolsky and N. Rosen, *Physical Review* **47**, 777–780; E. C. Kemble, *Physical Review* **47**, 973–974; N. Bohr, *Physical Review* **48**, 696–702; A. E. Ruark, *Physical Review* **48**, 466–467; H. Margenau, *Physical Review* **49**, 240–242; H. C. Wolfe, *Physical Review* **49**, 274; W. H. Furry, *Physical Review* **49**, 393–399; W. H. Furry, *Physical Review* **49**, 476. Copyright 1931, 1935, 1936 by the American Physical Society. These permissions will be displayed also in the text.

(n) By kind permission of the Cambridge Philosophical Society: Reprints of Schrödinger (1935e) and Schrödinger (1936).

(o) By kind permission of Bob Podolsky: Quotation from Boris Podolsky in 'Einstein Attacks Quantum Theory' (*The New York Times*, 4 May 1935); reprint of Boris Podolsky's 'States and Reality of Physical Systems'.

(p) By kind permission of Sebastian Born: Translation of the letter by Max Born to Erwin Schrödinger of 28 June 1935 (see also (j3)).

Every care has been taken to identify copyright owners of the original material and every effort has been made to ensure copyright information is accurate. Please contact the authors to signal omissions or inaccuracies.

Abbreviations and Editorial Conventions

Archives

AHQP	Archive for the History of Quantum Physics. Numberings within microfilm sections (where applicable) follow the cataloguing of the MPIWG
Alpbach	Erwin Schrödinger Nachlaß, Alpbach
BSCSupp	Bohr Scientific Correspondence Supplement, Niels Bohr Archive, Copenhagen
CERN-ARCH-PLC	Pauli Letter Collection, CERN Archive, Geneva
EA	Albert Einstein Archives, Jewish National and University Library, Hebrew University of Jerusalem
MPIWG	Max Planck Institute for the History of Science, Berlin
ZBP-Wien	Erwin Schrödinger Nachlaß, Zentralbibliothek für Physik, Vienna

Abbreviations used in the reprinted or translated sources

Berl. Ber.	Sitzungsberichte der Preussischen Akademie der Wissenschaften
Journ. of the Chem. Soc.	Journal of the Chemical Society
J. Frank. Institute	Journal of the Franklin Institute
J. Phil. Sci.	Philosophy of Science
Natur. Verden	Naturens Verden
Naturwiss., Naturwissenschaften	Die Naturwissenschaften
Phys. Rev.	Physical Review
Physik. Z. Sowjetunion	Physikalische Zeitschrift der Sowjetunion
Proc. Camb. Phil. Soc., Proc. Cambridge Phil. Soc.	Proceedings of the Cambridge Philosophical Society

Proc. Roy. Soc. Lond.	Proceedings of the Royal Society of London
Rev. Mod. Phys.	Reviews of Modern Physics
Z. f. Phys., Z. Physik	Zeitschrift für Physik

Editorial conventions and remarks

When reprinting sources, original spellings have been kept. We have uniformised typographic conventions such as order of punctuation and footnote marks, use of single and double quotes (except in the Ehrenfest quotations on pp. 14–16 and 118), etc.

Footnote numbering (arabic) starts anew in every chapter (and in every section in the edition of the correspondence in Part V). Endnotes (roman) are used for two of the translations.

Square brackets [ ] denote editorial amendments or (in the translations) original wordings. Editorial footnotes are also enclosed in square brackets and marked with (*eds.*).

References in the reprinted or translated materials have been only slightly modernised, but are included in full also in the references at the end of the volume. When double page references to papers reprinted or translated in our volume are given, the first refers to the original and the second to the present volume.

Finally, all translations are our own unless otherwise stated.

Part I

Main Contributions to the EPR Debate in 1935

1

Einstein on EPR*

1.1 Introduction

One of the most detailed studies of Einstein's dissatisfaction with quantum mechanics is certainly Don Howard's paper, '"Nicht sein kann was nicht sein darf"' (Howard 1990).[1] In it, Howard argues that Einstein was principally motivated by an *a priori* commitment to the separability of physical systems, and that this commitment arose from an even deeper commitment to describe interactions universally via field theories.

We shall not focus particularly on the aspect of Howard's argument regarding field theories as the path to a unified theory, but shall instead focus mainly on the degree to which Einstein's belief in the separability of systems drove his criticism of quantum mechanics in the years leading up to EPR. We agree with Howard that the failure of separability for Einstein was crucial in judging quantum mechanics was incomplete. Because nonseparability just is what we now call entanglement, one may justifiably state that Einstein's concerns related at least indirectly to the phenomenon of entanglement.

We also agree with Howard (as well as with Harvey Brown (1981) and Arthur Fine (1986, p. 28)) that Einstein's concerns were *not* primarily to do with the simultaneous measurability of conjugate variables. The issue of quantum mech-

* The material in this chapter relates to talks given at the CHPSTM workshop on *Quantum States: Ontic or Epistemic?*, Aberdeen (June 2011), the Philosophy of Physics Seminar at Oxford (February 2012), the 11th Biennial Meeting of IQSA, Cagliari (July 2012), the Purdue Wiener Memorial lectures 2014 (November 2014) and to Bacciagaluppi (2013, 2014, 2015, 2016b).

[1] But see also his 1985 paper, which is one of the first published discussions of the 1935 correspondence between Einstein and Schrödinger (Howard 1985).

anical completeness was, for him, a separate matter from Heisenberg's uncertainty relations, which he accepted as part of the correct formalism. Debates as to whether and when Einstein accepted the uncertainty relations in principle are typically grounded in studies of the various photon-box thought experiments. We add to these analyses by incorporating some new and some neglected primary sources, including the Einstein–Laue correspondence from 1936. We believe Einstein's recollections to Laue regarding the Rupp experiments show that already in 1926 Einstein was prepared to accept at least one direction of the energy–time uncertainty relation. We point out that the thought experiment Einstein cites in this correspondence as motivation for the Rupp collaboration is not only a precursor to the photon box, but also fits nicely with the views on the statistical nature of the quantum state that Einstein was to express in the context of EPR.

Given this earlier evidence of Einstein's thoughts on incompleteness, what was the real intended thrust of the 1930 photon-box thought experiment? We suggest it might have been whether a precise determination of energy implies a statistical spread in the aperture time – affirmation of which would impute uncertainty to a *macroscopic object* (the clockwork in the box).

The modified history of the evolution of the photon box and of Einstein's thinking proffered here will not only incorporate new materials (like the Einstein–Rupp proposal), but will also examine several moments in this story that have received less attention (e.g. the Einstein–Tolman–Podolsky paper from 1931), yet which we believe contribute importantly to the broader picture.

Following our re-examination of the years leading up to 1935, we present an updated narrative not only about the structure of the EPR paper, but also consider the logic of EPR in immediate contrast to Einstein's own views as expressed in his correspondence with Schrödinger in the summer and autumn of 1935.

A brief caveat on terminology. In his analysis of Einstein's critique of quantum mechanics, Howard provides the following definitions of separability and locality (Howard 1990, p. 64). Separability is 'the condition that spatio-temporally separated systems possess well-defined real states, such that the joint state of the composite system is wholly determined by these two separate states', while locality is 'the condition that such a real state is unaffected by events in regions of spacetime separated from it by a spacelike interval'. However, the concept Einstein christens the 'Separation Principle' in his letter of 19 June 1935 to Schrödinger (see p. 297) does not clearly distinguish between the concepts of locality and separability so defined. Indeed, Howard demonstrates that it is only in Einstein's correspondence with Epstein after World War II that the former makes this conceptual distinction explicit (Howard 1990, p. 64, note 7). Therefore, for the purpose of the present chapter we shall be satisfied to let the term 'separability' stand in for both separability and locality.

1.2 Einstein's Unpublished Hidden Variables Theory of 1927

On 5 May 1927, Einstein gave a report at the meeting of the Prussian Academy of Sciences in Berlin in which he described a hidden variables theory of his own devising.[2] This theory was also the subject of Einstein's brief note to Ehrenfest on that same day, wherein we read the following (EA 10-162 and Einstein 2018a, p. 815):

> I have now also done a little study of Schrödingery, in which I show that in a completely unique way the solutions [of Schrödinger's equation] can be assigned determinate motions that make any statistical interpretation unnecessary.

The paper was never published, for Einstein asked it to be retracted before printing. Various authors have variously interpreted this move; Howard (1990, pp. 89–91) makes a particularly convincing case. Based on examination of Einstein's addition in proof to the 1927 manuscript, Howard suggests that Einstein abandoned this hidden variables attempt upon realisation that its nonseparable features (implied by the non-vanishing cross-terms in the composite wavefunction of two subsystems) would violate 'a general requirement that must be imposed on a general law of motion for systems' (Einstein's 'Nachtrag', as quoted in Howard 1990, p. 90), namely separability of the subsystems.

In his treatment of Einstein's unpublished manuscript, Belousek suggests that Einstein's dissatisfaction with the work stemmed from its inability to provide sufficient causal explanations (Belousek 1996, p. 445; cf. Bacciagaluppi and Valentini 2009, p. 239). Belousek cites as evidence for his claim two 'sins' committed in the 1927 manuscript, the second of which is a sin of omission. Einstein's use of configuration space demonstrates his awareness of the possibility of waves living in a greater than three-dimensional space. And yet Einstein conspicuously avoids commenting on how quantum waves in such non-physical spaces could themselves be considered physical entities. A more recent paper by Holland (2005) describes several specific ways in which Einstein's theory would have failed to reproduce the statistical results already known to hold in quantum theory. Bacciagaluppi and Valentini (2009, p. 235) suggest that Einstein may have been planning to present his hidden variables theory at the Solvay conference that October until realisations such as those discussed by Howard led him to reconsider.

[2] EA 2-100 (unpublished manuscript, 6pp., in German). For more on the 1927 manuscript of Einstein, see Belousek (1996) and references therein.

1.3 The 1927 Solvay Conference

As is well known from the literature (e.g. Jammer 1974, Brown 1981, Howard 1990, Bacciagaluppi and Valentini 2009), during the October 1927 Solvay conference in Brussels, Einstein proposed one of the first in a long line of thought experiments intended to demonstrate shortcomings with the quantum theory. Einstein considers electrons passing through a single slit and impinging on a semi-spherical photographic film. Assigning wavefunctions to the electrons leaves open two interpretations of the event recorded by the photographic film, each associated with a cost. Brown summarises Einstein's reasoning here as follows (Brown 1981, p. 61):

> [Einstein] observed, using an extremely simple example, that a wave perturbation associated with a particle in wave mechanics seems to lead to an action-at-a-distance sort of effect during the process of locating the particle if we consider the wave to represent (as maintained by Bohr and Heisenberg) a complete description of the individual particle. On the other hand, the problem disappears if the [wave] description merely describes probabilities (following the Born rule) in relation to an 'ensemble' of independent particles with well-defined positions at every instant. In this second interpretation, wave mechanics represents an incomplete or 'statistical' description of a system only, leaving open the question of the possibility of a causal mechanism which governs the behaviour of the individual system. It was this position to which Einstein constantly adhered after 1927. [. . .] The crucial point here is that Einstein's comments constitute a criticism of the Copenhagen school, and they do not question the consistency of the formalism of quantum mechanics (and the criticism does not directly concern the indeterminacy relations). Moreover, the points raised by Einstein clearly already carry the germ of his more compelling and detailed 1935 arguments.[3]

Perhaps taking to heart the very problems encountered in his report to the Prussian Academy, Einstein concludes his Solvay comments by declaring dissatisfaction with any theory that relies on multi-dimensional configuration space, because permutations give rise to distinct configurations in this space 'which is not in accord with the new results in statistics', and because it leads to a 'less natural expression' of forces acting at short spatial distances than would a three- or four-dimensional description (Bacciagaluppi and Valentini 2009, p. 442).

There are two points in Brown's analysis we wish to emphasise. The first is the suggestion that Einstein's issues with quantum theory were *not* rooted in the uncertainty relations. Brown was one of the first authors to articulate this crucial point. Howard (1990, p. 94) would later corroborate this claim by proposing that if Einstein's concerns at Solvay in 1927 were tied to the uncertainty relations, it would be hard to explain his utter silence about them in correspondence and in print afterwards.

[3] Special thanks to Harvey Brown for sending us a copy of this paper, and even highlighting relevant passages! We also thank him for sharing relevant bits of his doctoral thesis, and for helpful discussions.

The second aspect we wish to emphasise is Brown's observation – correct in our view – that Einstein's comments at Solvay already contain the germ of EPR's arguments. This can be understood in several ways. To start, it is clear that Einstein already in 1927 is worrying about the tension between quantum mechanics and relativistic locality: how do spacelike separated detectors know when the other detector has fired already? One obvious difference with EPR is that the 1927 example does not rely on entanglement.[4]

But one can also argue for a closer parallel between 1927 and 1935, as Norsen (2015) has done, pointing out that Einstein's informal presentation of the EPR argument to Schrödinger at the start of his letter on 19 June 1935 (discussed here in Section 1.5.2) does not turn on entanglement, but rather employs a setup with just one 'particle' (a ball) and two 'locations' (boxes) in which it may be found. This version of the incompleteness argument *does* more closely parallel Einstein's 1927 considerations, wherein components of the wavefunction travel in different directions.

Even though not made explicit in his thought experiment during the 1927 Solvay conference, Einstein was at this time clearly worried about the nature of composite systems in quantum theory. This is evinced by his remark that an ensemble reading of the wavefunction may not comport with conservation laws at the individual level (see Howard 1990, pp. 94–97). So although there are noteworthy parallels between Einstein's commentary in 1927 and his incompleteness arguments in 1935, a more robust historical lineage for the EPR thought experiment is arguably to be traced through several transformations of another of Einstein's thought experiments: the photon box.

1.4 The Photon-Box Experiment

Einstein's famous photon-box thought experiment is usually dated to his debate with Bohr in the autumn of 1930 during the sixth Solvay conference in Brussels. However, in a letter to Laue in 1936 discussing their recently discredited collaborator Rupp, Einstein reveals that already in 1926 he had devised a thought experiment that is clearly a precursor to the 1930 photon box. The context in which this earlier thought experiment was devised is of course importantly distinct from its particular use as a criticism of quantum mechanics in 1930 and afterwards, but we nevertheless believe this report by Einstein underscores a continuous trend in his thinking about quantum phenomena. We begin this section therefore with a discussion of the 1926 thought experiment, and continue with a brief overview of the better-known history of the photon box.

[4] And, indeed, the explanation of this thought experiment provided by de Broglie's theory does *not* require action at a distance, whereas EPR's example does.

1.4.1 The Photon Box in 1926

In the spring of 1926, Einstein devised several experiments later (allegedly) carried out by Rupp as a means to probe the particle- or wavelike aspects of light (Van Dongen 2007a, b). Although in the interim period experiments by Compton, Bothe and Geiger, and solo work by Bothe, had overwhelmingly confirmed the quantised nature of light, Einstein was nevertheless motivated to seek further empirical evidence. Van Dongen writes: 'He did not have a straightforward answer to tough questions on how the wave train that was clearly exhibited in interference phenomena could be brought in line with an instantaneous emission process' (Van Dongen 2007b, p. 82; cf. Howard 1990, pp. 76ff.). As Einstein would colourfully write to Ehrenfest on 28 August 1926 – some time after Rupp had reported to Einstein that his experiments had yielded the expected wavelike results (as translated in Einstein 2018b, p. 359):

> I already wrote to you about this, that the canal ray experiments came out entirely in line with the wave theory. Here waves, here quanta! The reality of both is rock solid. Only the devil can make any *real* rhyme or reason out of it.[5]

Thanks to the careful work of Van Dongen we now know the details of the story of the Einstein–Rupp experiments, despite their having been largely forgotten after it transpired that Rupp's results for these tests (and many others besides) were fabricated. Van Dongen describes how although Rupp's status in the world of experimental physics became increasingly suspect into the 1930s, Laue and Einstein – both having published papers with Rupp – were not eager to accept the latter's discreditation. It seems Einstein and Laue only really dropped Rupp in 1935, after a damning paper by Gerlach and Rüchardt on the impossibility of Rupp's results appeared in the prominent journal *Annalen der Physik*, and after the German Physical Society had officially expelled Rupp from its membership for fabricating experimental results (Van Dongen 2007a, pp. 114–119).[6]

Thus it is a full decade after Einstein first suggesting wave-particle tests to Rupp and in the wake of Rupp's downfall that Laue initiates a brief exchange of letters with Einstein attempting to get clear on the physics at stake. The 1936 correspondence between Laue and Einstein comprises four letters: Laue to Einstein on 8 June and 29 July; a presumed but (alas) missing letter from Einstein to Laue between those dates, and a response to Laue on 29 August.[7] It is this last letter we find particularly interesting. Einstein writes:

[5] The last sentence displays an idiosyncratic mix of idioms rather characteristic of Einstein: 'Aber der Teufel macht einen Vers darauf (der sich *wirklich* reimt)'.

[6] See Van Dongen (2007a, b) for more on this fascinating episode.

[7] These letters are found, respectively, in EA 16-110 (typed, 2pp.), EA 16-111 (typed, 2pp.) and EA 16-113 (handwritten, 3pp., typed transcription EA 16-112, 3pp.), all letters in German.

No consideration at the time could have been a self-contained logical deduction [Keine damalige Überlegung konnte keine [*sic*] einheitliche logische Deduktion sein], because at the time we did not possess a theory that fit the entire complex of the phenomena, and still do not today. Thus it can only be a matter of using premises one believes are certain to obtain, yet without being able to provide a unified theoretical basis for them.

If we are farther along today than we were then, then in my opinion it is due to the following realisation: one largely does justice to the facts if one proceeds like this. One calculates consistently with waves and considers these as probability waves for the spatial occurrence of particles, or quanta. One must however add that in doing so one has to treat a plurality of particles as *one* particle in a multidimensional space. Neither of us, of course, believes that this method amounts to the final truth. But we are confident that, in applying this sort of consideration within our domain of interest, one does not arrive at gross inconsistencies. But at the time I wrote the short note, this insight was not yet achieved. At that time the question was: the quantum notion and the undulatory notion clash, but both contain something correct; how much of either is correct?

The simple example from which I was then proceeding (not in the publication) is as follows: in a mirror-lined box there is monochromatic radiation, and in fact a single quantum. A little door opens for a short time τ ($\tau = n\frac{\lambda}{c}$; λ wavelength, n finite). If the quantum escapes, is its frequency exactly that of the quantum initially located in the box, or does it deviate statistically from it? The naive quantum view leads to an expectation corresponding to the first case, the undulatory view instead to one corresponding to the second case. Today we are completely persuaded that the consequence drawn from the undulatory view corresponds to reality. Back then, however, I saw no direct possibility of deciding, and thus sought a sufficiently analogous case in which our knowledge enabled a decision even without conducting an experiment. So I was led to the case [kam ich auf den Fall] of a radiating particle which flies past *directly behind a slit* and radiates.

The decisive 'sufficiently analogous case' Einstein then details for Laue is close to the version Rupp adapted for the canal ray experiments he supposedly undertook in April and May of 1926. That is: the Einstein–Rupp experiment is a *transposition of the original photon box*.[8] After explaining the setup and detailing his argument for statistical spread in the measured frequency of the photon, Einstein asserts: 'Therefore I do not think that my erstwhile considerations are superfluous or false, in fact I even believe that they may claim some modest interest. For in my opinion, we still do not possess a serious theory even today.'

This letter is briefly referenced by Fine in support of his claim that 'right up to the end, Einstein believed that the conception of the ψ-function as corresponding to an ensemble of systems "was not logically refuted"' (Fine 1986, p. 42, note 3). The Einstein–Laue letters in 1936 are also invoked in order to corroborate Fine's thesis that at the time of EPR, Einstein was more worried about how to interpret

[8] Note, however, that Einstein had proposed Rupp-like experiments even as early as December 1921 (Einstein 1921), as discussed by Klein (1970, section 3) and by Tauschinsky and van Dongen (2008). Thanks to Jeroen van Dongen for alerting us to this fact.

the wavefunction than, say, the metaphysical status of the uncertainty relations. We certainly agree with this analysis, but wish to draw special attention to the thought experiment Einstein gives in his letter to Laue, reportedly from 1926.

In particular, examination of this thought experiment leads us to suppose that *already in 1926* Einstein had convinced himself that if the time window for the emission of the photon is short, then the frequency of the emitted photon will exhibit a statistical spread. Thus it seems we have direct evidence that Einstein was prepared to accept the time–energy uncertainty relation soon to be proposed by Heisenberg, at least in one direction: a precise time measurement implies a spread in energy. Indeed, we learn from Van Dongen (2015) that the Einstein–Rupp experiments were keenly seized upon by Bohr and Heisenberg, and contributed specifically to the development of the uncertainty relations. In the Chicago lectures (Heisenberg 1930b, pp. 79–80) Heisenberg cites the Einstein–Rupp tests as one of a handful of experiments seminal for the establishment of the new quantum theory.[9]

Einstein's phrasing in this letter (e.g. when he asks 'is its frequency exactly that of the quantum initially located in the box, or does it deviate statistically from it?') further suggests that he understood the uncertainty in frequency to be 'statistical' in the strict sense: the photon *does* possess a well-defined frequency (unlike a segment of a classical wave train!), but in an ensemble one will find a spread of frequencies rather than always the frequency of the original photon in the box. Of course, Einstein admits that this is far from being a straightforward picture ('[b]ut the devil knows how one should *really* understand this')!

Finally, Einstein's concluding remarks to Laue in this exchange confirm that in 19*36* he still believed his 19*26* argument for the statistical spread in the photon frequency to be correct (and perhaps better than arguments based on the uncertainty principle). So we have here not just the first instance of the photon box, but a likely smoking gun for the claim that the better-known 1930 version of this experiment was not intended by Einstein to demonstrate the *incorrectness* of quantum mechanics.

1.4.2 The Photon Box in 1930 and Beyond

The version of the thought experiment proposed by Einstein at the October 1930 Solvay conference involves a photon being released from a weighable box containing a clockwork mechanism for the release of the photon. The difference in weight of the box before and after the emission of light enables calculation of the photon's

[9] In addition to Bohr's and Heisenberg's initial interpretation of the Einstein–Rupp experiments, when re-analysed in the 1930s for the purpose of exposing Rupp as a fraud, these experiments were again interpreted explicitly as testing the validity of the uncertainty relations (Van Dongen 2015).

energy, and the clockwork fixes the time of release of the photon. Thus we appear to obtain precise knowledge both of the photon's energy and time of emission. In this version of the thought experiment no manipulation of the system of interest (the photon) is made, only inferences to its properties based on measurements of an ancillary system (the box).

One may be tempted to assume that Einstein's aim in this instance was to refute the energy–time uncertainty relation. This temptation is seemingly vindicated by Bohr's own recollection of the 1930 thought experiment – both in his contribution to the Schilpp volume (1949, pp. 225–226) and later when interviewed by Kuhn.[10] But if as argued here and elsewhere (with especial perspicuity in Howard (2007)) Einstein was not trying to cast doubt on the uncertainty relations, what *was* the intended thrust of this thought experiment? One might conjecture as follows: although Einstein had convinced himself that a short, precisely determinable time for the shutter's opening implied a spread in the photon's energy, he required further convincing that a precise measurement of energy of the *microscopic* system would likewise imply an uncertainty in the duration of the opening of the *macroscopic* shutter. The well-known analysis of the photon box in which weighing it changes its position in the gravitational field – thereby restoring uncertainty to the predictions – would indeed speak to this point (Bohr 1949, pp. 224–228[11]).

According to the more recent understanding of the 1930 photon box (which is not incompatible with the above), Einstein was arguing that quantum mechanics yielded a merely statistical description precisely *because* the photon must already know the answers to the predictions one can make by manipulating the box alone. In other words, while the uncertainty relations are correct, and while presumably Einstein had been inclined all along to read them epistemically, with the photon box Einstein is working towards an *argument* to the effect that the uncertainty relations are to be read as such – as limitations on what we can know or predict – rather than ontically as limitations on what is really there.

Bohr himself gives a near-contemporary description of the thought experiment – as well as a resolution of the 'paradox' posed by it – in a talk delivered in Bristol on 5 October 1931.[12] Specifically, Bohr says:

> These new principles, introduced by quantum mechanics, are of similar character as those of Einstein's theory of relativity and the question arises whether they are independent from each other, or one superseding the other, or contradictory to each other, or reconcilable. To my mind there is not the least difficulty, as the fundamental relations are relativistic[ally]

[10] AHQP 1419-001, section Bohr-006. Also available online at www.aip.org/history-programs/niels-bohr-library/oral-histories/4517-5.

[11] Bohr states (p. 226) that Einstein actively contributed to this analysis.

[12] AHQP-BMSS 12, section 5 (typescript, 14pp., in English). Transcription in Bohr (1985, pp. 367–369).

invariant and can be adapted to each Lorentz-frame of coordinates, but I shall enlarge on this subject in the published account of this lecture.[13]

I shall now proceed to discuss a problem designed by Einstein which seemed to show that the foregoing considerations are only valid as long as the view point of theory of general relativity is left out of the discussion, i.e. as long as gravitation is left out. Einstein proposed the following arrangement with which he thought it possible to determine the energy change of an object accurately and also the time at which this change took place accurately:

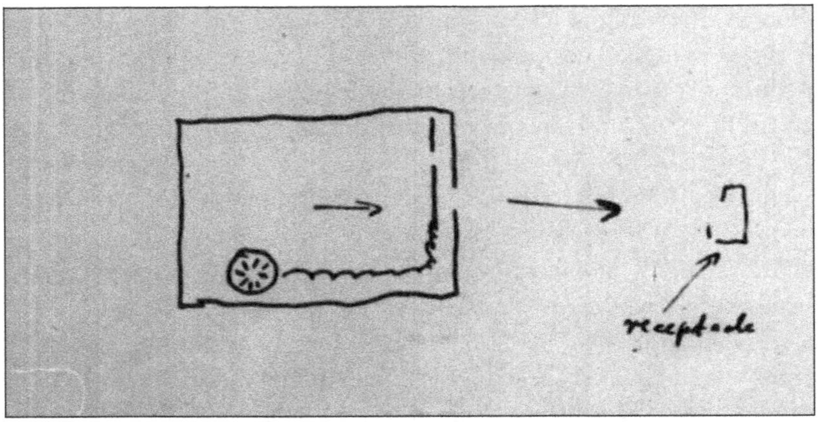

Fig. 1.1 Bohr's first figure: Einstein's photon box, with internal clockwork mechanism and external 'receptacle'.

The box contains a clock which, by some arrangement, can move a shutter, so as to open the box for a very short time ΔT sufficient to let pass out of the box just one light-quantum. We know, therefore, the time at which the light-quantum escaped with accuracy ΔT which can be made indefinitely small. In addition to this we will weigh the box in a gravitational field before and after the process to which there is also no limit of accuracy, so that time and energy of the emission of the light-quantum would both be known accurately. The solution of this paradox can be found by taking into account the principles of general relativity to a fuller extent, and by going into greater detail about the process of weighing. We will do the weighing by hanging up the box on a spring and reading the extension of the spring by means of a pointer attached to the box and running over a vertical scale.

Now the reading of the pointer cannot be done with infinite accuracy because this would involve infinite inaccuracy in our knowledge of the momentum of the whole box which would mean that the whole process of weighing was illusory. We must therefore be content with a finite accuracy in the determination of the mass. Let us say, for instance, that we do the experiment of reading the pointer by some arrangement which takes the time T. Then the uncertainty in mass ΔM will be connected with the gravitational force $\partial\varphi/\partial x$ and the uncertainty in the momentum by the equation:

[13] [Alas, Bohr did not publish the account of this lecture (*eds.*).]

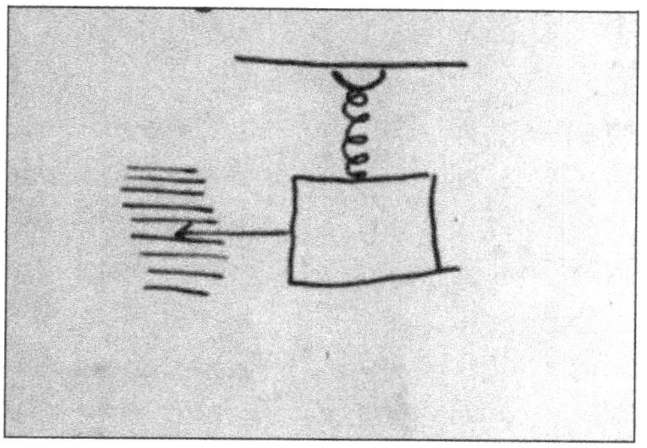

Fig. 1.2 Bohr's second figure: the weighable photon box with vertical scale.

$$\Delta M \cdot \frac{\partial \varphi}{\partial x} T = \Delta P_x = \frac{h}{\Delta x}$$

or

$$\Delta M \Delta \varphi T = h \qquad (1)$$

Such an arrangement would therefore involve also an uncertainty about the gravitational potential at which the whole box and the clock inside of it was placed. But this fact, according to a fundamental principle of general relativity would implicate uncertainty in the time scale indicated by the clock:

$$\Delta T = \Delta \varphi \cdot T/c^2$$

or, according to (1)

$$\Delta T = \frac{h}{c^2 \Delta M} \qquad (2)$$

Finally, we must remember that, also according to the relativity theory, mass and energy are in every case connected by the relation

$$Mc^2 = E$$

and this brings (2) back to the same form of uncertainty relation which we found valid in all those cases in which gravitation was not involved.[14]

The foregoing analysis of the paradox brought forward by Einstein shows that the phenomenon of gravitation which, up till now, in its mathematical treatment stands so far apart from the rest of modern physics does so by no means with respect to the primary physical concepts evolved by Einstein in his theory of general relativity in order to remove the obstacles which until then seemed to prevent the reconciliation of these two groups of physical phenomena.

This lecture together with his better-known recollections from 1949 present a consistent picture of what Bohr took away from discussions with Einstein at the

[14] [In the manuscript Bohr writes these last two equations $\Delta T = \frac{c^2 h}{\Delta M}$ and $M = c^2 E$ (*eds.*).]

1930 Solvay conference. What Einstein himself took away can be reconstructed from a number of sources from the early 1930s: a letter by Ehrenfest to Bohr in July of 1931, reports of talks given by Einstein in Berlin and Leiden in November of 1931, and some notes by Schrödinger from around the same time. Subsequent correspondence between Ehrenfest and Einstein in 1932 (and a related anecdote from Rosenfeld, to be mentioned in the next section) indicate some further developments.[15] Finally, while visiting Caltech in the winter of 1930, Einstein co-authored a paper with Tolman and Podolsky (he of EPR fame) on a version of the photon box. Despite its chronological precedence, we leave Einstein's Caltech collaboration (hereafter 'ETP') for the next subsection. This is because the ETP thought experiment (as confirmed in the report of Einstein's Berlin seminar) appears to be an extension of the photon-box experiment presented in the other sources. Indeed, we shall use the ETP experiment as a means of introducing some further aspects of both the EPR paper and the debates it instigated.

Ehrenfest's letter to Bohr on 9 July 1931[16] is prime evidence for Einstein's true intent with the photon box. Ehrenfest writes to Bohr in characteristic style (via Mrs Bohr and by way of Casimir as courier):

> I think now I understand again very well why he [Einstein], despite perfect familiarity with the uncertainty relation and the corresponding limited applicability of the classical concepts, still refuses to be put off the search for *micro*-field theories that can eventually prove fruitful for the observable phenomena. If one could just arrange it so that you two marvellous people do not speak PAST EACH OTHER.
>
> Of course I have also asked Einstein why he continually thinks up new "perpetua mobilia" = machines against the uncertainty relation. He told me that already since very long ago he absolutely no longer doubts the uncertainty relation and thus e.g. that he has not thought up the "weighable light-flash box" [...] AT ALL "contra the uncertainty relation", but for an entirely different purpose.
>
> [...]
>
> We want to think up a "machine" that ejects a "projectile". And AFTER the projectile has left the machine once and for all a "questioner" can, at his discretion, ask the machine operator to *predict* by means of a *subsequent examination of the machine*
>
> *either* which numerical value a for a quantity A
>
> *OR* which numerical value b for a quantity B
>
> the questioner will later find on the projectile (reflected and coming back after a frightfully long time) by performing either an A-measurement or a B-measurement.
>
> Here A and B shall be non-commuting quantities, so e.g. conjugate ones.

[15] Most of these sources (barring the Schrödinger notes, to which we shall return more extensively in Chapter 3 and which we translate as Chapter 8) are discussed in Jammer (1974), Fine (1986), Howard (1990) and Held (1998).

[16] AHQP-BSC 18, section 4 and BSC-EHR-310709t (typewritten, 2pp., in German).

> Thus for Einstein it is out of the question and beyond doubt that because of the uncertainty relation one must of course choose between either and or. But the questioner can choose between them AFTER the projectile is once and for all on its way.
>
> Can one think up such a machine? – And why is Einstein interested in it?
>
> *I can sympathise very well with Einstein about the latter*! But nevertheless I am unable to formulate it in a way as to be sure that he would be happy with my formulation. I would say: it is interesting to be clear about the fact that the projectile, which is already flying about in isolation "on its own", must be ready to satisfy very different ""non-commutative"" [*sic*] prophecies, "without even knowing" which one of these prophecies one will make (and check).
>
> [...]
>
> Now, his weighable light-flash box is meant to be such a machine.

Notice that Ehrenfest repeats *three times* (even in capitals and underlined) that the choice of which prediction to make is delayed until after the projectile has been emitted. Ehrenfest continues from here to describe a highly detailed version of the thought experiment: at 0 hours one sets up the clock to briefly open the shutter at 1,000 hours, and also weighs the box for 500 hours, after which one bolts it to the external frame. When the shutter opens, the light travels to a fixed mirror half a light-year away. One waits until 1,500 hours (in the external frame) and *then* makes the choice to predict either the time of return of the light front, or the colour of the returning light. In the first case, one leaves the photon box bolted to the external frame, opens it, and inspects the clock inside. One thereby determines what the gravitational lag of the clock has been and can correct for it, so as to know at what time in the external frame of reference the shutter had opened. In the second case instead, one unbolts the photon box from the external frame and re-weighs it.

Presumably, in the first case, opening the box and reading the time results in an uncontrollable exchange of energy with the photon box. And clearly in the second case, re-weighing the box allows one to determine the energy of the escaped photon, but disturbs once again the rate of the clock in the box. While at the end of the re-weighing (at 2,000 hours) it will also be possible to open the box and check the reading of the clock (and thus determine the *total* effect of gravitation on the clock) it will be impossible to reconstruct what external time corresponded to the intermediate 1,000-hours reading of the internal clock. Thus Ehrenfest not only clearly states that for Einstein these predictions are mutually exclusive, but also describes in considerable detail *how* they exclude one another.

Finally, notice in this version of the thought experiment that if one determines precisely the energy of the photon, it is the exact time *when the shutter first opens* that is uncertain – and thus the exact time of the arrival of the *wavefront* (one year later, since the distant mirror is half a light-year away). But now Ehrenfest highlights another paradoxical element: if the time window during which the shutter is open can be made quite small, only an extremely short wave train can escape, which is then 'very unmonochromatic' – that is, does not have a precise energy:

> I am worried about the following circumstance: even *if* one allowed the shutter to be open only FOR A VERY SHORT TIME, nevertheless one could determine through the double weighing the loss of weight very precisely, and thus also the energy of the emitted quantum; on the other hand it is clear that a very short light cork[17] is very unmonochromatic. I am deeply ashamed to be so stupid, but I cannot see what to think about this "contradiction". Of course it is only something stupid. Maybe Casimir, by whose hands I am letting this letter get to you, will sort out this question already.

This puzzling aspect was present already in the 1926 thought experiment: the opening of the shutter was presumed short, but Einstein concluded that the energy of the photon would yield a typically wavelike spread (which he interpreted statistically, however), and other versions of the photon-box experiment appear to lay emphasis on the duration of the opening.[18]

In any case, given the way Bohr described the thought experiment less than three months later, one might surmise that Ehrenfest's letter did not have the desired effect of clarifying Einstein's views to Bohr.[19]

Another source purporting to describe Einstein's version of the thought experiment is Schrödinger, in two pages of notes translated here as Chapter 8, and discussed more fully in Chapter 3. The description by Schrödinger closes with a verbatim attribution to Einstein (before Schrödinger goes on to express thoughts of his own):

> I have a box with automatic shutter, which tomorrow at midday shall briefly open a lid. I use the meantime to weigh the box very exactly.
>
> In the box there is very diluted monochromatic γ-radiation, the time during which the box is open is determined in *such* a way that it is probable that one quantum comes out. After the light quantum is gone, I can still choose whether I wish to transmit to observer B exactly the time of departure, so that he knows exactly the time of arrival, or whether I wish to weigh my box exactly and let him know the colour.
>
> Both together does not work, because through a *short* light signal to B, I change the weight uncontrollably. But already the possibility of an arbitrarily accurate communication of the colour is paradoxical, because the wave train *is* in fact short. I am forced to assume that for observer B the *temporal duration* of the lid opening is indeterminate.
>
> So Einstein said to me.

[17] [Ehrenfest refers to the emitted light quite consistently as a 'Lichtpropf' – a 'light stopper' – so to speak a cork popping out of Einstein's bottle (*eds.*).]

[18] See below for Casimir's own recollections of the photon box. He does not comment on this particular question, however.

[19] Indeed, one wonders if he even read it! For Ehrenfest recommended to Mrs Bohr: 'Do not pass on this letter at all to Niels if he is too tired', before sending his regards to 'all of you Bohrs and Bohrikins together'.

These notes are, admittedly, difficult to date: Einstein and Schrödinger had been colleagues in Berlin since October 1927, and Schrödinger's phrasing seems to refer elliptically to a conversation with Einstein. However, the version of the experiment described here by Schrödinger is slightly different than the one described by Ehrenfest: the emphasis is on the trade-off between the emitted energy and the *duration* of the opening of the shutter, rather than the time of the photon's emission. Exclusivity of measurements and delayed choice are again the salient features.

This is closer to a version of the experiment given by Einstein in a Berlin seminar in November 1931, a report of which was published in *Angewandte Chemie* early in 1932 (Einstein 1932; translated here as Chapter 7). The only significant difference from previous versions is that the report does not mention emission of a photon, but rather of a 'monochromatic light ray [...] of about 100 wave trains' (see p. 152). Details of the manipulations are not described in the brief report, but both the exclusivity of the two possible measurements and the option of delayed choice are again stressed: 'Only *one* measurement – of the time *or* the colour – can be carried out precisely, and in fact according to EINSTEIN one can still decide after the departure of the light ray which of the two predictions one wants to choose.' The report concludes by referring the reader to Tolman's extension of this argument in the ETP paper.

There are other fascinating materials from Schrödinger connected directly to the Berlin seminar: a set of notes on the discussion with further elaborations by Schrödinger (which will be discussed in forthcoming work by Jos Uffink and Christoph Lehner), and a letter from Schrödinger to Sommerfeld dated 11 December 1931,[20] containing an analogue of Einstein's photon box formulated in terms of the *position and momentum* of a photon and mirror, and where Schrödinger explicitly mentions – but then expresses reservations about – the idea that the unmanipulated photon possesses both a definite position and a definite momentum:

> [O]ur mirror is a 'universal instrument' for position *and* momentum of the light quantum in the following sense: if one 'applies it' to the light quantum in the above-described manner, then position *and* velocity of the light quantum leave a trace on the instrument, and in fact they are registered with a precision whose product can be pushed well below the limit h. Both traces really exist, because it depends on my free volition to make use of the one or of the other (whereby, however, the second gets smeared out). Since the light quantum no longer has anything to do with my manipulations at the mirror, one cannot well say that it 'receives' a sharp position or sharp momentum only through one of these operations, at any rate it is no 'receiving' in the real [sense], but at most in the mental sense. One might like to conclude therefore that the light quantum *possesses* at all times a fully definite position

[20] AHQP 84, section S-010 (handwritten, 5pp., in German), transcription in Meyenn (2011, vol. [illegible], pp. 488–492). Our thanks to Chris Fuchs for bringing this letter to our attention.

> *and* a fully definite momentum – a notion that we actually abandoned long ago as too harsh and paradoxical.

In Chapter 3 we investigate further Schrödinger's notes and his letter to Sommerfeld, and analyse what reservations (if any) Schrödinger might have had in fully embracing the conclusion of the incompleteness argument alluded to here.

There is yet one more piece of evidence on the photon box in 1931, coming from a talk Einstein gave in Leiden at about the same time as the Berlin seminar. In his 1982 biography of Einstein, Pais relates the following anecdote (Pais 1982, p. 449):

> My friend Casimir has written to me about a colloquium Einstein gave in Leiden, with Ehrenfest in the chair (this must have been in November 1930). In his talk, Einstein discussed several aspects of the clock-in-the-box experiments. In the subsequent discussion, it was mentioned that no conflict with quantum mechanics existed. Einstein reacted to this statement as follows: [...] I know, this business is free of contradictions, yet in my view it contains a certain unreasonableness.

There is a small historical discrepancy here: Einstein could not have been in Leiden either immediately preceding or following the 1930 Solvay conference that October. Prior to the Solvay meeting he spent time in southern Germany, and his official documents permitting entrance to Belgium for the Solvay conference are directly from Germany and do not mention the Netherlands. Einstein left the 1930 Solvay conference for the United Kingdom promptly at its conclusion, as he had an appointment on the evening of 30 October in London. Shortly thereafter Einstein set sail for the United States.

The date of the Leiden lecture is corrected in Casimir's recently reissued autobiography (Casimir 2010). There, an entire appendix is dedicated to 'The Discussions Between Bohr and Einstein on the Interpretation of Quantum Mechanics'. In it, Casimir distinguishes three phases of Einstein's complaints against quantum mechanics: the 'First Phase' corresponds to the 1927 Solvay conference and the 'Second Phase' to the conference of 1930. It is the 'Third Phase' that, by Casimir's lights, captures Einstein's stance from Solvay 1930 onwards to the publication of EPR and beyond (Casimir 2010, p. 316):

> Einstein conceded that the theory was free of contradictions, that it was consistent. He also admitted that it was extremely powerful and fertile. But he could not regard it as a complete description of nature and therefore he could not regard it as final. And here I can draw on my own personal experience. The scene is a colloquium at Leiden, almost certainly during the winter of 1931–32 when I was assistant to Ehrenfest. Ehrenfest, of course, was in the chair, but more silent than usual, and Einstein was giving a lecture. What exactly did he discuss? I think – but I have to admit that I am not completely certain – that he started with a box, containing a clockwork and a shutter and placed on a sensitive balance, as in his 1930 paradox. But this time he no longer tried to beat Heisenberg's principle. He stressed that

after the light quantum had escaped we can still choose whether we want to read the clock or alternatively determine the weight of the box. So, without touching the light quantum in any way, we can either determine its energy or the time at which it will return from a distant mirror. As I have said, I am not completely certain this was the experiment he discussed. I am certain that it was a case where we can still choose whether we want to know one quantity or the other (energy or time, place or momentum), although the particle itself is already out of reach; and that was also the gist of the later paper by Einstein, Rosen, and Podolsky. Ehrenfest had entrusted me with the task of opening the discussion and I explained to the best of my ability the Copenhagen views on such questions. Einstein listened, perhaps slightly impatiently, and then he said – and I believe now I can vouch for the exact words: 'Ich weiss es, widerspruchsfrei ist die Sache schon, aber sie enthält meines Erachtens doch eine gewisse Härte'. Pais, who also mentions this episode, translates this as: 'I know, this business is free of contradictions, yet in my view it contains a certain unreasonableness'. I think 'a certain unpalatability' comes closer to what Einstein had in mind.

We know that Einstein visited the University of Leiden some time during the last week of November in 1931 (leaving Berlin on 23 November and arriving in Antwerp in order to sail on 1 December). Thus, it was mostly likely in this time period that Einstein gave the talk remembered by Casimir.[21] The latter's report can thus be considered a good barometer of Einstein's thinking at the time, and it corroborates other sources mentioned here, especially regarding emphasis on free choice of measuring one or another quantity after interaction has (presumably) ceased.

In the spring following Einstein's Berlin colloquium, we hear from the man himself: in a missive dated 5 April 1932 from Rotterdam Harbour, Einstein confesses to Ehrenfest that the latter has provoked him to think more about the thought experiment. Relevant bits from this letter read as follows (EA 10-231):

Dear Ehrenfest!

You piqued me yesterday about altering the 'box experiment' so that it uses concepts that are less foreign to the wave theorist. I do this in what follows, in which I only use idealisations which I know you find harmless.

It employs a schematised Compton effect.

Let the mass m follow a straight line. A radiation pulse is sent from Q through the lens with 'precisely' known time of emission. It is collimated by the lens and reaches the straight line at right angles to it. Let the light pulse be diffracted by m, and let it likewise travel along the straight line (this is unclean but unimportant). Let the segment AB be long enough that we may assume that constraining the initial position of m on AB is compatible enough with the assumption that the mass is initially at rest.

One inquires about location, resp. momentum, of the mass m after the Compton effect has occurred, whereby experiments may be carried out with the diffracted light quantum.

[21] Thanks to Barbara Wolff (formerly at the Albert Einstein Archives) for investigating this point with one of us (EC).

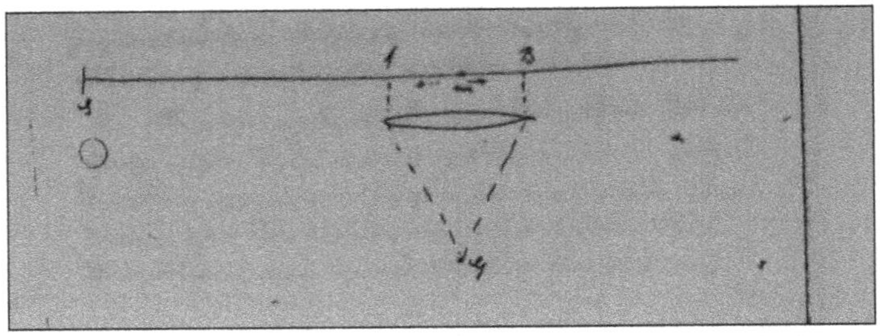

Fig. 1.3 Einstein's sketch of a wave-theoretic thought experiment.

Let the mirror S be somewhere far away, which sends the light quantum back toward the right. Let the experimenter sit somewhere near A. While the quantum is en route, he can still freely choose whether he wants to prophesy what the momentum [may be] or whether he wants to prophesy [what] the location of m may be (after the completed Compton process), and in fact without doing anything with the mass m.

Indeed, according to the law of momentum conservation, the momentum I of m is given by

$$I = \frac{h\nu}{c}$$

where ν is the frequency of the diffracted and reflected quantum. Thus if I measure ν I learn I.

But I can also determine the time of arrival of the quantum. Then I know exactly the time elapsed since the Compton action, thus also the location of m immediately after the Compton action. The greater the mass m is, the more exactly this location coincides with that which can be determined at A as time of arrival after the return of the quantum.

Thus it is possible to prophesy at will *either* the momentum *or* the location of m with arbitrary precision without an experiment on m.

This is the reason why I feel compelled to ascribe objective reality to *both*. It is not logically necessary, however, that I admit.

As has been remarked in the literature, this letter contains the first evidence of Einstein himself couching the thought experiment in terms of position and momentum, thus a setup (in line with Ehrenfest's urging) amenable to a description in terms of wavefunctions – even though Einstein does not provide then such an explication.[22]

[22] Jammer (1974, pp. 173–174) describes what appears to be an intermediate stage of development of the example taken from Einstein's recollections in his 1945 correspondence with Epstein: measuring the position of the box at the time of release of the photon allows one to predict the time of arrival (and hence the position) of the photon, while measuring the momentum recoil on the box allows one to predict the energy or momentum of the photon.

In his 1990 paper, Howard provides a reading of the photon-box experiment *qua* incompleteness argument, undergirding this reading with evidence from Ehrenfest's above-quoted letter to Bohr. Howard also remarks upon the striking similarity between Einstein's critique of quantum theory in 1931 and in 1935 (Howard 1990, p. 99):

> In fact, the logic of the 1931 version of the photon-box *Gedankenexperiment* is almost exactly that of Einstein's own 1935 incompleteness argument. The whole point of placing the reflector $\frac{1}{2}$ light-year away is to assure a spacelike separation between the inspection of the photon-box and the projectile. The argument works as a criticism of the quantum theory only if one assumes that the projectile, when thus separated from the box, is, in virtue of that separation and its therefore possessing its own independent reality, wholly unaffected by what we do to the box when we inspect it. But quantum mechanics makes a different assumption. [. . .] [Q]uantum mechanics says that the state we ascribe to the photon depends crucially on what we do to the box. What Einstein is thus arguing is that classical assumptions about the separability of previously interacting systems lead to different results than the quantum mechanical account of interactions.

We agree, with one small qualification. While the separation between the box and the photon is spacelike at the time we *perform* our freely chosen manipulation of the box, in Einstein's presentations of the experiment, by the time we later *verify* our prophecy, the photon has been invariably reflected by a (fixed) mirror. Thus the separation between the prediction and its verification is always timelike. This may seem a minor point, but we shall argue in Chapter 5 that it is of some significance in the context of Bohr's reply to EPR.

1.4.3 The ETP Paper

Einstein visited Caltech in the winter of 1930–31. The fruits of his labour there include a paper co-authored with Tolman and Podolsky, published in February 1931 (Einstein, Tolman and Podolsky 1931, reprinted herein as Chapter 6). Not quite two pages in length, ETP nevertheless manage to describe and assess a thought experiment closely resembling the photon box – yet with notable differences.

What most readily springs to the eye from a technical vantage point is that the ETP paper considers *two* particles being emitted in different directions from a two-holed box (and also that the particles are not necessarily photons, but merely 'particles in thermal agitation'). One might acknowledge here a superficial similarity with the EPR arrangement, but we wish to point out that there is even closer similarity with *Bohr's* version of the EPR example (see Chapter 5).

Indeed, in his reply Bohr considers two particles, each passing through a different slit in a screen. The initial total momentum of the particles and the screen is known, as is the distance between the slits. By measuring the momentum of the screen immediately after the passage of the particles, we know the sum of the particles' momenta and we know the difference in their positions. The ETP thought experiment, then, is just the energy–time analogue of Bohr's version of the EPR example. In the ETP experiment, the initial total energy of the particles plus the box is known (via a first weighing), and later the particles are released simultaneously. After the particles are emitted, we weigh the box a second time and so learn the sum of the particle energies; we also know the time difference of their escape (zero). The rest of the story is analogous to the single-photon box: measurements on one particle can be used to make predictions about the other particle (which travels very far in order to allow sufficient time to accurately re-weigh the box); these predictions can be verified when the second particle at last returns from being 'elastically reflected'.

Yet the thought experiment described by ETP is more than just another Einsteinian photon box (apart from details in setup). In the case of the latter, the original version is often taken as suggesting that simultaneous measurements on one system (the box) would enable simultaneous predictions of the other (the photon). But the uncertainty relations forbid this consequence, and careful analysis reveals that measurements on the macroscopic box must also be subject to the constraints of uncertainty. Hence the photon box illustrated the correctness of quantum mechanics at the macroscopic as well as microscopic scale, while at the same time suggesting it is incomplete by virtue of the possibility of a delayed, arbitrarily chosen measurement.

The ETP paper instead initially suggests a *different* loophole for obtaining simultaneous predictions on the distant particle. Then – as with Einstein's 1931 photon boxes – from grounds of uncertainty, ETP argue that no attempt to exploit this loophole can succeed. Thus in ETP quantum mechanics is *explicitly assumed* to be correct, whatever else may follow.

What is this new apparent loophole? Before delving into ETP's argument, it will be useful to distinguish between two aspects of a measurement: (1) determining the value of a quantity at a time immediately *before* the measurement; and (2) preparing a state at a time immediately *after* the measurement. The standard reading of the uncertainty relations – and the one ETP presuppose – is as a limitation on simultaneous state preparation: there are no measurements such that the prepared state allows one to exactly predict the results of further measurements of either of two conjugate quantities, *despite* the fact that scenarios like the photon box suggest that a system simultaneously possesses values for such quantities. However, what was generally supposed – an assumption ETP aimed to correct – was that retrodictions of values of conjugate quantities were possible, even though physically meaningless because devoid of predictive power. This is the line taken by Heisenberg in the Chicago lectures, for instance (Heisenberg 1930b, p. 20). The classic example is of two successive position measurements taken at definite times: the (average) momentum of a particle can be reconstructed for the past, while the second position measurement makes the momentum of the particle uncertain for the purpose of future measurements.

While Heisenberg took it as a 'matter of taste' whether one wanted to attribute physical reality to the past position and momentum of a particle, certainly Einstein would have had no qualms attributing reality to retrodicted values. To Bohr might be attributed yet a third stance: retrodiction is in fact what happens in *all* measurements, but it is noticeably limited by complementarity only in quantum mechanical cases. Indeed, in any measurement a system of interest interacts with a measuring apparatus. We then manipulate the apparatus in such a way as to be able to reconstruct *certain features* of the previous interaction, while other features are uncontrollably lost. This allows us to make certain further predictions on the system, while also explaining why we have cut ourselves off from the possibility of making certain other predictions.[23]

Though the further details of ETP's thought experiment are slightly involved, the reasoning is clear enough. Because the box gets re-weighed with great precision, the time of the particles's escape cannot be read off the clockwork mechanism internal to the box (this realisation came out of Bohr and Einstein's tussle over the photon box in 1930[24]). It must instead be reconstructed after the fact by conducting measurements on the first particle. ETP imagine that while the first particle is in flight, one can measure its momentum without disturbing it (this action also yields a measurement of the particle's energy at the time of escape from the box). Measuring the time of arrival of the first particle then also allows one to reconstruct the time of release of both particles, and therefore (given that the box gets re-weighed), enables the prediction of both the energy and time of arrival of the second particle.

Being able to predict both the energy and time of arrival of the second particle, however, is precluded by the (standard reading of the) uncertainty relations, which ETP assume to be correct. 'It is hence to be concluded', they write, 'that the principles of the quantum mechanics must involve an uncertainty in the description of past events which is analogous to the uncertainty in the prediction of future

[23] This take on Bohr is very close to Howard's (1994) analysis of Bohr's doctrine of classical concepts. A similar take was articulated in considerable detail by Grete Hermann (1935a, b), who emphasises the idea that the context of observation determines which features of the interaction are causally relevant for explaining a given measurement result, so that quantum mechanics is in fact causally complete *within* a measurement context (though not *across* all possible contexts). We shall return to the views of Hermann at some length in Chapter 4, and to the views of Bohr at considerable length in Chapter 5, where we also discuss Pauli's arguably related discussion of measurements in his 1933 handbook article (Pauli 1933, 1990, section A.9).

[24] ETP note that 'it is of special interest to emphasize the remarkable conclusion that the principles of quantum mechanics would actually impose limitations on the localization in time of a macroscopic phenomenon such as the opening and closing of a shutter' (ETP, p. 781; see p. 151). See also the end of Schrödinger's notes (Chapter 8).

events' (ETP, p. 781; see p. 151). It follows that a momentum measurement which does not disturb the momentum of the particle is impossible to carry out, and time-of-flight measurements enabling the reconstruction of the particle's trajectory are impossible. This last point is presumably a veiled criticism of Heisenberg's discussion of retrodiction. A similar (but not in the least veiled) negative judgement on Heisenberg's treatment of time-of-flight measurements is made by Hermann (1935a) in regards to her thesis about the relative context of observation. (Hermann is doing precisely the kind of limited retrodiction that ETP are envisaging and sanctioning here.)

In a way, then, the ETP argument (so to speak) sides with Bohr and Hermann against Heisenberg: it shows that if we were able to reconstruct the past of one particle in a way that violates the constraints of uncertainty, we would be able to make simultaneous predictions about the other particle that also violate uncertainty. This is the new loophole ETP find – and then close off – using their thought experiment. The paper does not *really* side with Bohr (nor with Hermann), in that ETP tacitly assume throughout (e.g. by talking about disturbing the momentum of the particle) that the limitations imposed by uncertainty – and also on the possibility of retrodicting the energy and time of release of the particles – are purely *epistemic* limitations. The paper is even titled 'Knowledge of Past and Future in Quantum Mechanics'. Had they instead argued for an ontic reading of uncertainty based on their thought experiment, they would have anticipated Bohr's reply to EPR!

1.4.4 Thinking Outside the (Photon) Box

In 1933 Rosenfeld gave a lecture in Brussels based on a joint paper with Bohr, and Einstein was in attendance. After the lecture, Einstein described to Rosenfeld a thought experiment involving no boxes, only two interacting particles – that is, an experiment nearly identical to that of EPR. Here is Rosenfeld's recollection of Einstein's words on that occasion (Rosenfeld 1967, pp. 127–128):

> What would you say of the following situation? Suppose two particles are set in motion towards each other with the same, very large, momentum, and that they interact with each other for a very short time when they pass at known positions. Consider now an observer who gets hold of one of the particles, far away from the region of interaction, and measures its momentum; then from the conditions of the experiment, he will obviously be able to deduce the momentum of the other particle. If, however, he chooses to measure the position of the first particle, he will be able to tell where the other particle is. This is a perfectly correct and straightforward deduction from the principles of quantum mechanics; but is it not very paradoxical? How can the final state of the second particle be influenced by a measurement performed on the first, after all physical interaction has ceased between them?

Jammer (1974) has suggested that Einstein's inspiration for removing the box from his quantum mechanical thought experiments was a 1934 paper by Karl Popper, 'Zur Kritik der Ungenauigkeitsrelationen' ['A Critique of the Uncertainty Relations'] (Popper 1934) which contained, according to Popper's own later assessment, a grievous mistake. Jammer writes, 'It is not impossible that it was precisely this "mistake" which prompted Einstein (who immediately recognized the error) to publish, together with Podolsky and Rosen, the argument against the completeness of quantum mechanics' (Jammer 1974, p. 174). We have just seen that Einstein started thinking in terms of examples that did not involve the photon box prior to 1934, but Popper's paper (which was published together with a critical reply by Weizsäcker) might well have contributed to motivate Einstein to publish his ideas on the subject.

Indeed, Popper's claim *sounds* like Einstein's: that quantum mechanics is correct but incomplete, in the sense that the uncertainty relations are empirically correct but need to be interpreted epistemically. But Popper is doubly mistaken: his argument to establish the incompleteness of quantum mechanics is incorrect (as spelled out by Weizsäcker), and even had the argument been correct, it would have established that the limits imposed by the uncertainty relations *can* be overcome. Einstein may well have been unhappy to let the published record of the question end on such a note.

Popper's thought experiment involves a ray of electrons and a ray of photons (with known momenta) crossing each other, giving rise to Compton scattering for individual electron–photon pairs. Popper then claims that one can measure the momentum a particle has on the journey to a detector together with its time of arrival. From this one can then reconstruct both the time and place of the collision, and the momentum of the other particle; thereby one can predict both the momentum and the time of arrival of that particle. This scenario is uncannily similar to that of the ETP paper, and so Einstein was very well aware that a measurement of the kind Popper proposed was impossible. Indeed, both ETP and Weizsäcker use the same example of a momentum measurement (employing the Doppler effect) to illustrate that such a procedure does not provide the desired information. And ETP clearly recognise that if such a procedure were to yield the desired information, this would constitute an empirical violation of the uncertainty relations; hence they conclude to its general impossibility.

Popper's second mistake was to think that because he was considering *rays* of photons and electrons, the information gathered through the measurement would tell us that there was *a* particle in the other ray with such and such properties, but it would not allow us to tell *which* particle that was, ergo it would not allow us to select a sub-ensemble of particles from the ray that was more informative than a quantum mechanical pure case. But nothing, at least in principle, prevents one from using rays of such low intensity that individual particles can be tracked.

Even though Popper was trying to establish that quantum mechanics was correct but incomplete, he failed to realise that for his argument he should have applied ETP's *modus tollens* instead.[25]

1.5 The EPR Paradox

We now turn to the EPR paper itself (Einstein, Podolsky and Rosen 1935; see Chapter 9), beginning with a summary of its contents and continuing with an analysis of its logic and an examination of Einstein's own stance – different from the published version.

When Einstein arrived at the Princeton Institute for Advanced Study in late 1933 he was quickly joined by Podolsky (with whom he had already collaborated on the ETP paper) and also by Rosen, who had been at MIT as a graduate student and, among other things, had published on two-body problems in quantum mechanics.[26] Specifically, in December 1931 Rosen had published a paper in the *Physical Review* (Rosen 1931), which gives a calculation using the variational method (as opposed to more common perturbation approach) of dissociation energy, moment of inertia and fundamental vibrational frequency of a two-body system, in particular the hydrogen molecule. In such a bound two-particle system, the distance between the two particles is approximately fixed, so an eigenstate of momentum of the hydrogen molecule could be thought of as an EPR state. Still, such a state does not appear explicitly in Rosen's paper.[27] It further appears that the idea of a collaborative paper came from Rosen (Rosen 1985; see also Pais 1982, p. 494).

1.5.1 Structure and Logic of the EPR Paper

EPR begin by providing distinct definitions for the correctness and completeness of a physical theory: correctness is meant in the empirical sense, while completeness is defined as the condition that '*every element of the physical reality must have a counterpart in the physical theory*'; they then formulate their (in)famous sufficient

[25] Popper gives a more detailed discussion of the thought experiment, embedded within a more extensive discussion of quantum mechanics, in *Logik der Forschung* (1935, chapter 7 and appendices V–VII), which he sent to Einstein along with a reprint of his paper. In appendix VI he gives an explicit argument for why such selective measurements cannot be used to prepare 'overpure' states, but the argument equivocates between the inefficiency of such a procedure and its impossibility. N.B. Grete Hermann (1935a) also criticises Popper's arguments, at the same time as she criticises Heisenberg's agnostic reading of 'time-of-flight' measurements. In the English version of *The Logic of Scientific Discovery* (1959), Popper retracted his original experiment, but reiterated his critique of quantum mechanics, invoking the EPR thought experiment instead. See Frappier (2017) for a detailed discussion of Popper and Hermann on these issues.

[26] Details of how the three authors came together at the Princeton Institute for Advanced Study are given in Jammer (1974, pp. 180–181).

[27] See Uffink (2018) and his expected new work with Lehner for some crucial new discoveries about a different possible origin of the EPR state.

criterion for establishing physical reality (EPR, p. 777; see p. 158, all emphases in the quotations are original):

> *If, without in any way disturbing a system, we can predict with certainty (i.e., with probability equal to unity) the value of a physical quantity, then there exists an element of physical reality corresponding to this physical quantity.*

This criterion also applies to quantum mechanics: if a particle is in an eigenstate of momentum (thus allowing for the prediction of its momentum with probability 1), the corresponding value of momentum is an element of reality. In quantum mechanics, however, one also interprets the fact that for this same state there is a uniform probability for the conjugate variable of position as implying that the position '*has no physical reality*' (p. 778/159). More generally, if one assumes that the wavefunction provides a complete description of physical reality, then quantities corresponding to non-commuting operators '*cannot have simultaneous reality*' (p. 778/159).

EPR next consider a state of the form

$$\Psi(x_1, x_2) = \sum_{n=1}^{\infty} \psi_n(x_1)u_n(x_2) = \sum_{s=1}^{\infty} \varphi_s(x_1)v_s(x_2) \tag{1}$$

(their equations (7)–(8)), where the $u_n(x_2)$ are eigenfunctions of a quantity A, and the $v_s(x_2)$ of a quantity B. According to the standard '*reduction of the wave packet*', a measurement of A on the second system leaves the first system in a state $\psi_k(x_1)$, a measurement of B in a state $\varphi_r(x_1)$. They write (p. 779/161):

> On the other hand, since at the time of measurement the two systems no longer interact, no real change can take place in the second system in consequence of anything that may be done to the first system. This is, of course, merely a statement of what is meant by the absence of an interaction between the two systems. Thus, *it is possible to assign two different wave functions* (in our example ψ_k and φ_r) *to the same reality* (the second system after the interaction with the first).

They then introduce the EPR state as a special case, and conclude that 'by measuring either A or B we are in a position to predict with certainty, and without in any way disturbing the second system, either the value of the quantity P (that is p_k) or the value of Q (that is q_r)' (p. 780/162). EPR apply their criterion of reality to this situation, concluding that p_k and q_r are elements of reality. Thus, they claim, from the assumption of completeness it also follows that 'two physical quantities, with non-commuting operators, can have simultaneous reality' (p. 780/162). Since the assumption that the wavefunction is a complete description of the quantum state leads to contradictory statements in this way, it must be false.

EPR concede that one might object along the following line: the criterion of reality ought to be applied to physical quantities '*only when they can be simultaneously measured or predicted*' (p. 780/162). But were this so, the reality of P and Q for one system would depend on the measurement process on the other system, which does not disturb the first. And of course this would be unacceptable. Nevertheless, EPR end on an optimistic note, expressing their belief that a theory providing a complete description of physical reality is possible.

Much has been said already on how to read the argument, and how Einstein himself was unsatisfied with the published version. Starting with the latter, as famously pointed out by Fine (1981, 1986) and Howard (1985), in his letter of 19 June 1935 to Schrödinger – which we here translate in full for the first time (see p. 297) – Einstein complained that the article had been redacted by Podolsky and that the argument had been 'buried in erudition'. This is evident regarding the point made in the preceding paragraph: while in the paper we saw that EPR use (7)–(8) to show that '*it is possible to assign two different wave functions* [...] *to the same reality*', according to Einstein this general case *already establishes incompleteness* without appeal to the specific EPR state, nor indeed to the controversial criterion of reality. Einstein writes (see p. 299; notation adapted):[28]

> What is essential now is only that $[\psi_k]$ and $[\varphi_r]$ are at all different from each other. I claim that this difference is irreconcilable with the hypothesis that the ψ-description corresponds bijectively to physical reality (to the real state).
>
> Indeed, after the collision the real state of [(I,II)] consists of the real state of [I] and the real state of [II], two states that have nothing to do with each other. *The real state of* [I] *now cannot depend on what measurement I perform on* [II]. ('Separation principle' from above.) Then, however, corresponding to the same state of [I] there are two (in fact arbitrarily many) equally legitimate [wavefunctions for I], which contradicts the hypothesis of a bijective or complete description of the real states.

An argument very similar to this is nicely expressed by Schrödinger in his cat paper (see p. 196; again, notation adapted):

> Suppose now that in this way [by carrying out a measurement plan for II] I derived a [I]-catalog in a particular case. Then I can reflect and consider whether I might perhaps have found a *different one* if I had put into action a *different* measuring plan for [II]. But since after all I neither have actually touched the system [I], nor in the imagined other case would have touched it, the statements of the other catalog, whatever it might be, must all *also* be correct. They must therefore be entirely contained within the first, since the first is maximal. But so is the second. So it must be identical with the first.
>
> Strangely enough, [...] examples can be set up where the requirement is necessarily violated.

[28] Note that in the letter, Einstein is *not* careful to point out that the different wavefunctions in fact depend on the result k or r one obtains on the other system, a point emphasised instead by Schrödinger (see Section 3.4.1). For detailed discussion of this issue, see Uffink (2020).

Neither Einstein nor Schrödinger mention elements of reality in these passages, yet both directly infer violation of completeness (or 'maximality'). Nor in fact does either passage make use of a particular observable pair. Indeed, the states of system I (while each of them will be the eigenstate of *some* observable) need not form an orthogonal family, and will in general form one only for a single measurement on system II (uniqueness of biorthogonal decompositions), as Einstein presumably appreciates (see p. 299):

Note: I don't care a *fiddle* [ist mir *wurst*] whether the [ψ_n] and [φ_r] can be considered eigenfunctions of observables [A], [B].

In Born's letter to Schrödinger of 28 June 1935 commenting on how disappointed he is about the EPR paper, he also focuses on the generic non-maximally entangled example rather than using the EPR state (see p. 300).

Let us now turn to the interpretation of the logic in the published paper.[29]

First of all, it is unclear where completeness actually enters into the demonstration that p_n and q_r are elements of reality. We shall presently discuss this in some detail.

Second, the repeated use of 'can' in the formulations has raised questions regarding the modal aspects of the argument – for example, whether in their premises EPR mean the *possibility* of making predictions and in their conclusions mean the *possibility* of simultaneous existence, or whether they mean to conclude from possible measurements to actual simultaneous reality, and so on.[30] We shall not discuss this at length, but it seems to us that the argument runs from the separate possibility of certain measurements to the simultaneous actuality of certain elements of reality.[31]

Third, it seems obvious that EPR are criticising an ontic reading of uncertainty (e.g. that position does not always have a physical reality).

Fourth, the EPR state is not strictly speaking a special case of their equations (7)–(8) because it is not a normalisable wavefunction, but it is well known that one can run the argument using Bohm's example of the spin singlet state (Bohm 1951, pp. 614–615). Neither does Einstein's preferred version of the argument rely on the EPR state, instead using the more general case of (7)–(8).

Returning to the role of completeness, it seems EPR do not carry out their argument as initially declared. They propose to argue from the assumption of completeness (from which it follows that non-commuting quantities do not have simultaneous reality) to a contradictory conclusion: that non-commuting quantities do have simultaneous reality after all. However, their derivation of the simultaneous

[29] On this subject, see in particular also Whitaker (2004) and Fine (2017).

[30] Cf. van Fraassen (1974) and Dickson (2002a).

[31] This seems to be the case in Schrödinger's reading of the argument in the cat paper, and in Section 5.3 we shall argue that this is also how Bohr understands EPR's argument.

reality of position and momentum is often read as calling upon the reality criterion and not the completeness condition. Through the former one establishes directly that position and momentum of the distant particle are elements of reality, and this already contradicts completeness. There is a clear consensus in the secondary literature on this point. (Even Einstein's preferred version of the argument just needs the separation principle.)

One author who begs to differ, however, is *Schrödinger*; he does so in his letter to Born of 29 June 1935 (see p. 304):

> Note the very cautious form of the conclusion of that paper: if one accepts the present interpretation with the addendum that the description (in the pure case) is complete, then one arrives at the following contradiction: 1) conjugates are not determinable simultaneously; 2) conjugates must be determinable simultaneously. Thus, what one has accepted is inconsistent. Statements 1) and 2) are deduced only to bring out this contradiction.

Incidentally, Schrödinger is making an extraordinary claim here – one not mentioned by EPR: that the assumption of completeness implies (absurdly, of course) that the position and momentum of the distant particle must be simultaneously determinable.[32] Be that as it may, he is clearly reading the overarching logic of the EPR argument as proceeding *from the assumption of completeness* to two contradictory conclusions.

But what role can completeness be playing in deriving the simultaneous existence (or possibly even determinability) of position and momentum? Einstein may be giving us a clue in his 19 June letter (which, allowing 10 days for the letter to reach Schrödinger in England, was received in time to conceivably influence Schrödinger's comments to Born).In it, Einstein appears to draw a conclusion from completeness that is not explicit in the EPR paper (see p. 298):

> One would now like to say the following: ψ corresponds bijectively to the real state of the real system. The statistical character of the measurement outcomes is to be imputed exclusively to the measurement apparatus, or the process of measurement. If this works, I talk of a complete description of reality through the theory. If, however, such an interpretation is not feasible, I call the theoretical description 'incomplete'.

The remark about the statistical character is presumably to be understood in the light of a parallel passage in the immediately preceding description of the two-box experiment: 'The statistics only come about because the observation introduces insufficiently known factors foreign to the described system' (see p. 297).

In the case of classical mechanics, a complete state description is already enough to provide full explanations for the results of all possible measurements (with no

[32] On the question of why Schrödinger might be making this further extraordinary claim, see our discussion in Section 3.4.1.

detailed analysis of the measurement process needed). But assuming that the wavefunction is a complete description of the state of a quantum system is clearly *insufficient* for explaining individual measurement results. Thus unless quantum theory can provide a successful explanation of individual measurement results via an explicit analysis of the *process of measurement*, the only way to explain measurement results will be to admit that the wavefunction is in fact an incomplete description of the state of the system.

This reading of completeness fits nicely with the ostensive argumentative strategy in the EPR paper: assume quantum mechanics is complete. Then, not only is the state description assumed complete, but also some appropriate analysis of the 'foreign factors' brought in through the measurement process is assumed, which will provide a full explanation for individual measurement results. Apply this to a measurement (say, of position) on one of the systems in the EPR example. In principle, knowledge of the 'foreign factors' brought in by the act of measuring one particle's position should allow us to predict with certainty the result of such a measurement. But due to the perfect correlations exhibited by the EPR state, we should also be able to 'predict with certainty (i.e. with probability equal to unity) the value of a physical quantity' of the second particle. And assuming a separation principle, we can make such predictions 'without in any way disturbing' the unmeasured particle. In other words: assuming quantum mechanics is complete, the argument suggested by Einstein's remarks allows one to establish the *antecedent* of the criterion of reality, and in this way completeness *does* feature as a premise in the published version of the argument.

Our conjecture is that during their many conversations Einstein, Podolsky and Rosen discussed this version of the argument and that Podolsky had this in mind when writing the initial paragraphs announcing the logical strategy to be used, but failed to bring it out clearly later in the body of the paper. Of course, as we mentioned above, once the separability principle is assumed, incompleteness of the state description can be shown without any appeal to the criterion of reality. On this view, the core of the incompleteness argument stands independently of the faintly baroque logic within which it lies buried in the published version.

Nonetheless, some good may be borne of this roundabout argument for incompleteness of the state description. For EPR establish along the way that granting the assumption of state-description completeness, a purely quantum mechanical analysis of the measurement process *cannot explain the results of individual measurements* – otherwise known as the measurement problem.[33]

[33] More precisely, in the literature such a result is known as an 'insolubility theorem': it is not possible to *solve* the measurement problem simply by postulating some appropriate unitary interaction of the system to be measured with an imperfectly known state of a macroscopic apparatus; see Brown (1986) and Bacciagaluppi (2013) for discussion and further references, and Bassi and Ghirardi (2000) for the simplest and most general

Let us now turn to a deeper examination of Einstein's arguments as given in his correspondence with Schrödinger in the latter half of 1935.

1.5.2 Einstein's 1935 Correspondence with Schrödinger

The correspondence between Einstein and Schrödinger germane to EPR comprises the following eight letters, translated into English in full for the first time in this volume:

7 June: Schrödinger to Einstein (p. 281)
17 June: Einstein to Schrödinger (p. 295)
19 June: Einstein to Schrödinger (p. 297)
13 July: Schrödinger to Einstein (p. 317)
8 Aug.: Einstein to Schrödinger (p. 327)
19 Aug.: Schrödinger to Einstein (p. 332)
4 Sept.: Einstein to Schrödinger (p. 335)
4 Oct.: Schrödinger to Einstein (p. 343).

We shall be discussing some further aspects of this correspondence in our chapter dedicated to Schrödinger (Chapter 3). For now we focus on Einstein.

One of the most weighty contributions to this correspondence is the aforementioned letter of 19 June. In it, Einstein articulates *his* central complaint against quantum mechanics as opposed to the argument 'buried in erudition' in EPR, and presents a characterisation of interpretations of the wavefunction in a manner strongly reminiscent of his comments at the 1927 Solvay conference. This portion of the letter, perhaps more frequently quoted than any other part of this correspondence, runs as follows (see p. 297):

> Before me stand two boxes with hinged lids into which I can look if they are opened; the latter is called 'making an observation'. In addition, there is a ball, which is always found in one box or the other when one makes an observation.
>
> Now I describe a state thus: *the probability that the ball is in the first box is* $\frac{1}{2}$. – Is this a complete description?
>
> *No*: A complete statement is: the ball *is* in the first box (or it is not). This then is how the characterisation of the state should look like in a complete description.
>
> *Yes*: Before I lift the lid of the box, the ball is not at all in *one* of the two boxes. This being in a definite box only comes about through my lifting of the lid. Only through this does

proof. The first version of the theorem was proved already by von Neumann in his book, because such a process would of necessity violate unitarity (von Neumann 1932, p. 233; 1955, pp. 437–439). Our preferred construal of EPR's logic can then be seen as an alternative proof of the insolubility theorem – which we might call the 'EPR Theorem'. A preliminary reading of Einstein along these lines is proposed in Bacciagaluppi (2013), and a related proof of the insolubility theorem is spelled out in Bacciagaluppi (2014).

the statistical character of the world of experience, or rather of its empirical regularities, come about. The state before lifting the lid is *completely* characterised by the number $\frac{1}{2}$, whose meaning, however, when carrying out observations manifests itself only as a statistical result. The statistics only come about because the observation introduces insufficiently known factors foreign to the described system.

According to Einstein, the 'Schrödingerian' view that the wavefunction is a complete description commits one, at least in principle, to providing an explanation of definite experimental results in terms of external factors brought in by the measurement interaction. (It is perhaps because no such detailed explanation is provided that Einstein calls this view 'spiritualist' – or maybe because the only way this view can work is by invoking 'spooky' action at a distance!) Under a 'Bornian' interpretation a more complete description is available in principle, and a measurement simply reveals the elements of this more complete description, rendering unnecessary any appeal to external factors introduced through measurement.

A wee note is needed on Einstein's use of the term 'Bornian' here. Is Born not one of the champions of the idea that the quantum mechanical description is complete? Is Einstein equivocating on Born's use of the term 'statistical' in characterising his own approach to quantum mechanics?[34] As a matter of fact, it appears that in the papers in which Born first introduced and discussed the 'statistical interpretation' (Born 1926a, b, c), Born was applying it *exclusively* to the statistics of stationary states – consistent with the assumption held at that time that a system is *always* in a stationary state and performs quantum jumps between them. Thus Born's interpretation of quantum mechanics is 'statistical' in the Einsteinian sense that the wavefunction does not describe the state of the system completely, but only the statistical distribution of stationary states. And this despite Born's own pronouncements about the completeness of the theory.[35]

Returning now to the letter of 19 June: Einstein remarks that the 'Talmudist' (Bohr) refuses to take seriously talk about reality and conflates the Bornian and Schrödingerian interpretations. But insisting on the separability principle exposes this conflation. Why? Because if one insists on the separability principle, one realises that a mechanism explaining why one gets a certain result when one

[34] As early as November 1928, in a speech delivered to the Gesellschaft der Wissenschaften zu Göttingen [Scientific Society of Göttingen] Born states that it can be shown in a mathematically exact way that the established formalism of quantum mechanics allows for no completion. Thus if one wants to retain the hope that determinism will someday be recovered, one must consider the present theory '*contentually false*' (Born 1929, p. 118). This is an echo of the views that von Neumann expressed in the paper wherein he first proved what became known as his 'no-go' theorem against hidden variables – a paper that was presented by Born to the very same society almost exactly one year earlier (von Neumann 1927). On von Neumann's paper, see also Bub (2010, 2011), Crull and Bacciagaluppi (2017b) and Bacciagaluppi (2022).

[35] For a more detailed discussion of Born's statistical interpretation, see Bacciagaluppi (2008), Bacciagaluppi and Valentini (2009, chapter 3), Bacciagaluppi, Crull and Maroney (2017), and Bacciagaluppi (2022).

performs a measurement cannot explain the result (whether positive or negative[36]) obtained *at a distance*. Thus the Schrödingerian view and the Bornian view are not only very different, but the former is *ruled out* because it fails in general to explain nonlocal results. The only view that remains is the Bornian one. Einstein then proceeds to give his preferred version of the quantum mechanical thought experiment, as already discussed above.

Also revealing is Einstein's letter of 8 August, in which he presents an entirely different argument against the 'Schrödingerian' view (see p. 327):

> Let the system be a substance in a chemically unstable equilibrium – say, a pile of gunpowder that can ignite itself by internal forces, where the average lifetime is of the order of one year. In principle this is very easy to represent quantum mechanically. Initially the ψ-function characterises a sufficiently definite macroscopic state. Your equation, however, ensures that this is no longer the case after a year has passed. The ψ-function describes then rather a kind of mixture of a not-yet and an already-exploded system. Through no interpretational artistry can this ψ-function be made into an adequate description of a real state of affairs; in truth there just is no middle thing between exploded and unexploded. Thus your equation certainly cannot provide the description of the actual process as you in fact imagine. (On the other hand, in the statistical sense it can certainly reproduce correctly the changes in a collection of systems.) With this example I want to suggest that your attempt at an interpretation fails with respect to what we know with certainty from coarse macroscopic experience.

Schrödinger quickly replies with his famous cat example, and in his 9 September letter Einstein concedes that in this matter he and Schrödinger are very much in agreement. What is worth emphasising is the difference between this argument and the EPR argument, and even Einstein's preferred version of EPR: the gunpowder thought experiment dispatches the 'Schrödingerian' view directly, without any need for an additional separation principle. This makes sense if one recalls that Einstein invoked the separation principle in the 19 June letter specifically in order to beat the 'Talmudist'– who might not otherwise care – and in order to argue for a distinct 'Bornian' view.

We know from various sources that Einstein remained convinced of the incompleteness of quantum mechanics until his death. Among these sources are a paper written less than a year following EPR, entitled 'Physik und Realität' ['Physics and Reality'] (Einstein 1936), which contains a clear albeit informal statement of Einstein's preferred version of the incompleteness argument.[37] Another source is a

[36] For a striking (and possibly the earliest) discussion of 'negative result' or 'null result' experiments, see Schrödinger (1934) and our related remarks in Chapter 3.

[37] For a translation and brief discussion of the relevant portion of this article, see Howard (1985, pp. 184–185). Note also the separate argument in Einstein's footnote 3: even when described by a superposition of energy eigenstates, a quantum state must have a definite energy, because of $E = mc^2$ and because the *inertia* of a system must have a well-defined value. Applied to the photon box, this argument of course again shows that Einstein thought of the spread in energy as purely statistical.

paper written as a contribution to the *Festschrift* occasioned by Born's retirement from the University of Edinburgh (Einstein 1953). The latter makes an explicit plea for the 'Bornian' view of quantum mechanics using a gunpowder-like argument, this time with a little macroscopic ball bouncing between two walls in a box: quantum mechanics describes this by a standing wave and predicts correctly the statistics of position and momentum, but a macroscopic ball surely has a definite position and momentum at all times.[38]

Before going on to discuss Schrödinger's point of view in Chapter 3, we shall give a brief survey of reactions to the EPR paper in the coming chapter, including the 'minor' responses to EPR. This will also provide a wider context in which to situate our detailed discussion in subsequent chapters.

[38] For discussions of Einstein's 1953 paper (where Einstein uses his example also to criticise Bohm's (1952) new theory), see Myrvold (2003) and Bacciagaluppi (2016b).

2

Others on EPR

2.1 Introduction

In his letter to Einstein of 13 July 1935, Schrödinger writes disparagingly of the responses to EPR he has seen up to that point (see p. 317):

> What I have seen so far in printed reactions is not very clever. I understand E. C. Kemble least, in *Phys. Rev.* 15th June – indeed, he does not even mention the *one* case that gives us headaches. It is as if one says: in Chicago it is bitterly cold, and another answers: that is a fallacy, it is very hot in Florida. – Then in Nature on 22nd June there was an equally rather vacuous report, I think by Flint (it was only signed with initials). I foamed with rage about N. Bohr's letter to *Nature* of 13th July. He *just* makes one curious, reveals not a word of what he thinks, and refers to a forthcoming article in *Phys. Rev.*

The fact that numerous physicists attempted a response to EPR, in tandem with the fact that each seems to have misunderstood EPR in their own way, is revealing about the state of interpretations of quantum mechanics in this time period. We shall not provide comprehensive analyses of each response here, but we do wish to provide a bit of context for each author and response, as they can be interestingly compared to the responses serving as the focal point of this book.[1]

One way in which the minor respondents might be interestingly contrasted with the group of physicists corresponding with Schrödinger about EPR has to do with the criterion of reality. Whereas physicists in the latter group make scant mention of EPR's criterion, it serves as the focal point for the former. Those with whom Schrödinger corresponded about EPR are those in closer contact with Einstein, and central players in these debates. Perhaps it is unsurprising that from this position in the thick of the debate, these physicists were better prepared to appreciate the true

[1] Of course many know that in the 1940s Einstein revisited questions raised by EPR in his correspondence with Paul Epstein. Insights from these letters into Einstein's interpretation of quantum mechanics have already been distilled by Jammer (1974, pp. 231–235) and Howard (1990). Since we have nothing to add, and since this correspondence takes place in a later time period, we shall not discuss the Epstein correspondence further.

thrust of Einstein's concerns as somewhat independent from the criterion of reality as Podolsky articulated it.

We deal in somewhat more detail with a couple of these minor replies because, while not managing to directly address the concerns of the EPR paper, they do introduce some interesting ideas. Kemble's is a case in point, since he appears to subscribe to an 'ensemble interpretation' of quantum mechanics, which in turn is briefly criticised in Margenau's own reply to EPR. Margenau is another author we discuss somewhat more thoroughly, because he provides a discussion and criticism of the collapse postulate (in the mistaken belief that it plays a crucial role in EPR's argument). Finally, Furry's two articles are major contributions of their own to the foundations of quantum mechanics. Again, the focus is mainly on the collapse postulate, which is somehow assimilated to EPR's position, but Furry's papers provide a detailed argument for why the collapse postulate should be considered only a device apt for all practical purposes, and not a fundamental postulate (Furry's first paper supports this by giving an extensive mathematical discussion of entangled systems and bipartite measurements).

In addition to our goal of providing robust context for participants in Schrödinger's 1935 correspondence, a complete, annotated list of EPR responses in 1935 is nowhere else compiled. Although Jammer discusses earlier and later reactions to EPR (1974, sections 6.4–6.7), we supplement his list and update his analysis. We present the responses chronologically, according to date of composition.

2.2 *The New York Times*

'Einstein Attacks Quantum Theory', 'Raises Point of Doubt' and 'Conservative Einstein' – three articles in the issue of 4 May 1935, pp. 11, 11, and 12, respectively. And: 'Statement by Einstein' – issue of 7 May 1935, p. 21.

The first of these is the famous article in which the news of the EPR paper was first publicly broken. The article itself is somewhat wooden, and mistakenly reads the main target of EPR's criticism to be the uncertainty relations. The second half, however, gives a long quotation from Podolsky himself, describing the requirements of correctness and completeness, emphasising how successful quantum mechanics has been (so that its correctness is not in doubt), and finally giving the following explanation of how EPR reached their conclusion that quantum mechanics is incomplete:

> In quantum mechanics the condition of any physical system, such as an electron, an atom, &c., is supposed to be completely described by a formula known as a 'wave function'. Suppose that we know the wave function for each of two physical systems, and that these two systems come together, interact, and again separate (as when two particles collide and

move apart). Quantum mechanics, although giving us considerable information about such a process, does not enable us to calculate the wave function of each physical system after the separation. This fact is made use of in showing that the wave function does not give a complete description of physical reality. Since, however, description of physical systems by wave functions is an essential step of quantum mechanics, this means that quantum mechanics is not a complete theory.

Even though this is a popular exposition of the reasoning in the paper, it is curious that the way Podolsky presents it here is neither the version using the criterion of reality, nor Einstein's preferred version (in which quantum mechanics appears to permit the assignment of more than one state to a physical system). Rather, Podolsky seems to be assuming that after an interaction ceases, two systems each ought to be described separately by a pure state – which of course is generally false in quantum mechanics. As we shall see below, this reading may be what both Margenau and Furry adopt in their own published responses to EPR.

The second, very short, article reports the opinion of Edward U. Condon (of Princeton), who had been asked to comment. Condon may (or may not?) have had only the report by Podolsky quoted by the *Times* to comment on, but he states that much 'hinges on just what meaning is to be attached to the word "reality" in connection with physics', noting that EPR 'have certainly discussed an interesting point in connection with the theory'. He remarks that Einstein had never been satisfied with the 'statistical causality' of the theory (even quoting 'Der liebe Gott wuerfelt nicht [the good Lord does not throw dice]') and that in recent years he had (unsuccessfully) 'subjected the quantum mechanical theories to very searching criticism from this standpoint'.

Jammer (1974, pp. 189–190) reports on both of these articles but does not discuss the opinion piece appearing on the next page of the *Times*, which expresses a *rather* popular assessment of quantum theory ('Natural laws? There are none. We make them and then attribute them to nature. [...] Experiments mean nothing.'). More interestingly for our purposes, this anonymous piece presents us with a description (the first in the general press?) of Einstein as 'out of touch' with modern physics:

> EINSTEIN expresses his dissatisfaction with the quantum theory, for the development of which he is in part responsible. [...] EINSTEIN is not alone in quarreling with this point of view. Even MAX PLANCK, inventor of the original quantum theory, sighs for cause and effect. To the younger school the two are classicists for all their once revolutionary views. Both veterans experience an intellectual need for causality. Whereupon they enter the mire of metaphysics. Which shall it be – determinism and a machine universe, or free will? The old theological controversy still rages.

To these two articles (and perhaps also the third!), Einstein reacted with a curt statement included in a précis on 7 May. By now well-known and already quoted in Jammer (1974, p. 190), Einstein's statement reads:

Any information upon which the article 'Einstein Attacks Quantum Theory' in your issue of May 6 [*sic*] is based was given to you without my authority. It is my invariable practice to discuss scientific matters only in the appropriate forum and I deprecate advance publication of any announcement in regard to such matters in the secular press.

2.3 Kemble

'The Correlation of Wave Functions with the States of Physical Systems' – written 25 May 1935 and published as a Letter to the Editor of Physical Review *on 15 June 1935. Reprinted herein as Chapter 17. Also 'Physical Reality and Quantum Mechanics' – written between mid-October and 9 November 1935 but unpublished at the time. Published here as Chapter 23.*

As Schrödinger points out acerbically with his 'Chicago and Florida' remark, Kemble's reply seems to be talking past EPR by taking completeness to mean something like the most complete knowledge one can have within the bounds circumscribed by the uncertainty relations. This interpretation leads Kemble to conclude somewhat trivially that '[t]here seems no reason to doubt the completeness of the quantum-mechanical description of atomic systems within the frame of our present experimental knowledge' (p. 974; see p. 259). Kemble complains that whereas EPR assume that sans disturbance 'system β' must remain in the same 'state', we can make different predictions about its subsequent behaviour depending on the measurement we perform on 'system α': '[t]hus each possible wave function for system β gives a correct description of the corresponding state of that system' (p. 974/258). This of course is just EPR's *correctness* condition, which is not in dispute.

The most interesting aspect of the reply (which Kemble explicitly mentions is crucial to add to the above remarks) is that he subscribes to an 'ensemble interpretation' of quantum mechanics (or as he puts it, 'the interpretation of quantum mechanics as a statistical mechanics of assemblages of like systems'), a view he attributes to Slater (1929). According to Kemble, what one means by the 'state' of a system according to such a view is that 'we cannot know more about an individual electron than the fact that it belongs to a suitably chosen potential assemblage' (p. 974/258). What is more, Kemble appears to associate coherent superpositions with mixtures of sub-ensembles ('Every analysis of a wave function into a linear combination of orthogonal functions can be interpreted as a resolution of the complete assemblage into a collection of sub-assemblages' (p. 974/258)), and to interpret wave function collapse as a purely epistemic process ('The "reduction of the wave packet" is the process of substituting the wave function of the sub-assemblage for that of the original complete assemblage which must accompany the transfer of one's attention from an arbitrary member or members of the original assemblage to systems known to belong to said sub-assemblage' (p. 974/258)). It is unclear to us how much of this view is in fact attributable to Slater.[2]

In any case, even if there *were* some feature that distinguished among systems belonging to different sub-ensembles of a given ensemble, although it is not included in the quantum mechanical description of the original ensemble – its existence already concedes EPR's point about incompleteness. Kemble eventually sees this, confessing as much in a letter to Einstein on 15 October 1935. The letter contains a reprint of Kemble's *Physical Review* note along with the following apology (AHQP 53, section 7-034):

> May I take this occasion to add that after some discussion with Ruark and my colleagues here [at Harvard] I have reluctantly come to the realization that my own letter was based on a misconstruction of your article on the 'Quantum-mechanical Description of Physical Reality'. My own philosophic bias is quite different from yours and in the attempt to interpret your letter from my own angle I seem unconsciously to have twisted the plain meaning of your words. To be more specific I assumed that your definition of physical reality was intended to be equivalent to the following.
>
> 'If we know the value of a physical quantity so that we can without disturbing the system predict with certainty the result of a measurement of that quantity, the value in question is physically real. The values of two physical quantities have simultaneous physical reality if we know both of them so that we can predict with certainty the result of a measurement of either one'.
>
> As a reexamination of your paper convinces me that you meant to define two physical quantities as simultaneously real even when the values of both cannot be simultaneously predicted, my argument is beside the point. Personally I prefer the definition given in the quotation marks above, which I believe to be in harmony with quantum mechanics. However, as Ruark would say, 'That is a matter of taste'.

His own position is perhaps made clearer in a letter to Margenau of 9 November 1935, shortly after Bohr's reply had appeared in print (AHQP 54, section 3-017):

> It seems to me that Bohr's paper properly understood is a complete and satisfactory answer to their [EPR's] discussion. I do not see that the resulting physical situation is awkward unless one adopts a realistic point of view which I have, for one, cast over-board some years back. To put it briefly, we say that a dynamical variable is defined by the method for measuring it. [. . .] The method of making the measurement and the result of the measurement, i.e. the eigenvalue, are the only necessary elements in the situation. It would be better to say that we make measurements Type A, Type B, etc., and never talk about measuring things as if dynamical variables were sitting around waiting to be measured.

[2] Slater's (1929) is a short paper whose stated aim is to present the physical meaning of wave mechanics. Slater writes that wave mechanics is not a theory like Newtonian mechanics but rather like classical statistical mechanics, in that it deals with ensembles. What ensembles represent in wave mechanics are repetitions of measurements, while wavefunctions represent the distributions of those measurement results in terms of the squares of the coefficients in an appropriate basis.

In the same letter, Margenau also included a typescript of a retraction he had 'seriously considered sending to the *Physical Review*', adding: 'Some of my colleagues' (possibly Ruark and Furry) 'consider this hardly necessary, but I am still not quite sure that I ought not to send it in.' In the end, Ruark did not send in his retraction, and we publish it here for the first time as Chapter 23.

2.4 Podolsky

'States and Reality of Physical Systems' – dated 6 July 1935, reproduced herein as Chapter 18.

Jammer (1974, pp. 192–194) includes a transcription of a riposte to Kemble composed by Podolsky. Although Podolsky's response remained unpublished at the time, Jammer showed it to Kemble some time later, causing Kemble to lament the fact that it was not published, as he felt it put into nice contrast the very point upon which they disagreed. In this note, Podolsky writes that he had not been in a position to consult with Einstein and Rosen on the content (he had in the meantime moved to the University of Cincinnati).

2.5 Flint

'Quantum Mechanics as a Physical Theory' – date of writing unknown, but Schrödinger received page proofs for it on 14 June 1935 (it was published in Nature *on 22 June). Reprinted herein as Chapter 19.*

The rather short note by Flint begins with a fairly accurate outline of the argument in the EPR paper, and Flint seems to accept the argument as valid. His only doubt is as to the validity of the criterion of reality, even though he does not provide any criticism of it. He goes so far as to agree with the spirit of EPR in writing that instrumentalism should not be taken for granted, and that a more direct description of reality than the one currently provided by quantum mechanics might be desirable. Overall, however, the note is inconclusive as an assessment of the EPR paper itself.

2.6 Schrödinger

'The Completeness of Quantum Mechanical Description' and 'The Observation of Canonically Conjugates' – respectively, 12 June 1935 and undated; the latter reproduced here as Chapter 20. Also: 'Discussion of Probability Relations between Separated Systems' and 'Die gegenwärtige Situation in der Quantenmechanik'

[*'The Present Situation in Quantum Mechanics'*] *– both completed mid-August 1935, and published in November/December in the* Proceedings of the Cambridge Philosophical Society *and* Naturwissenschaften, *respectively; reprinted herein as Chapters 10 and 11.*

The first of these was a short note Schrödinger had outlined and suggested as a contribution to *Nature*, but which does not seem to exist in completed form (see the extract from the relevant letter in Chapter 3). The second is an unpublished note Schrödinger had included in the folder containing his correspondence about EPR. It references Flint's reply, and is related to Schrödinger's letter to Einstein on 13 July 1935. We briefly discuss it in Chapter 3. The last two are well-known major papers by Schrödinger: they are extensively discussed in Chapter 3 and reprinted as Chapters 10 and 11 (the latter in the standard translation by John Trimmer).

2.7 Bohr

'Quantum Mechanics and Physical Reality' – a Letter to the Editor of Nature *dated 29 June and published 13 July. And: 'Can Quantum-Mechanical Description of Physical Reality be Considered Complete?' – published in* Physical Review *on 15 October 1935. These are reprinted herein as Chapters 21 and 16, respectively.*

The *Physical Review* article is Bohr's full-length response to EPR, which we discuss in detail in Chapter 5. The manuscript was circulated in early July 1935, and the paper was received 13 July. At the same time, Bohr published the Letter to the Editor of *Nature*. A mere three paragraphs in length, Bohr's letter is a concise and accurate statement (despite Schrödinger finding it infuriating) of the strategy Bohr uses in his main reply in the *Physical Review*. Bohr had made further corrections to the latter by mid-September, and it appeared a month later.

2.8 Ruark

'Is the Quantum-Mechanical Description of Physical Reality Complete?' – written 2 July 1935 and published in Physical Review *on 1 September 1935, this response to EPR is reprinted herein as Chapter 22.*

In essence, Ruark's reply merely points out that the disagreement between EPR and many other physicists lies in the adoption of different criteria of reality. According to Ruark, the generally accepted criterion is that 'a physical property of a given system has reality only when it is actually measured' (Ruark 1935, p. 466/268). A proponent of such a view, he writes, will consider an indirect measurement like the one sketched by EPR equivalent to a direct one, and in particular will 'feel that the ingenious method of measurement discussed by Einstein, Podolsky and Rosen suffers from all the essential difficulties common to measurements which result in disturbing system II' (p. 466/269). No justification is given for preferring this purportedly majority view. In fact, Ruark ends by claiming that the arguments for both positions are inconclusive, thus it all boils down to 'a matter of personal choice or definition' (p. 467/269).

We know from Kemble's correspondence that Ruark (then at UNC Chapel Hill) had conversed with him about EPR.

2.9 Heisenberg

'Ist eine deterministische Ergänzung der Quantenmechanik möglich?' ['Is a Deterministic Completion of Quantum Mechanics Possible?'] – July–August 1935. Translated herein as Chapter 14.

This manuscript was composed throughout the summer of 1935 and then sent to Bohr for comments. Unpublished at the time, it only appeared 50 years later – in the second volume of Pauli's correspondence, in German (Pauli 1985, pp. 409–418). The first English publication of Heisenberg's response to EPR is provided here as Chapter 14. Discussion of this response can be found in Chapter 4.

2.10 Hermann

'Die naturphilosophischen Grundlagen der Quantenmechanik (Auszug)' ['The Natural-philosophical Foundations of Quantum Mechanics (Excerpt)'] – abridged version prepared for the 18 October issue of Naturwissenschaften *of an essay that had appeared in March 1935. This version is translated here as Chapter 15.*

Published as it was in March of 1935, the original, full-length essay by Hermann (Hermann 1935a) was obviously not a reply to EPR. However, in Heisenberg's own unpublished response to EPR he cites this earlier version of Hermann's essay as a presentation of the same position he takes. It is tempting to speculate that Heisenberg encouraged Hermann to publish the shorter version of her work partly as a contribution to the EPR debate. We discuss Hermann's views in Chapter 4 together with Heisenberg's, and translate the short version of Hermann's essay as Chapter 15.[3]

[3] We have translated the full version as chapter 15 of Crull and Bacciagaluppi (2017a).

2.11 Margenau

'Quantum-mechanical Description' – received 16 November 1935 and published on 1 February 1936 in Physical Review*; reprinted herein as Chapter 24.*

Margenau's is one of the most original responses, and additionally interesting because he corresponded with Einstein about it prior to its publication. Although Jammer (1974, pp. 226–229) provides a summary of the germane Einstein–Margenau correspondence, we revisit these letters briefly in order to highlight a few points of relevance.

A letter was sent on 13 November 1935 to Einstein, then at Princeton, from Henry Margenau, then at Yale: Margenau had written a short response to EPR which he hoped to publish in *Physical Review* – and did – and on which he desired Einstein's comments. Margenau writes: 'It is my impression that the replies which so far have been made to your paper are not dealing at all with the fundamental issue involved, but attempt to eliminate the difficulty by superficial means' (EA 17-65).

Margenau's strategy in the response to EPR is to argue in the first place against the standard collapse postulate of quantum mechanics, suggesting that the EPR argument only applies to quantum mechanics with this postulate added. Enclosed in Margenau's letter to Einstein is a typescript of the article, which differs from the published version in a few interesting respects.

For one, Margenau's typescript contains a sentence near the beginning indicating that while Bohr's arguments in response to EPR on the basis of complementarity were interesting, they nevertheless 'hardly relieved' the philosophical consequences of quantum theory as stated by EPR. This criticism of Bohr is omitted in the final version, and Bohr's reply (about which Kemble had also written to Margenau on 9 November) is mentioned only in a footnote, alongside citations of the original EPR paper and the replies of Kemble and Ruark.

The main difference between the typescript and the published version concerns Margenau's rejection of two claims he attributes (both in the typescript and in the published article) to Kemble: that the quantum state should be interpreted statistically, and that the collapse of the wavefunction is a process that selects a sub-ensemble from the whole.

In the typescript, after claiming that state preparations can be dispensed with and referring to a longer paper in the works, Margenau has stricken this additional comment:

> At this particular point two possibilities of interpretation arise: (1) quantum mechanical description, though explicitly referring to a single system, tacitly involves reference to a large class of similar systems. This is apparently the view of Kemble. Its rigorous persecution reduces quantum mechanics to a statistical discipline, operating with a notion similar

to Gibbsian ensembles; (2) quantum mechanical description refers to individual systems, the statistical feature being introduced into the theory by an essentially a-causal character of the results of observations (*not* of states!). It is difficult to decide at present between these two possibilities.

We know that Margenau and Kemble corresponded on these points,[4] but Margenau's remarks might also be read in connection with Einstein's own suggestion that quantum states should be interpreted statistically, clearly expressed in his response to Margenau on 3 January (EA 17-66):

I want to be clear here that our [EPR] remarks are not meant to be construed as an actual argument against quantum theory as such. We only wanted to show that the description using a ψ-function cannot be considered a complete physical description of an individual case.

It seems to me the whole difficulty disappears with the view that the ψ-function does not describe a single system but rather a statistical ensemble. Making a measurement then brings with it a passage to a new ensemble. But this is just a more detailed way to admit that the description by the ψ-function is an incomplete description of a single matter of fact.

All of this is not to say that a more complete description would allow the formulation of definite laws. Most physicists today are convinced this is not the case. But on this question at present there is no knowledge, only belief.

Perhaps as a result of these comments, the published version of Margenau's essay contains early on a brief but explicit argument against a statistical ensemble reading of the projection postulate. The publication makes two new points: (1) if ψ were an ensemble with a greater number of (identical) systems than the sub-ensemble picked out by measurement, then one would have to attribute configuration spaces to the two ensembles which differ in dimensionality, and thus 'it would be difficult to give mathematical meaning' to the equation describing the projection postulate; (2) Margenau argues that '[i]t is certainly proper to make a measurement on a single system', and this would render meaningless the selection of a sub-ensemble.[5] 'We thus conclude', he writes, 'that [the postulate], if it has significance at all, must imply the actual transformation of the state function for a single system into an eigenstate function for the same single system' (Margenau 1936, p. 241/273).

This short discussion in the published paper seems intended to clear away any misunderstandings regarding the collapse postulate, which then serves as the main object of Margenau's criticism.

4 AHQP 54, sections 3-016 to 3-020: Margenau to Kemble, 6 November (handwritten, 1p.), Kemble to Margenau, 9 November 1935 (typed, 6pp.), Margenau to Kemble, 13 November (typed, 4pp.), Kemble to Margenau, 20 November (typed, 4pp.), and Margenau to Kemble, 27 November 1935 (typed, 1p.), all letters in English.

5 In the longer paper, he even discusses electron diffraction with a single electron at a time (Margenau 1937, p. 354).

With regard to the collapse postulate, Margenau remarks in passing that it is time-asymmetric, but his main bone of contention is that it is inconsistent with the Schrödinger equation. He thus proposes to drop the collapse postulate. Indeed, Margenau suggests that a quantum state is essentially a probability distribution; as such, it can only be determined by a large number of measurements, and not a single one. Margenau is aware that measurements in quantum mechanics are commonly thought of as preparing a state, and that dropping the projection postulate 'renders impossible the *preparation* of states in the manner advocated by Dirac and others' (p. 242/274, emphasis original). However, using measurements to ascertain the quantum state of a system is still possible (even though Margenau mentions 'difficulties'). Instead, he claims that the preparatory aspect of measurements is dispensable, referring for further discussion to the planned longer paper in *J. Phil. Sci.* [*sic*], that is, *Philosophy of Science* (Margenau 1937). The final section of Margenau's response claims that in the EPR case one therefore needs to consider not single measurements but a 'great number' of measurements. These would lead to the determination of all the coefficients in EPR's equations (7) and (8), and a difficulty could arise only if (contrary – he says – to any indication we have) this determination yielded two different answers for the entangled state of the two particles.

Even though Margenau's note is vague on some points (including spelling out the role of the collapse postulate in the EPR paper), it is clear that his main conclusion amounts to rejection of the standard collapse postulate, which Einstein does indeed defend in his letter of 3 January 1936 (EA 17-66):

> Current quantum mechanics is adapted to the following postulate, which considering the facts of experience appears unavoidable:
>
> If a measurement on a system yields the value W, then the same measurement immediately thereafter yields again the value W with certainty.
>
> Example: if a light quantum has passed a polariser P_1, then I know for certain that it will also pass a second polariser P_2 oriented parallel to the first.
>
> This is valid independently of how the quantum is produced, thus also in the case in which before the passage through the first polariser (P_1) the probability for the direction perpendicular to the polarisation direction of the polariser was not zero (e.g. in which the light quantum comes from a polariser P_0, whose direction of polarisation forms an acute angle with that of P_1).—
>
> Therefore in my opinion one cannot avoid the assumption that the measurement changes the probability coefficients of a state, that is it brings about a new state in the sense of quantum mechanics, which is an eigenstate with respect to the variables to which the measurement refers.

The later paper (Margenau 1937) announced in his shorter November 1935 publication is both more rigorous and more illuminating than its predecessor on a number of points, including how dropping the collapse postulate is meant to assuage EPR's worries. In the 1935 note, for instance, Margenau is unrigorous in that he limits himself to talking about the quantum state in terms of the probability distribution for the values of a single observable, while in the 1937 paper he considers the set of probability distributions defined for arbitrary observables. Again, he is more careful in the later paper with regard to how to ascertain a quantum state: he shows that in general the statistics of a single observable does not yield enough information to reconstruct the state, but he points out that for any state there is some observable of which the state is an eigenstate, and measuring that observable will reliably confirm what the state is.

With regard to why the preparatory aspect of a measurement can be dispensed with in quantum mechanics, and specifically in the case raised by EPR, the following passage is revealing (Margenau 1937, pp. 367–368, original emphasis):

> When a measurement on one particle is made, the state function corresponding to that particle is known: it represents the eigenstate belonging to the measured value. But this function appears in the original sum as multiplied by a coefficient which contains only the coordinates of the other particle, and which can therefore be regarded as its state function after the measurement. Thus, it is contended, a measurement causes the originally mixed function to reduce itself suddenly to a simple form, thus 'resolving the wave packet'.
>
> On the present view, no such event happens; it can only be shown [. . .] that there exists a peculiar correlation between the probabilities for the various results of measurements. It is this: When suitable measurements on the two particles are made, *the probability for the occurrence of a pair of values is zero unless the pair is the one which corresponds to the two eigenfunctions of the reduced package*. The physical consequences are the same as on the customary view, but the result appears now in less metaphysical language. The method of the reduction of the wave package remains in good practical standing, only its interpretation requires greater care.

What Margenau appears to be suggesting is that the quantum state never collapses (although he refers consistently to the state of microscopic systems and not, say, to a quasi-Everettian 'wavefunction of the universe'), and that all the probabilistic predictions of quantum mechanics – including the correlations in an EPR scenario – are already encoded in the non-collapsing state.

The 1937 paper moreover contains an explicit distinction between 'measurements' (which provide statistical information about the original state of the system) and 'preparations', including a discussion of preparations via polarisers, which Jammer reads as a direct reply to Einstein's example from the correspondence quoted above.[6] Margenau suggests that passage through a polariser is a preparation,

[6] Note, however, that in the meantime Kemble had also written to Margenau insisting on the importance of preparations (Kemble to Margenau, 20 November 1935, AHQP 54, section 3-019): '[I]t seems to me very evident that we cannot get along with a completely homogeneous class of measurements. Retrospective measurements and predictive measurements perform very different functions, and although the same apparatus may suffice for both, clarity of thought is assisted by treating these two things as distinct.'

in the sense that *if* an appropriate measurement were then performed, one could predict the result of that measurement with certainty. But he stresses that mere passage through a polariser cannot itself count as a measurement, for at this stage one has not yet established whether the photon was actually transmitted.

With regard to the EPR paradox, what is confusing is that one could use Margenau's own remarks about ascertaining a state to argue that the state of EPR's distant particle *has* collapsed: by performing measurements of the relevant observable on an ensemble of suitably prepared particles, one could ascertain that the state is in fact the corresponding eigenstate. Hence what Margenau must be suggesting is that physical quantities are not *real* until they are actually *measured*, while a mere preparation does not guarantee that a corresponding quantity is real. Understood in this sense, Margenau is denying the validity not only of the projection postulate, but also of EPR's criterion of reality; presumably he takes the *latter* to be an expression of the *former*.

2.12 Furry

'Note on the Quantum-Mechanical Theory of Measurement' – written 12 November 1935, but only published in Physical Review *on 1 March 1936. Reprinted herein as Chapter 12. Furry also contributed: 'Remarks on Measurements in Quantum Theory' – written 2 March 1936 and published in* Physical Review *very soon thereafter, 15 March 1936. Reprinted herein as Chapter 13.*

Unlike many replies to EPR, Furry's first contribution to the debate is substantial – comparable in length to Bohr's article rather than a letter to the editor. And although with Furry's two papers (of which the second is an addendum pursuant to the publication of Schrödinger's entanglement paper and cat paper (Schrödinger 1935e, f)) we see again a contribution to the debate that fails somewhat to address EPR's true concerns, still we find a rather masterly contribution to the foundations of quantum mechanics.[7] Furry argues that although the usual collapse postulate is certainly utilisable for all practical purposes, it nevertheless cannot be *fundamental*. He believes that the main literature on the theory of measurement (Heisenberg's Chicago lectures, Pauli's handbook article and von Neumann's monograph) is as a whole susceptible to misinterpretation on this point.

Thus, Furry's main focus (like Margenau's) is the collapse postulate. And (again like Margenau) Furry appears to assimilate the collapse postulate to EPR's position. More precisely, Furry takes EPR to be suggesting the following: after two systems

[7] Furry's work receives little mention in the foundations literature, two exceptions being the ever-comprehensive Jammer (1974, pp. 221–225) and Brown (1978).

have ceased to interact, they both have separate 'real' attributes in the sense that if the interaction has produced the entangled state

$$\Psi(x_1, x_2) = \sum_k (w_k)^{\frac{1}{2}} \varphi_k(x_1) \xi_k(x_2) \tag{1}$$

(Furry's eq. (3) in slightly adapted notation), then the two systems must have transitioned stochastically to one of the pairs of states $(\varphi_k(x_1), \xi_k(x_2))$, with probability w_k. Put another way: in keeping with the standard literature on measurement, since (1) is the form of the wavefunction of a system and an apparatus post-interaction, and since the measured system behaves as if it 'really' is described by a (proper) mixture of the states $\varphi_k(x_1)$, EPR – like many physicists – allegedly interpret these measurements simply as ascribing separate realities to separate systems. Furry will argue that this interpretation, though perhaps the most obvious one, is nonetheless mistaken.

Furry calls the assumption of such a transition upon interaction (specifically a measurement interaction) *Assumption A*, and *Method A* the calculation of probabilities for results of measurements on system II using the corresponding mixed state. In contrast, he calls *Method B* the calculation of such probabilities from the entangled state $\Psi(x_1, x_2)$ which, according to the Schrödinger equation, arises from the same interaction. He then discusses four different cases:

a. No measurement is performed on system I and an arbitrary observable S is measured on system II.
b. The observable L associated with the basis $\varphi_k(x_1)$ in the biorthogonal decomposition of the state is measured on I and an arbitrary observable S is measured on II.
c. An arbitrary observable M is measured on I and the observable R associated with the basis $\xi_k(x_2)$ in the biorthogonal decomposition is measured on II.
d. Neither of the observables measured on systems I and II commutes with L or R, respectively.

One then asks what the probabilities are for measurement results on II (conditional on any results on I), and finds that Methods A and B give the *same* answers for cases (a), (b) and (c), but generally *different* answers for case (d).

Even if one considers L a 'pointer observable' on system I, so that all predictions regarding II conditional on the results of measurements of L can indeed be derived using Method A, there are yet other measurements one can in principle make such that Method A yields answers which disagree with those obtained via a strict application of quantum mechanics (i.e. via Method B). Thus, Method A is not exact and Assumption A is wrong, despite the fact that 'it is possible for a reader to form the impression that the theory is consistent with Assumption A' (pp. 396–397/214).

After this mathematical discussion, Furry presents a reasonably realistic physical example. What he needs is 'a case in which one system (I) is used as a measuring instrument in observing another system (II), and in which, after interaction has ceased, some other observable besides the one suited to serve as a "pointer reading" can be measured on system I' (p. 397/215). But rather than detailing an example of this kind along the lines of his case (d), he chooses a slightly different strategy: 'There is, however, a particularly neat and striking special type of example, to which EPR have directed attention; and the one we shall describe is of this type' (p. 397/215); it is in discussing this example that Furry comes nearest to EPR's intended mark.

Furry starts by setting up the experiment, essentially along the lines of Heisenberg's γ-ray microscope (where one reconstructs the momentum or position of an electron via measurements on a photon; see Chapter 4). In order to give a more realistic example, Furry considers instead the probing of a proton using an *electron* microscope – an instrument which was actually under development at the time. Furry writes (1936a, p. 398/216):

> The 'Assumption A' which corresponds to the first experiment is: During the collision between the electron and the proton, the proton made a transition into some state with well-defined momentum; *which* state this is can be determined by a measurement made on the electron. Corresponding to the second experiment one has an 'Assumption A' which reads exactly the same except that the word 'momentum' is replaced by the word 'coordinate'.

He now makes use of the *delayed choice* feature of this setup (p. 398/216):

> Since one is still free to choose which experiment is to be performed *after* the electron has been scattered into the microscope [...] and ceased to interact with the proton, these assumptions must both be true at once, if one accepts the point of view on which they are based.

What Furry is saying here sounds paradoxical: if 'Assumption A' is taken strictly to mean assigning a system a quantum state from a biorthogonal decomposition, then assigning quantum states from two different biorthogonal decompositions is simply inconsistent, and there is no need to proceed further in a demonstration of the assumption's untenability. So suppose we take 'Assumption A' to mean assigning *some* quantum state to the system, and assume the state whose existence follows from the first application of Assumption A to be the *same* as the state whose existence follows from the second application of Assumption A. Or to put it in EPR's words, 'since at the time of measurement the two systems no longer interact, no real change can take place in the second system in consequence of anything that may be done to the first system' (EPR 1935, p. 779/161).

And now, what Furry goes on to show through detailed analysis of the electron microscope experiment is that the uncertainty in the momentum and the position of the proton resulting from both measurement predictions has a product far below the bound stipulated by the uncertainty relations. Thus we must jettison the assumption that there is some single quantum state we can attribute to the proton after interaction has ceased.

If, like Furry, one is investigating the tenability of assigning a particular *quantum* state to a system after it has ceased interacting (the view Furry seems to attribute to EPR), then certainly the analysis of the electron microscope reveals this to be untenable. If instead one assumes (or even *infers*) that a system has *some* real state after interaction – not necessarily a quantum state – then the fact that no single quantum state can play that role in the electron microscope experiment shows that the quantum mechanical description is incomplete.

Were our more charitable reading of 'Assumption A' Furry's own, he would have gotten appreciably closer to the core of EPR than most. But the lesson Furry himself takes away here (and which is made even more explicit in his second note) is that the standard collapse postulate, though fine for all practical purposes, is false when interpreted literally.

2.13 Wolfe

'Quantum Mechanics and Physical Reality' – written on 30 December 1935 and published in Physical Review *on 1 February 1936. Reprinted herein as Chapter 25.*

In his short letter to the editor, Wolfe insists that quantum mechanical wavefunctions embody our knowledge of a system. Under a merely epistemic interpretation of wavefunctions such as this, the EPR paradox 'simply dissolves', since '[d]ifferent measurements on the first system give us different information about the second and therefore different wave functions and different predictions as to the results of measurement on the second system' (p. 274/277).

After some truly substantial contributions, this final word to be written on EPR in 1935 – and the last of the immediate flurry of responses elicited by that paper – is, alas, rather an anticlimax.

3

Schrödinger on EPR*

3.1 Introduction

This chapter will try to do partial justice to the complex story of Schrödinger's struggles with the foundations of quantum mechanics, in particular the concept of entanglement, in the period leading up to and including the golden year of 1935.

The background of this story (which we discuss in Section 3.2) is of course Schrödinger's development of wave mechanics as a consciously 'anschaulich' alternative to Heisenberg's matrix mechanics, and Schrödinger's realisation that wave mechanics had interpretational problems of its own. Some of this is well-known, and will be summarised here only briefly.[1] By the time of the fifth Solvay conference in 1927, Schrödinger had been publicly and energetically promoting the point of view of wave mechanics, while also openly admitting that it left certain problems unsolved (e.g. by providing only a partial description of radiation). As shown by Joas and Lehner (2009), however, behind the confident public façade, Schrödinger was struggling in his notebooks with even deeper worries – indeed, partly to do with entanglement. We shall follow their lead and look in some detail at how Schrödinger was beginning to realise just how mystifying quantum mechanics was when applied to composite systems.[2]

* Part of the material in this chapter relates to talks we gave at the History of Quantum Physics satellite meeting, Sydney (May 2008), the History of Quantum Physics 2 meeting, Utrecht (July 2008) and the 16th UK and European Meeting on the Foundations of Physics, Aberdeen (July 2010), as well as to Bacciagaluppi (2008) and Bacciagaluppi and Crull (2009).

[1] For a more detailed summary, see e.g. Bacciagaluppi and Valentini (2009, chapters 3 and especially 4).

[2] While Joas and Lehner (2009) mention entanglement only very briefly, we were made aware of Schrödinger's very specific worries about it through their two talks at New Directions 2011 (Joas 2011, Lehner 2011). Numerous discussions with them have greatly clarified and complemented our understanding of Schrödinger. We give them very special thanks also for sharing their commented partial transcriptions of Schrödinger's research notebooks AHQP 40-5-001 to 42-1-002, which we have used extensively.

The further development of Schrödinger's views has perhaps been most thoroughly discussed by Michel Bitbol (1996), who sees the 1930s as a time for reflection in which Schrödinger's earlier ideas got transformed into his views of the 1950s, and who provides a beautiful analysis of the latter (which did not completely satisfy Schrödinger – essentially because of the measurement problem). The folklore is that the EPR paper had revitalised in Schrödinger his latent criticism of the 'orthodox' quantum theory, which (so the story goes) had abated after the stormy confrontations of 1926–27, ending with the 1927 Solvay conference. Like Bitbol, we understand Schrödinger's views in the period 1927–35 as more interesting than the folklore has it. While Schrödinger did talk about wave aspects and particle aspects of matter, that was by no means an embracing of Bohr's views on complementarity: for Bohr complementarity was the *answer* to the puzzles posed by quantum mechanics, while for Schrödinger the apparent dual nature of matter was a *symptom* that we do not yet understand quantum mechanics. Schrödinger no longer believed that his earlier ideas would be able to provide an adequate picture of quantum mechanics, but in this period he became if possible even more dissatisfied with the orthodox view. To put it briefly, Schrödinger did not have an interpretation of quantum mechanics in this period *because no satisfactory interpretation was available*.

Part of this bewilderment with quantum mechanics derived from Schrödinger's engagement with the proto-EPR arguments, as mentioned already in Section 1.4.2. Fuller details of the role played by Schrödinger in the story of proto-EPR are currently emerging through the work of Jos Uffink and Christoph Lehner, who have recently discovered additional source material, in particular Schrödinger's notes and calculations after Einstein's Berlin seminar of November 1931.[3]

In our final section we turn to Schrödinger's reactions prompted by the EPR paper itself: a thick correspondence with Einstein and others, and two famous papers – one on entanglement and famed for that (Schrödinger 1935e, reprinted here as Chapter 10), and one on 'The Present Situation in Quantum Mechanics' and famed above all for its feline thought experiment (Schrödinger 1935f, reprinted here following essentially the standard translation by John Trimmer as Chapter 11). The correspondence is very rich and provides a wealth of clues for the views on the topic of several of Schrödinger's correspondents, as well as direct insights into the further development of his own views. As we shall see (and as pointed out already in the excellent study by Fine (1994)), Schrödinger's considered view of the EPR argument is that it is truly a paradox: a dilemma between some kind of 'spooky action

[3] We would like to thank Jos and Christoph very much for discussions and advance snippets of material from their work on Schrödinger and EPR (on which more in Section 3.3). This is very exciting work which will importantly complement what we have to say in this chapter. Some details have already transpired in fascinating talks by Uffink (2017, 2018).

at a distance' and a bewildering mess of 'elements of reality' violating the expected algebraic relations – bringing Schrödinger no nearer to a satisfactory understanding of quantum mechanics. The provisional conclusions that Schrödinger draws in 1935 are beautifully summarised in his cat paper, and we close with a brief discussion of the EPR argument as presented there.

3.2 Anschaulichkeit and Early Worries about Entanglement

A crucial element in Schrödinger's approach to the foundations of quantum mechanics was shaped by the conflict between wave and matrix mechanics. Beyond the clash of personalities, the role in this conflict played by Schrödinger's philosophical inclinations has become clear in recent years, in particular the *methodological* role played by the notion of Anschaulichkeit (i.e. the use of detailed 'physical pictures' to supplement the phenomena, a legacy of Boltzmann's work on the kinetic theory of gases). Despite the close formal correspondence discovered between the two theories soon after they were developed, the participants in the debate clearly believed that wave mechanics and matrix mechanics were inequivalent research programmes, based on fundamentally different physical concepts and having different potential for further development and generalisation. This is where Schrödinger saw the specific methodological value of Anschaulichkeit.[4]

One of the main conceptual differences between wave mechanics and matrix mechanics was the notion of a quantum state. While in matrix mechanics *there is no object* corresponding to the state of a system, wave mechanics for Schrödinger was from the start about the *individual states* of systems, described by wavefunctions. Wavefunctions, or some closely related concepts, appeared to be candidates for representing not only Bohr's stationary states, in a very concrete and anschaulich way, but also non-stationary states and thus potentially the transitions between the stationary states. Matrix mechanics instead was understood as describing directly only the totality of stationary states, between which random transitions occurred (the quantum jumps). Intuitions somewhat differed as to whether stationary states themselves had an anschaulich meaning (Born retaining a geometrical intuition for them? Heisenberg rejecting it?), but quantum jumps were universally thought of as fundamentally unanschaulich.[5] Thus, in his 'equivalence' paper on the relation of

[4] For analyses of this conflict, see in particular Beller (1997), Bacciagaluppi and Valentini (2009, section 4.6) and Jähnert and Lehner (2012). Specifically on the concept of Anschaulichkeit, see de Regt (1997). For a discussion of how even the formal equivalence depended on mutual borrowings between the two theories, see Muller (1997, 1999a).

[5] Cf. Bacciagaluppi (2008). While explicit statements as to this understanding of matrix mechanics can be found only from around October/November 1926 (cf. Born (1926c) – which incidentally includes very explicit statements about geometrical trajectories for free particles – Heisenberg (1927b) and Jordan (1927)), we have argued together with Owen Maroney (Bacciagaluppi, Crull and Maroney 2017) that it is a crucial element

wave and matrix mechanics (Schrödinger 1926b, submitted March 1926) Schrödinger identifies the issues of scattering and of radiation as ones for which he believes wave mechanics may provide a more fruitful approach than matrix mechanics. Heisenberg, on the other hand, was eager to emphasise that corpuscular and discontinuous phenomena seemed naturally to support matrix mechanics.[6]

However, as Joas and Lehner (2009) point out, from the spring of 1926 onward Schrödinger's public and private research programmes start to diverge. In his published work, Schrödinger goes from strength to strength working out successful applications of wave mechanics, while in his notebooks he is often struggling with fundamental questions, in particular about the meaning of the wavefunction. Central to these preoccupations are the fact that the wavefunction is defined in configuration space and the question of how to couple material particles to the electric field – that is, the mechanism of radiation. By the fourth paper on quantisation (Schrödinger 1926d, submitted 21 June 1926) we have Schrödinger settling on a picture of the wavefunction in which the expression $|\psi|^2$ represents a smeared-out charge density, where the density contributed by each particle should be obtained by integrating over the other particles.[7] From this, Schrödinger develops a theory of radiation that is still used today as a highly successful semi-classical treatment, in which this charge density and the corresponding current are used as classical sources of electromagnetic radiation. As Schrödinger was to emphasise repeatedly over the next few years, however, this semi-classical theory only deals with emission: electromagnetic fields do not *act* in the usual way on the thus introduced charge densities.[8]

On the question of collisions, Schrödinger conceivably had in mind that both loosely bound and free electrons could be represented by localised wavepackets; however, already in May 1926 Lorentz had pointed out to him in correspondence that electrons in high quantum orbits would be spread out (Przibram 1967, p. 47). Ironically, and much to Schrödinger's dislike, the issue of scattering led to Born's introduction of the 'statistical interpretation' of the wavefunction, in his two papers

already in Jordan's derivation of Einstein's blackbody fluctuation formula in the Dreimännerarbeit of late 1925 (Born, Heisenberg and Jordan 1926).

[6] E.g. Heisenberg to Pauli 28 July 1926, where he chides Schrödinger for throwing overboard 'everything quantum theoretical: namely photoelectric effect, Franck collisions, Stern–Gerlach effect etc.' (Pauli 1979, p. 338). Heisenberg also reported later on that Bohr had insisted to Schrödinger in their discussions in Copenhagen in the autumn of 1926 that quantum jumps were necessary for the derivation of Planck's radiation law (Heisenberg 1969, 1971, chapter 6). See also Heisenberg (1926).

[7] Schrödinger had initially taken $\psi \frac{\partial \psi^*}{\partial t}$ as the charge density of the electron. In an added footnote to the French translation of his paper on the equivalence of wave and matrix mechanics, he clarifies the original motivation as 'the desire to allow a relativistic generalisation of the theory, in which this spatial density would appear as the *time component* of a four-vector' (Schrödinger 1933, p. 97).

[8] Cf. e.g. Schrödinger's comments at the fifth Solvay conference (Schrödinger 1928a), Marcel Brillouin's Preface to the French translation of Schrödinger's papers on wave mechanics (Brillouin 1933) and other explicit references below.

on collisions and (in much more detail) in his subsequent paper on the adiabatic theorem (Born 1926a, b, c).

In the collision papers, Born considers an asymptotically incoming plane wave $e^{\frac{2\pi}{\lambda}z}$ and an atom initially in a stationary state $\psi_n(q)$, and solves the Schrödinger equation perturbatively to obtain a final asymptotic state of the form

$$\sum_m \int_{\alpha x+\beta y+\gamma z>0} \Phi^{\tau}_{nm}(\alpha\beta\gamma)\psi_m(q)e^{k^{\tau}_{nm}(\alpha x+\beta y+\gamma z+\delta)}d\omega, \tag{1}$$

where the wave number k^{τ}_{nm} corresponds to the energy $E^{\tau}_{nm} = h\nu_{nm} + \tau$ (the ν_{nm} being the transition frequencies of the atom, and $\tau = \frac{h^2}{2m\lambda^2}$ the initial energy of the incoming particle). This is an entangled state, but in line with the physical interpretation of matrix mechanics, Born believes that the final state after the collision must be one of the products $\psi_m(q)e^{k^{\tau}_{nm}(\alpha x+\beta y+\gamma z+\delta)}$, so that the atom and the scattered particle have jumped (in an unanschaulich process) from their respective initial stationary states to the final ones (possibly with a resulting exchange of energy). He is thus naturally led to interpret $|\Phi^{\tau}_{nm}(\alpha\beta\gamma)|^2$ as the probability for this transition, or equivalently as the probability of occurrence of the state $\psi_m(q)e^{k^{\tau}_{nm}(\alpha x+\beta y+\gamma z+\delta)}$ after the collision.

Heisenberg was equally irked by Born's embracing of Schrödinger's formalism (albeit interpreted 'in the sense of Heisenberg'[9]), but at Pauli's suggestion managed to reformulate matrix mechanically the calculation of transition probabilities *à la* Born, in a short paper on 'Fluctuation Phenomena and Quantum Mechanics' (Heisenberg 1927b) that used the example of two systems in resonance (i.e. with an energy difference in common).[10]

As for Schrödinger, Joas and Lehner point out that, after Born's introduction of the 'statistical interpretation', his efforts to physically interpret the wavefunction became increasingly febrile. Of particular interest are two of Schrödinger's research notebooks from this period.[11] The main public result from the notebooks is Schrödinger's paper on energy exchange (Schrödinger 1927, submitted 10 June 1927), in which Schrödinger turns the tables on Heisenberg, reinterpreting wave

[9] As Born states in the paper on the adiabatic theorem: 'the fundamental ideas of the matrix form of the theory initiated by Heisenberg [...] have grown directly out of the natural description of atomic processes in terms of "quantum jumps" and emphasise the classically-geometrically incomprehensible nature of these phenomena. It is definite that both forms of the theory arrive at the same results for stationary states; the question is only how one should treat non-stationary processes. Here, Schrödinger's formalism turns out to be substantially more convenient, provided one interprets it in the sense of Heisenberg. I therefore wish to support a merging of both points of view, in which each fulfills a very particular role' (Born 1926c, p. 168).

[10] For further details of how the 'statistical interpretation' developed and of Heisenberg's continued deflationary view of the wavefunction, see Bacciagaluppi (2008), Bacciagaluppi and Valentini (2009, chapter 3), Bacciagaluppi, Crull and Maroney (2017, section 4), and Bacciagaluppi (2022).

[11] AHQP 41-1-002 ('Wave statistics I. General remarks on Born's probability hypothesis. The collision formulas as perturbations of degenerate systems') and AHQP 41-1-003 ('Wave statistics II').

mechanically the results of the note on fluctuations, and ending with the suggestion that blackbody radiation should be amenable to the same kind of treatment. The ultimate aim of much of these notebooks appears to be to discuss blackbody radiation from a wave mechanical point of view, and Schrödinger spends much effort on whether one could construct a wave mechanical parallel of Boltzmann's approach to statistical mechanics (e.g. he is wondering about the analogue of Liouville's theorem), as well as on coupled systems. This is where Schrödinger reacts to Born's statistical interpretation and to Heisenberg's short paper on fluctuations, and this is where he first truly stumbles upon entanglement.[12]

Let us look at these two notebooks in more detail, starting with 'Wave statistics I'. The notebook is organised in three main parts, as specified in the title.[13]

To begin with, Schrödinger discusses coupled systems with an eye towards blackbody statistics. He seems to want to investigate whether a Boltzmannian treatment can go through, with a crucial ingredient being averaging over the phases of the blackbody modes ('The averaging over the phases must replace completely the statistical obscenity [Schweinerei] from Göttingen').

Then, Schrödinger makes some general comments about Born's ideas. First, one must distinguish the hypothesis that $|c|^2$ gives the probabilities for stationary states and the hypothesis that $|\psi|^2$ is the probability density for positions. Second, Schrödinger is critical of the assumption that particles might have point-like positions (even trajectories in a Wilson chamber or particle detections do not indicate there are point-like particles). If one grants there are such point-like particles, then the wave equation must play the role of a diffusion equation as in Brownian motion (a point he then rather inconclusively elaborates on – not without noting, however, that a δ-function will spread out instantaneously, threatening 'instantaneous action at a distance').[14]

[12] Note that Jordan's derivation of the blackbody fluctuation formula (which he himself considered to be one of the great successes of matrix mechanics, and his finest contribution to it) was also carried out within a Boltzmannian framework, and that Heisenberg thought of Jordan's derivation as a special case of the point of view he sketched in his note on fluctuations. For details, see Bacciagaluppi, Crull and Maroney (2017).

[13] For the dating of this notebook: Born's first paper (Born 1926a) was received 25 June 1926, but a letter from Schrödinger to Wien of 25 August (quoted in Moore 1989, pp. 225–226) suggests he knew of it through Born's second paper, received 21 July (Born 1926b). In the notebook, he also comments on Born's interpretation of $|c|^2$ as a probability, which also appears only in the second paper. In the third part of the notebook, Schrödinger references his own third paper (Schrödinger 1926c, published 13 July 1926), refers to his fourth (Schrödinger 1926d, submitted 21 June 1926 and published 5 September), refers implicitly to a paper by Wentzel (probably Wentzel (1926), received 19 November), and mentions Heisenberg's ' "as-if" discontinuities', which presumably refers to Heisenberg's note on fluctuations (Heisenberg 1927b, received November 1926, published July 1927). Schrödinger tended to use different notebooks for different topics, so each notebook could have been in use for a considerable time. It is thus conceivable that at least the second part of the notebook could date from the summer of 1926, while the third part maybe even from as late as the middle of 1927.

[14] He then (mistakenly) also states that this would not happen relativistically. Nowadays, there are a number of results about superluminal propagation of localised states in relativistic quantum theory, collectively known as 'Hegerfeldt theorems'. See e.g. Malament (1996) or Muller (1999b) for discussion and references.

Finally, we have the discussion of collisions. Schrödinger notes the analogy with coupled systems in resonance, and states that it is the resonance condition that is 'the true sense in which "systems exchange energy only in quanta"'. He also states this is the way to describe statistics ('small subsystem of a very large total system'). After some calculations about collisions, Schrödinger goes on to the systems in resonance. He asks himself explicitly how Heisenberg can get his strange result about 'as-if' discontinuities, and answers that it is because the vibrational states of the joint system that result from the interaction are not resolvable into states of the single systems.[15] Even worse: such 'non-resolvable' states generally persist even after the interaction has ceased.

Schrödinger notes that this situation is one with no unambiguous result (i.e. no definite final states of energy for the subsystems), and tries to separate the final state into factors in which both systems are each in a superposition of both stationary states. He admits this is not strictly correct, but convinces himself that from the point of view of his fourth paper (i.e. in the charge density interpretation), if he integrates over the other system this will have no effect on the '*real* state' of the system.[16] He was soon to realise (in the second notebook) that this is not a harmless procedure. And indeed, he (obviously later) footnoted this remark with 'Fehlschluß!' ['Fallacy!']. There follows a failed attempt to calculate the fluctuations wave mechanically.

The notebook concludes with a series of remarks, notably to the effect that non-factorisability is a step further in the direction of having to renounce specifying a classical position for an individual particle; that the expression 'electrical density' has its limitations, in particular can only be used to describe sources of the electric field, but not its action on physical systems (it is the wave equation that determines the evolution of the particles, not the semi-classical field); and that the apparent corpuscular nature of the electron and perhaps the trajectories in a Wilson chamber are problematic for his interpretation.

In the second notebook,[17] Schrödinger begins by trying to describe the coupling with the electromagnetic field as a separate quantum system, and continues his discussion of statistics (among other things enquiring about an analogue of Liouville's theorem, needed for Boltzmann's formalism to apply). Then, in a two-page

[15] In his own words: 'precisely because the oscillatory states of the composite [die gemeinsamen [?] Schwingungszustände] that are established immediately upon interaction – in fact established even when before the start of the interaction each of the individual systems was excited only through one eigenoscillation – are such that they can no longer be resolved into states of the individual systems'.

[16] As he writes, '[o]ne sees that the mere adjoining [der einfache Hinzutritt] of orthonormal functions of a foreign system to the eigenfunctions of our system will have no effect whatsoever. One can thus leave them out.'

[17] Regarding dating: in the middle of the notebook Schrödinger refers to a discussion on 27 May 1927 with Fues, Heitler and London; parts of the notebook are related to his paper on energy exchange (Schrödinger 1927, submitted 10 June 1927); and in a passage in shorthand towards the end he inserts the date 18 June 1927.

section headed 'Meaning of the amplitudes' – the most significant for our purposes – Schrödinger retracts his attempted interpretation of entangled subsystems, in which integrating over the other system is allegedly harmless. The main reason for this retraction is that when we consider the total system in the degenerate state

$$c_1 lm' + c_2 l'm \tag{2}$$

and calculate the charge density for one or the other subsystem, it does *not* contain (oscillating) terms of the form ll' (or mm'),[18] because they disappear when integrating over the other system. The total state is stationary, and such a state always remains stationary: merely thinking of the two systems as separate cannot change that.[19] He says that one system of eigenfunctions 'screens the other off', in the sense that it prevents the formation of a density distribution that would radiate.

Schrödinger compares this to the situation of a many-electron atom (which he treats in his talk at the fifth Solvay conference of October 1927), where the different electrons screen each other off, preventing the radiation that would take place in a one-electron atom. He seems inclined, however, to think that there is a breakdown of the non-relativistic quantum mechanical formalism, as a consequence of looking at separated systems using forces operating instantly at a distance:

> The key point seems to me this (discussion in the evening of 27 May with Fues, Heitler, London at the Franziskaner[20]). During the separation of the two systems there is evidently a moment when one may no longer describe their mutual influence according to the framework assumed, because this has grown out of classical mechanics with its *instantaneous* action at a distance.

This is a thought to which he returns at the 1927 Solvay conference, and in a number of instances in the following years (even at the end of his second Cambridge paper (Schrödinger 1936), where he hypothesises that entanglement might decay with distance – see Section 3.4.3 and the Appendix to Chapter 10). More quantum statistics and a variety of other material follow, and the notebook concludes with a series of scattered remarks.

In his public persona, Schrödinger publishes his paper on energy exchange in August 1927, in which he reinterprets wave mechanically the case treated by Heisenberg in his note on fluctuations, thereby (in public) turning the tables on

[18] This is *unlike* the case of a state $c_1 l + c_2 l'$ of a single system.

[19] As he puts it, '[i]t is of course absurd to think that the *conceptual* process of separation will suddenly give rise to this oscillating density distribution'.

[20] Erwin Fues, Walter Heitler and Fritz London were all three Rockefeller Fellows in Zürich with Schrödinger – Fues since November 1925, the others since April 1927. Fues was presumably about to leave for Copenhagen, where he spent the last few months of his fellowship, while when Schrödinger left for Berlin later that year, Heitler went to Born in Göttingen, and London followed Schrödinger to Berlin. The Franziskaner – now Café Henrici – is a well-known café on Stüssihofstatt in the Zürich old town.

Heisenberg by arguing that one could thus eliminate quantum jumps from the picture. Schrödinger also includes remarks on reinterpreting the idea of quantised energy in terms of wave frequency, suggesting that resonance phenomena are the key to the 'quantum postulates'. From statistical considerations about the distributions of the amplitudes of the two systems in resonance, Schrödinger goes on to the idea of the squares of the amplitudes as measures of the strength of excitation of an eigenstate. He concludes with hopeful remarks on the case of a system coupled to a heat bath, where resonance considerations would (so he conjectures) in principle suffice for the derivation of Planck's radiation formula.

However, Schrödinger's realisation about at least some of the troubling nature of entanglement must have privately worried him, and it is perhaps significant that at the fifth Solvay conference Schrödinger did not present these upbeat results, but a seemingly lower-key paper on the multi-electron atom (with hints at its relevance for a spacetime picture of the wavefunction), reiterating his smeared-out charge density interpretation, but also emphasising the limitations of this picture as only apt to describe the emission of radiation. Indeed, although Schrödinger clearly appears to think that some of the problems might be an artefact of the non-relativistic formalism, the difficulty in interpreting entangled states must have been a major stumbling block for Schrödinger's hopes of making sense wave mechanically of blackbody statistics, and even more of scattering: not only might the scattered particles not remain localised as wavepackets (as suggested already by Lorentz's remarks), but the final state after the interaction was now 'non-resolvable' in a way that did not seem to provide a sensible description of the individual subsystems. If Schrödinger had had hopes of treating collisions in terms of a statistical ensemble of wavepackets scattering off in different directions, there was clearly no scope for that now.

3.3 Schrödinger's Evolving Ideas on Quantum Mechanics

The usual view of the development of Schrödinger's ideas on the interpretation of quantum mechanics is well summarised by Bitbol (1996): a first phase in which the wavefunction is interpreted literally as a field in ordinary space soon gives way to a second phase in which waves in configuration space give rise to a charge density in ordinary space; then after the 1927 Solvay conference Schrödinger appears to be satisfied with the Copenhagen interpretation, before reverting to a variant of his second wavelike interpretation in the 1950s. Bitbol gives a very nuanced reading of the period from 1927 onwards, which he sees as a time for reflection in which Schrödinger's earlier ideas get transformed into his mature views of the 1950s. In this section, we shall have a closer look at this transitional period (up to the climax of 1935), during which Schrödinger is more critical of his earlier views but also develops new lines of criticism of the 'dogmatic quantum mechanics'.

One can multiply the examples in this period where Schrödinger expresses his dissatisfaction with the current situation in quantum mechanics, sometimes in passing but very explicitly,[21] including very publicly in his Nobel lecture of 10 December 1933, which Schrödinger concludes with the words (Schrödinger 1935a, p. 154):

> I would describe the present state of our knowledge as follows: The light ray, or track of the particle, corresponds to a *longitudinal* continuity of the propagating process (that is to say, *in the* direction of the spreading); the wave-front, on the other hand, to a *transversal* one, that is to say, perpendicular to the direction of spreading. *Both* continuities are undoubtedly real. The one has been proved by photographing the particle tracks, and the other by interference experiments. As yet we have not been able to bring the two together into a uniform scheme. It is only in extreme cases that the transversal – the spherical – continuity or the longitudinal – the ray – continuity shows itself so predominantly that we *believe* we can avail ourselves either of the wave scheme or of the particle scheme alone.

As we shall see, throughout this period Schrödinger still upholds the ideal of Anschaulichkeit, stating time and again that physical theorising is about filling the gaps between the actual observations (think of the cloud chamber!) by providing a picture of physical reality as serviceable as the one we have of tables and chairs. In many places he still emphasises that continuous wave-like concepts have not exhausted their potential for providing a fruitful understanding of quantum phenomena, but he often distances himself from *both* classical wave concepts and classical particle concepts, judging that his earlier views on wave mechanics and Heisenberg's views on matrix mechanics *both erred in opposite directions*.

Indeed, Schrödinger consistently sees quantum mechanics as advocated by Bohr and by Heisenberg as clinging to classical concepts as to crutches (both when it emphasises particulate concepts, quantum jumps, discontinuities and uncertainty, and when it invokes duality between particles and waves). To Bohr himself, Schrödinger addresses pressing questions about his insistence on the use of classical concepts. For Schrödinger, *new concepts* will have to be developed in order to provide the requisite physical pictures. These are all ideas that will recur in his 1935 correspondence and publications.

As to what these new concepts might be, Schrödinger is open to a variety of options. Often he revisits the ideas of his teacher Franz Exner about all physical regularities being ultimately statistical in nature and fundamentally indeterministic, and expressing scepticism that either determinism or the idea of probabilistic laws might be part of the physical picture to be developed (even suggesting that they

[21] Cf. e.g. Schrödinger (1929b, 1929e and 1932d).

may be conventional elements *à la* Poincaré). One occasionally sees him experimenting with novel concepts, for instance critiques of the applicability of geometry, or even forwards- and backwards-in-time equations. But none of these ideas ultimately provide him with a satisfactory answer to the problems posed by quantum mechanics.

Indeed, sometime in this period Schrödinger gives up for good the idea that the wavefunction provides a direct representation of a quantum system,[22] and in some ways, perhaps unwittingly, his views do get close to some aspects of Bohr's thinking. Schrödinger comes to believe that the ascription of quantum states to systems is *conditioned by the choice of our observations*, in such a way that quantum mechanical wavefunctions in fact encode the *relation* between subject and object. This perhaps crucial element in the evolution of Schrödinger's views will reappear in connection with Schrödinger's involvement with proto-EPR arguments. But, unlike for Bohr, for Schrödinger this relational aspect of the quantum state is an indication of the *limitations* of the quantum mechanical description and of our inability thus far to construct a satisfactory underlying picture of physical reality.

3.3.1 Schrödinger's Old and New Ideas

One of the earliest examples of Schrödinger's transitional attitudes is contained in the last of four lectures on wave mechanics that Schrödinger delivered between 5 and 14 March 1928 at the Royal Institution in London (Schrödinger 1928d). In the final lecture, after discussing some corrections to the treatment of the hydrogen atom and some perturbation theory, Schrödinger goes on to discuss resonance between two systems, concluding by reiterating his own interpretation of the phenomenon.[23] Then, in the concluding section, after explaining once more his charge density interpretation of the wavefunction, Schrödinger appears to distance himself from it, but goes on to criticise the rival statistical interpretation (in ways related to his report at the 1927 Solvay conference and the attendant discussion; see Schrödinger 1928a), as well as the orthodox notion of *measurement*. This criticism is not the later 'measurement problem', but the question of *what kind of quantities are supposed to be 'measured'* in the first place, and how, in the absence of any proper

[22] In his correspondence with Schrödinger, Einstein calls this the 'Schrödingerian view', to which Schrödinger responds on 19 August 1935: 'I am long past the phase when I thought one can consider the ψ-function somehow directly as a description of reality' (see p. 333).

[23] '[T]he amplitudes belonging to $E + F'$ and to $E' + F$ will show a slow secular interchange, all the others remaining essentially constant. The sum of the squares of the two amplitudes in question is also constant. Interpreted in the single systems, this cannot very well have any other meaning but that e.g. the amplitude of F increases at the expense of that of F' and, so to speak, to compensate for the amplitude of E' increasing at the expense of that of E. This seems to be the appropriate wave-mechanical description of what in the older form of the quantum theory [!] was called the transfer of a quantum of energy $E - E'$ $(= F - F')$ from one system to the other' (Schrödinger 1928d, pp. 50–51).

understanding of this question, orthodoxy hangs on to classical quantities as a lean-to. (As we shall see, the specific sense in which Schrödinger thinks or will come to think of this is very important for some aspects of Schrödinger's attitude in the EPR debate.)

Here is the final section of the lecture in its entirety, entitled 'The Physical Meaning of the Generalized Wavefunction' (Schrödinger 1928d, pp. 51–53):

Perhaps the latter conclusions are obscured by the fact that we have hitherto avoided putting forward any definite assumption as to the physical interpretation of the function $\psi(q_1, q_2, \ldots, q_n, t)$ relating to a system whose configuration in terms of ordinary mechanics is described by the generalized co-ordinates $q_1, q_2, \ldots, q_n$. This interpretation *is* a very delicate question. As an obvious generalization of the procedure of spreading out the electronic charge according to a relative density function $\psi\overline{\psi}$ (which furnished satisfactory results in the one-electron problem [...]), the following view would present itself in the case of a general mechanical system: the real natural system does not behave like the picture which ordinary mechanics forms of it (e.g. a system of point-charges in a definite configuration), but rather behaves like what would be the result of spreading out the system, described by $q_1, q_2, \ldots, q_n$, *throughout its configuration space* in accordance with a relative density function $\psi\overline{\psi}$. This would mean that, if the ordinary mechanical picture is to be made use of at all, the actual system behaves like the ordinary mechanical picture, present in all its possible configurations at the same time, though 'stronger' in some of them than in others.

I maintained this view for some time. The fact that it proves *very* useful can be seen from the one-electron problem [...]. *No* other interpretation of the ψ-function is capable of making us *understand* the large amount of information which the constants a_{kl} [appearing in Schrödinger's discussion of the one-electron atom] furnish about the intensity and polarization of the radiation. Yet this way of putting the matter is surely not quite satisfactory. For what does the expression 'to behave like' mean in the preceding sentences? The 'behaviour' of the ψ-function, i.e. its development in time, is governed by nothing like the laws of classical mechanics; it is governed by the wave-equation. –

An obvious *statistical* interpretation of the ψ-function has been put forward, viz. that it does not relate to a single system at all but to an assemblage of systems, $\psi\overline{\psi}$ determining the fraction of the systems which happen to be in a definite configuration. This view is a little unsatisfactory, since it offers no explanation whatever why the quantities a_{kl} yield all the information which they do yield. In connexion with the statistical interpretation it has been said that to any physical quantity which would have a definite physical meaning and be in principle (*principiell*) measurable according to the classical picture of the atom, there belong definite proper values (just as e.g. the proper values E_k belong to the energy); and it has been said that the results of measuring such a quantity will always be one or the other of these proper values, but never anything intermediate. It seems to me that this statement contains a rather vague conception, namely that of *measuring* a quantity (e.g. energy or moment of momentum), which relates to the classical picture of the atom, i.e. to an obviously wrong one. Is it not rather bold to interpret measurements according to a picture which we know to be wrong? May they not have quite another meaning according to the picture which will finally be forced upon our mind? For example: let a beam of electronic rays pass through a layer of mercury vapour, and measure the deflection of the beam in an electric and in a magnetic field before and after the beam has traversed the vapour. According to the older conceptions this is interpreted as a measurement of differences of energy-levels in the mercury atom. The wave-picture furnishes another interpretation, namely, that the frequency of part of the electronic waves has been diminished by an amount equal to the difference of two proper frequencies of the mercury. Is it quite certain that these two interpretations do not interfere with one another, and that the old one can be maintained together with the new one? Is it quite certain that the conception of energy, indispensable as it is in macroscopic phenomena, has *any* other meaning in micro-mechanical phenomena than the number of vibrations in h seconds?

Later in 1928 we find explicit statements of Schrödinger's views on the role of Anschaulichkeit and suggestions that new concepts are needed (connected to hints that framing the debate in terms of the old concepts might in fact be meaningless), specifically in a talk of 8 December 1928 ('Der erkenntnistheoretische Wert physikalischer Modellvorstellungen', Schrödinger (1928e) [translated as 'Conceptual Models in Physics and their Philosophical Value' in Schrödinger (1935d)]). In this talk, Schrödinger discusses the idea of 'virtual perceptions' supplementing actual perceptions in forming both everyday and physical reality. He remarks that electron trajectories do not exist, since we know that if they existed they would lead to electromagnetic fields quite different from the ones that we observe. Thus, talk of 'orbits' in Bohr's model is misleading.[24] Regarding the uncertainty relations, he comments that if energy and time are not simultaneously well defined, then 'the problem which ardently interested us before, namely, whether energy actually passed by jumps or in a steady flow from one atom to another, naturally becomes illusory'. And he concludes: 'I definitely believe that the elimination [of spatiotemporal detail] ought to be possible without leading to the consequence that no visualisable scheme of the physical universe whatever will prove feasible. The situation will turn out similarly to that of the "colour of the electron". The clearness of the idea of the electron was not seriously interfered with by the fact that the property of possessing this or that definite colour, though common to all perceptible objects, could not be attributed to the electron. In the same way it will be necessary to acquire a definite sense of what is *irrelevant* in our new models and schemes, before we can trust to their guidance with more equanimity and confidence' (p. 132).

As regards specifically the wave concepts, one gets occasional glimpses of the way Schrödinger thought they were still useful (possibly more than the matrix

[24] 'Despite the *immeasurable* progress which we owe to Bohr's theory, I consider it very regrettable that the long and successful handling of its models has blunted our theoretical delicacy of feeling with reference to such questions. We must not hesitate to sharpen it again, lest we may be in too great haste to content ourselves with the new theories which are supplanting Bohr's theory, and believe that we have reached the goal which indeed is still far away' (p. 124). (Note that the argument on the non-existence of trajectories is not dissimilar to Heisenberg's argument in his June 1927 letter to Einstein, discussed in Section 4.2.1.)

concepts), but would have to be modified. One example is 'Die Erfassung der Quantengesetze durch kontinuierliche Funktionen' (['Capturing the Quantum Laws with Continuous Functions'], Schrödinger 1929c). Schrödinger starts from the photoelectric and photochemical effects, where it is the Fourier coefficients of the wave (which are obtained through integrations over the whole of space) that are relevant to the description of the phenomena. Similarly in quantum mechanics, to obtain the quantities that are relevant in the Heisenberg–Born formulation of the theory, integrals of wavefunctions over the whole of space are required. Thus, the wave mechanical picture is not classical, in that it is not *local* properties of the waves that matter.

A little-known source (because not reprinted in Schrödinger's collected works) are the added notes to the French edition of the papers on wave mechanics (Schrödinger 1933).[25]

The longest of these notes (Schrödinger 1933, pp. 98–99) is a postscript to Schrödinger's paper on the equivalence (more accurately, the relation) between wave mechanics and matrix mechanics (Schrödinger 1926b). Appearing in the contents of added material as 'Comparative advantages of the wave hypothesis and of Heisenberg's postulate about transition probabilities', this footnote emphasises the notable success of the theory of emission based on interpreting $\psi^*\psi$ as a charge density,[26] while hinting at difficulties with the Raman effect, discussed in a footnote (Schrödinger 1933, p. 176) to the translation of Schrödinger's fourth paper on quantisation (Schrödinger 1926d).[27] He then expresses in no uncertain terms his conviction in the superiority of wave mechanics in terms of its potential for relativistic generalisation (p. 99):

[25] The collected works only include the introduction by Schrödinger, in which he emphasises that the reader should not get the impression from these early papers that wave mechanics provides in fact a three-dimensional picture of quantum phenomena, adding that 'the internal difficulties of the new theory are much more clearly apparent to us today than in the beginning. They pile up in front of our eyes and culminate in the irreducible antinomy between *waves and particles* (both pictures that we are forced to keep because we do not know yet how to get rid of them) – as well as in the contrast between the evolution of the wave phenomenon, which takes place in a perfectly determined way, and the observable behaviour of the particles, which, in all appearances, is determined only statistically' (Schrödinger 1933, p. xiv). (While one might read this as an embracing of wave–particle complementarity on Schrödinger's part, the parenthetical remark indicates rather that he thinks of these classical pictures as mere crutches.)

[26] 'This hypothesis has received a sensational confirmation through the results obtained in the analysis of the intensity of the X-rays reflected by the different lattice planes of a crystal, an analysis that allows us to determine with sufficient precision how the "diffracting matter" is distributed around the points of the lattice; at present, one can say that the interpretation of these experimental results using a different hypothesis would be, if not impossible, at least very complicated (see M. von Laue, *Die dynamische Theorie der Röntgenstrahlinterferenz in neuer Form*, in the *Ergebnisse der exakten Naturwissenschaften*, 1931, p. 133, Berlin, Springer) [Laue 1931]' (pp. 98–99).

[27] In that footnote, Schrödinger indeed comments on the difficulty for wave mechanics in predicting the Raman effect (cf. Bacciagaluppi and Valentini 2009, pp. 128–131), criticising at the same time also the matrix mechanical calculation, and stating that a (provisional) solution has been achieved only through second quantisation.

Heisenberg's postulate regarding transition probabilities is equivalent to my wave hypothesis about radiation *only to the first approximation*. If one does not stop at this first approximation, one sees that the wave hypothesis turns out to be superior to the said postulate, at least in one point of capital importance, namely the following. According to this hypothesis, the integrals taken *over the whole of space* (keeping time fixed!), which according to Heisenberg give the transition probabilities, determine the emission of radiation only if the dimensions of the emitting system are *small* with respect to the wavelength. The rigorous calculation of emission needs of course to involve the *retarded potentials*, which are given by integrals taken over the *light cone* (to be more precise over the *forward light cone*) rather than taken over the whole of space with time fixed. Now, the notion of light cone is an invariant notion, both in special and general relativity, while the notion of 'the whole of space at a certain time' is not. In these circumstances, there can no longer be any doubt as to the superiority of our point of view with respect to its effectiveness and the possibilities of further development.

What Schrödinger is saying is that the matrix mechanical quantities correspond to spatial integrals of wave mechanical quantities, and that taking such integrals presupposes an absolute notion of simultaneity, while the wave mechanical quantities themselves will be easier to generalise to the relativistic case.

Somewhere during this period, however, Schrödinger appears to have definitively given up the idea of the wavefunction as a direct description of reality.

A striking example of Schrödinger's new attitude, especially interesting partly because it is quoted as an example of the alleged softening of Schrödinger's criticism of orthodoxy, is his talk of 6 May 1930 at the Deutsches Museum in Munich (Schrödinger 1962). It is discussed by Moore (1989, pp. 250–251), who characterises it as 'surprisingly favorable to the Copenhagen–Göttingen interpretations, more so than most of his subsequent statements'. Indeed, Schrödinger states (as quoted in Moore 1989, p. 251):

The wave functions do not describe Nature in itself, but the *knowledge* that we possess at any given time of the observations actually carried out. They allow us to predict the results of future observations not with certainty and precision but with just that degree of unsharpness and probability with which observations actually made on the object permit predictions about it. The wave description that is presently accepted [...] is based on the fact that observations mutually disturb one another – a circumstance that in one respect increases our knowledge of the object, in another respect decreases it.

Most of us today feel that this necessary abandonment of a purely objective description of Nature is a profound change in the physical concept of the world. We feel it as a painful limitation of our right to truth and clarity, that our symbols and formulas and the pictures connected with them do not represent an object independent of the observer but only the relation of subject to object. But is this relation not basically the one true reality that we know?

As Moore puts it here, 'Schrödinger seems to have met Bohr and Heisenberg more than half way', and indeed these passages clearly show Schrödinger going back on his interpretation of the wavefunction as an objective representation of a quantum system. However, this is not so much a concession to orthodoxy, but again an idea for making sense of the limitations of *both* the continuous and the discontinuous approaches. Indeed, the passages leading into the preceding quotation begin with a criticism of the discontinuous jumps picture, or as Schrödinger characterises this picture, 'natura facit nil nisi saltus – Nature does nothing at all except discontinuous jumps' (Schrödinger 1962, pp. 24–25):

> One should however be very careful, indeed, with this interpretation. As I have taken licence to explain at the beginning of this talk, at any rate something discontinuous, jump-like permeates *our observations*; indeed, each single observation fundamentally and always, however one may multiply their number. Out of a certain finite, limited constitution of our mind we are utterly incapable of asking Nature a question that allows for a continuous sequence of answers.

We shall see these thoughts resurface when Schrödinger gets involved with proto-EPR considerations, in fact tempering the more radical conclusions that one might otherwise extract from them.

Another line of thought that Schrödinger starts developing further, beginning in early 1929, is the idea that determinism and indeterminism may be – epistemologically and perhaps ontologically – on the same footing. He had already expressed such views in his inaugural lecture at the University of Zürich of 9 December 1922, quoting his teacher Exner as their originator.[28] In January 1929 Schrödinger published this lecture unaltered in *Naturwissenschaften* (Schrödinger 1929a). In it, Schrödinger states that large-scale regularities are statistical (as revealed by the analysis of fluctuations, as in Brownian motion), and can be interpreted as arising from *either* causal *or* acausal mechanisms. The former interpretation thus requires an additional principle of causality that appears to be redundant (he even intriguingly states that Einstein's gravitation need not be an elementary theory, but may also be merely statistical in nature).

Schrödinger clearly considered these views to be representative of his current point of view, for he now reiterates them and develops them further on a number of occasions. In his Inaugural Lecture at the Prussian Academy of Sciences (4 July 1929, published 24 July as Schrödinger (1929f)), after expressing yet again the view that wave and matrix mechanics both fail to shed light clearly on quantum phenomena,[29] he quotes again Exner's views on acausality, then states his belief that the issue cannot be decided empirically, and that it is presumably a matter of convention and convenience, as Poincaré has argued in the case of geometry.

[28] As such, these ideas predated the Heisenbergian version of fundamental indeterminism(!). For a review of this tradition of 'Vienna indeterminism', see Stöltzner (1999).

[29] '[W]e are still rather far from a real understanding of the view of nature hinted at on the one hand through wave mechanics, on the other hand through HEISENBERG's quantum mechanics' (p. C).

Indeed, we are far from even determining which of the two approaches *is* the most convenient.

Similar views are expressed in the popular article 'Das Gesetz der Zufälle' (Schrödinger (1929d) [published in translation as 'The Law of Chance' in Schrödinger (1935d)]). Schrödinger again imputes to statistical regularities (and fluctuations!) as in thermodynamics our shaky belief in the law of causality. (He notes: 'We shall be especially inclined to sacrifice the causal principle if we follow Hume in recognizing it is not a necessary feature of our thought, but only a convenient habit, generated by the observation of that regularity in the course of actual occurrences the merely statistical character of which is now clearly perceived', pp. 40–41.) Alternatively, we can see causality as fundamental and chance as arising from our inability to discover detailed causal mechanisms. Deciding between these two options is not going to be empirically possible.

These thoughts are repeated also in a number of talks and essays from the early 1930s. In 'Anmerkungen zum Kausalproblem' (['Remarks on the Problem of Causality'], Schrödinger 1932b), Schrödinger again publishes earlier material, namely extracts from a letter to Reichenbach from January 1924. He reiterates both his belief in fundamental acausality à la Exner, and the idea that determinism and indeterminism are experimentally indistinguishable. Schrödinger is specifically commenting on Reichenbach's own ideas, but he concludes that probabilistic laws are equally as mysterious as the causal law.

In some of these writings of the 1930s Schrödinger introduces new aspects to this discussion. In 'Physical Science and the Temper of the Age' (Schrödinger 1935c[30]), he yet again reiterates that causality cannot be proved or disproved experimentally and is just a way we regard nature ('as Hume pointed out long ago'), and makes the parallel with Poincaré's conventionalism in geometry. But he further states that a discarded principle (like that of causality) may be re-adopted at a later date (pp. 93–94). This is a thought that he shares with Laue, to whom we shall presently return.[31]

We see important new aspects perhaps more directly in 'Indeterminismus in der Physik', a talk given in Berlin on 16 June 1931 (Schrödinger (1932c) [translated as

[30] Schrödinger (1935b) and Schrödinger (1935c) are both (very) free but approved translations of, respectively, sections I–II and III of Schrödinger (1932d), itself an expanded version of a talk at the Prussian Academy of Sciences of 18 February 1932.

[31] This talk is particularly memorable, however, for the humorous comparison of quantum mechanics with the [*Neue*] *Sachlichkeit* movement in art (maybe providing some circumstantial support for the Forman thesis! – cf. Forman (1973)). Indeed, Schrödinger compares our no longer shunning 'broad empty spaces in our furniture or on our walls' to the quantum mechanical prohibition of ascribing meaning to unmeasured quantities: 'Our world picture has to remain bare and empty in this respect – we are not afraid of the empty space on our canvas. I have been giving you the current view without criticizing it. You may call it the scientific fashion of the day, if you like, for that is what we are interested in for the purpose of the present discussion' (pp. 90–91).

'Indeterminism in Physics' in Schrödinger (1935d)]). Again it contains the Exnerian views,[32] but also states that positions and velocities on extremely small scales are unobservable in principle because the corresponding *measuring instruments* are not available, 'probably [...] as a matter, not of contingency, but as an essential feature of the physical world' (p. 62). It even 'may be possible that the very ideas of position and trajectory may be seen to be inapplicable, when used with reference to such extremely small spatial and temporal dimensions. This is the present-day attitude of physics; time will show whether it is right or wrong' (p. 63). Maybe this last point is just a paraphrase of orthodox pronouncements, but in this period Schrödinger starts showing an increased interest in the issue of the applicability and definability of geometrical (and temporal) concepts in the light of quantum mechanics.

Indeed, in 'Spezielle Relativitätstheorie und Quantenmechanik' (['Special Relativity and Quantum Mechanics'], Schrödinger 1931b), Schrödinger points out that quantum mechanics poses limitations on the operations required for setting up inertial coordinate systems (e.g. synchronisation of clocks), so that the requirement of Lorentz-invariance is ill-defined (at least if we disregard the possibility of using instruments of infinite mass). He further criticises in some detail the possibility of constructing a quantum mechanical clock.[33] Schrödinger's comments about geometry will find their fullest expression in a paper of 1934 (Schrödinger 1934), to which we shall return.

Many of the ideas that Schrödinger explores in these years are collected and developed further in his lectures at the Institut Henri Poincaré (May 1931, published as Schrödinger (1932a)), where he explores more fully yet another new idea, first sketched in 'Über die Umkehrung der Naturgesetze' ['On the Reversal of Natural Laws'] (Schrödinger 1931a). Here he hints at the quantum mechanical probability density being the product of the solutions of a forwards-in-time and a backwards-in-time equation. This idea is accompanied by a quote from Eddington's Gifford Lectures (Eddington 1928, footnote on p. 216):

The whole interpretation is very obscure, but it seems to depend on whether you are considering the probability *after you know what has happened* or the probability for the purposes of prediction. The $\psi\psi^*$ is obtained by introducing two symmetrical systems of ψ waves

[32] Schrödinger (1932c) and Schrödinger (1932d) were published together as Schrödinger (1932e) and dedicated to Exner.

[33] He had already suggested at the 1927 Solvay conference that it was inconsistent to treat time as continuous in a discontinuity view, citing work by Campbell (1921, 1926; see Bacciagaluppi and Valentini 2009, p. 409). In connection with quantum and spacetime concepts, Schrödinger (1930) also interestingly concludes by suggesting quantum mechanics and unified field theory are two research programmes (both relevant to clarifying the nature of electromagnetism) that might hopefully be some day integrated with each other.

traveling in opposite directions in time; one of these must presumably correspond to probable inference from what is known (or is stated) to have been the condition at a later time.

Schrödinger goes on to discuss the case of classical Brownian motion in one dimension, in which it is supposed that one knows the probability distribution at two times t_0 and t_1, and asks for the probability distribution at an *intermediate* time t. He then shows that the solution is a *product of solutions of the forwards and the backwards Fokker–Planck equations*. In the simple case of δ-distributions as boundary conditions, the solution is a Gaussian that spreads out during half of the time, then contracts again, and is centred along the inertial trajectory connecting the initial and final conditions.

Schrödinger comments on the perfect time symmetry of the resulting equations (reducing to the forwards and backwards Fokker–Planck equations if one chooses uniform distributions as final or initial constraints), and noting that, should one observe a sizeable fluctuation out of equilibrium, then the *most probable* way it has come about will be through the *temporal reversal* of the usual diffusion equation.

He then concludes his Poincaré lectures with speculation about the analogy with quantum mechanics: how this approach restores temporal reversibility to the classical equations (!), while lacking the undulatory aspect (because the equations are real), and wondering whether instead of considering the value of the complex function ψ at a single time as the natural 'initial' condition, one might not instead consider the values of $\psi^*\psi$ (or of $\psi^*\varphi$ with one future-evolving and one past-evolving wavefunction!) at two times as the natural boundary conditions for quantum evolution.[34]

Schrödinger's final criticism of orthodox quantum mechanics in this period is a paper of 1934 on the non-applicability of geometry to small scales (Schrödinger 1934). Besides EPR and together with a handful of papers by Laue, this paper seems to have been one of the most successfully irritating public expressions of quantum dissent.[35]

[34] Schrödinger's continued interest in this approach is indicated by a manuscript note on his own offprint of 'Über die Umkehrung der Naturgesetze': 'This is still a challenge to mathematicians, to solve a certain boundary value problem. 14. III. 52. E.S.' (Schrödinger 1984, p. 412). For contemporary time-symmetric approaches to classical stochastic equations and specifically in classical statistical mechanics, see Bacciagaluppi (2010) and Uffink (2007, section 7), respectively. For explicit consideration of forwards and backwards Schrödinger equations (two-vector quantum mechanics), see e.g. Aharonov and Vaidman (2007). For an approach to quantum mechanics based on time-symmetric diffusion processes see Nelson (1966, 1985); see also the review article by Bacciagaluppi (1999), and the modification of the theory by Derakhshani (2017), which solves the notorious 'Wallstrom problem' (Wallstrom 1994).

[35] Heisenberg in his draft reply to EPR (Chapter 14) cites both Schrödinger's and Laue's papers explicitly. Pauli, in his letter to Heisenberg of 15 June 1935 (see p. 293), complains about both Laue and Einstein being 'older gentlemen' who were 'haunted by the idea that quantum mechanics should indeed be *correct* but *incomplete*'. Pauli expresses similar thoughts to Schrödinger himself in his letter of 9 July 1935 (see p. 312). Grete Hermann quotes Laue's (1932) in her manuscript 'Determinismus und Quantenmechanik' ['Determinism and

The earliest of these papers is by Laue in *Naturwissenschaften* (Laue 1932), and is a critique of the doubts raised by quantum mechanics with regard to the notion of causality, in particular the claim (based on the uncertainty relations) that future events cannot be predicted because the present state of a system cannot be defined with sufficient precision. For Laue, this reasoning is invalid, because it essentially uses the classical conception of a point-mass: 'if there is no point-mass, there is no need to worry about its position and momentum' (p. 916). Statistical methods should be used as long as they are fruitful, but Laue compares the situation with causality in quantum mechanics to that with the principle of contiguous action, which Newtonian gravitation appeared to have to give up (at the unease even of a Newton). He states that Newton's law of gravitation based on action at a distance was incredibly fruitful for the development of physics, and remains to this day of fundamental importance as an approximation. But relativity has restored the principle of contiguous action to its rightful place, even though such a development would not have been possible in Newton's time. No difficulties in physics, Laue concludes, even if they underscore the importance of epistemological issues, can force a change in epistemological standpoint (a point briefly repeated at the end of his (Laue 1933), where he attributes it also to Schrödinger in his (1932b), as we have indeed seen above).

In 1934, Laue adds a critique of the disturbance idea behind the Heisenberg uncertainty relations (also in *Naturwissenschaften*, Laue (1934)). He concedes that the finiteness of the quantum of action imposes limitations on the way we currently understand measurements, in fact limitations that would carry over to new experimental techniques. However, he takes this conclusion to depend on our current way of understanding measurements. A *reconceptualisation* of existing evidence that might lead to re-assessing the claims about uncertainty and causality is not thereby ruled out. Laue concludes by reiterating his idea that it is the conceptualisation in terms of corpuscular concepts (position and momentum in space and time) that is fundamentally limited.

Besides detailed presentation of Schrödinger's arguments about geometry (briefly sketched above), the critique of the disturbance theory of measurement is also one of the main topics of Schrödinger's (1934). Schrödinger provides a number of short but incisive criticisms of the disturbance idea (energy levels do not always have finite spacings, repetition of an ideal measurement apparently does not

Quantum Mechanics'] from 1933, her original critique of completeness in quantum mechanics (Dirac Archives, DRAC 3/11; translated into English in Crull and Bacciagaluppi (2017a, chapter 14); published in Herrmann (2019, pp. 185–203)) – before she turned against the idea of incompleteness in her (Hermann 1935a), which quotes both Laue (1934) and Schrödinger (1934). And Schrödinger himself refers to Einstein, Laue and himself in his letter to Born of 29 June 1935, stating: 'Pardon me for the arrogant line-up, but these happen to be the most renitent obstructionists' (see p. 303).

involve disturbance), and in particular he discusses (to our knowledge for the first time) the idea of a negative-result measurement (p. 519):

> By the way, it is also actually nothing like obvious that the measurement instrument can make no statement about the object without an interaction. A target that is *not* hit at least excludes certain trajectories of the projectile. And if the target encloses the marksman in the shape of a hollow sphere that has only a very small hole, then if not hit it provides very exact information about the trajectory.

Schrödinger then suggests that the orthodox theory is using the wrong concepts for the analysis of measurements: 'the just-mentioned result of the thought experiments indeed speaks to the internal consistency of the theory, but not that it is meaningful to interpret any experiment as a measurement at all of a quantity that *has* no definite value. Does one really still want to call it a *measurement* [...]?' (p. 519), suggesting the term 'Procrustean' for the practice of forcing a system to produce a result in such a 'measurement'.[36] This is the case also for allegedly geometrical measurements of position. As Schrödinger had already suggested, the definition of a rigid body and thus of a frame of reference fails in quantum mechanics (whether relativistically or not), and thus Heisenberg errs in saying position and momentum are separately definable.

3.3.2 Schrödinger and Proto-EPR

By the autumn of 1927, Schrödinger had moved from Zürich to Berlin, where he joined, among others, Laue and Einstein, and this is where Schrödinger's worries about entanglement intersect the prehistory of EPR.

There is a prelude, however, in that Schrödinger may have played a role in Einstein's thinking about the Einstein–Rupp experiment – itself (as we have seen) a prelude to the 1930s version of Einstein's photon box. Schrödinger in fact corresponded with Einstein about the Einstein–Rupp collaboration, and in a letter of 28 April 1926 made suggestions as to why one might after all expect the result of the experiment to conform to the wave aspect of light. Comparing dates, this appears to be the exact same week when Einstein definitively made up his mind that this was indeed the correct expectation (Van Dongen 2007a, p. 84).[37]

[36] The mythological reference is to the highway robber Procrustes who stretched or cut off the limbs of his victims until they matched the length of his iron bed. He was disposed of by Theseus using the same method.

[37] A fragment of this letter is included in Meyenn's edition of Schrödinger's correspondence, with a footnote explaining that the letter is missing from the Albert Einstein Archives and wondering where it might be (Meyenn 2011, vol. 1, p. 218). It is to be found among Sommerfeld's correspondence in AHQP 84, section S-001 (3pp., carbon copy, in German). Schrödinger had evidently sent Sommerfeld a copy of the letter, which primarily discusses Schrödinger's (1926a) paper on Einstein's gas theory about which Sommerfeld and Schrödinger were also corresponding.

It is well-known that, in his 7 June 1935 letter to Einstein, Schrödinger refers to EPR issues as having been much discussed already in Berlin (see p. 281). While Einstein and Schrödinger may well have discussed proto-EPR issues in private at any time during this period, there is direct evidence that Schrödinger was heavily involved in discussions after the seminar that Einstein gave in Berlin in November 1931 (the report of which is translated herein as Chapter 7). As discussed in Chapter 1, the seminar described the Bohr–Einstein version of the photon box, in which either the frequency of the photon or its time of emission can be determined by manipulating the box, with the small addition of a fixed far-away mirror to reflect back the photon, as in Ehrenfest's letter to Bohr of July 1931. (The report describes both determinations as 'predictions', which can be independently checked by measuring the frequency or the time of arrival of the photon,[38] with an emphasis on the possibility of a delayed choice of which prediction to make *after* the photon has already left.)

Indeed, in his letter to Sommerfeld of 11 December 1931, Schrödinger describes in quite some detail a very similar example:[39]

In the seminar we have debated for hours about the following, actually very simple case (conceived *one*-dimensionally):

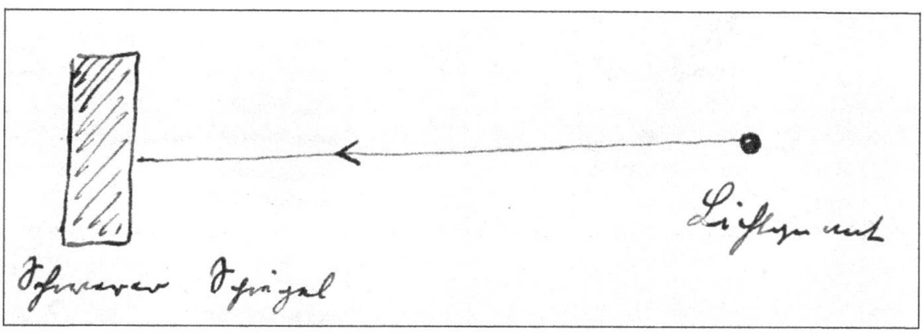

Fig. 3.1 Labels read 'Heavy mirror' and 'Light quantum'. Thanks to Anna Krutsch for providing a digital scan.

We measure the *momentum* (= 0) of a heavy mirror *very* exactly and simultaneously the *position* of a light quantum *very* exactly. (The colour of the light quantum and the position of the mirror are thus very uncertain.) Then I let the light quantum be reflected by the mirror. It thereby absorbs double the momentum of the light quantum. After this has happened and the light quantum has moved away a bit, I have the *choice* through measurement *on*

[38] Cf. Schrödinger's description of the photon-box experiment in the two pages of notes to be presently discussed, and translated as Chapter 8.

[39] The letter to Sommerfeld is AHQP 84, section S-010, transcription in Meyenn (2011, vol. 1, pp. 489–492). We would like to thank Chris Fuchs for attracting our attention to this letter and to the published report of the seminar.

the mirror to ascertain *either* the *position* or the *colour* of the light quantum. Indeed, if I measure the *position* of the mirror, then (with a sufficiently heavy mirror) this still agrees with *that* position where the collision occurred, sufficiently precisely to make a very precise position statement about the light quantum, whose initial position was in fact known. If instead I decide to make a momentum measurement on the mirror (giving up on the mirror's *position*), then since its initial momentum was known, I can infer the *colour* of the light quantum.

In its structure, the example corresponds to Einstein's photon box, but it is notable for involving position and momentum like the EPR example.[40]

The next paragraph in Schrödinger's letter is if anything even more revealing (emphases original):

So on the one hand I can *not* come to know position *and* colour of the light quantum simultaneously. But *without intervening on the light quantum* I can come to know either the one or the other, as I please. Still more crassly: our mirror is a 'universal instrument' for position *and* momentum of the light quantum in the following sense: if one 'applies it' to the light quantum in the above-described manner, then position *and* velocity[41] of the light quantum leave a trace on the instrument, and in fact they are registered with a precision whose product can be pushed well below the limit h.[42] Both traces really exist, because it depends on my free volition to make use of the one or of the other (whereby, however, the second gets smeared out). Since the light quantum no longer has anything to do with my manipulations at the mirror, one cannot well say that it 'receives' a sharp position or sharp momentum only through one of these operations, at any rate it is no 'receiving' in the real [sense], but at most in the mental sense. One might like to conclude therefore that the light quantum *possesses* at all times a fully definite position *and* a fully definite momentum – a notion that we actually abandoned long ago as too harsh and paradoxical. – Just about the only solution seems to be the following 'emergency measure': in quantum mechanics, statements about what is 'real' – statements about the *object* – are forbidden; they *only* deal with the object-subject *relation*, and in fact apparently in a much more radical sense than is ultimately true of *any* natural description.

40 In fact it is rather closely analogous to Bohr's example in his reply to EPR of a particle passing through a single slit in a screen that is movable (like the mirror in Schrödinger's example): while a momentum measurement can be made at any time after reflection (passage through the slit), the position of the photon (particle) can be ascertained only a very short time after the reflection at the mirror (passage through the slit), by measuring the position of the mirror (slit).

41 In the original, this word has a wavy underlining, and there is the marginal annotation 'Impuls' [momentum].

42 In the original, the words starting at 'whose product' have a wavy underlining, and there is the marginal annotation '?'. Sommerfeld's reply to Schrödinger (ZBP-Wien B33-377, handwritten, 1p., in German) refers to this passage: 'About the other problem Bethe thinks: you should talk consistently about the position and *momentum* (not veloc[ity]) of the light quantum. With your universal instr[ument] you can measure the position of the light quantum up to amounts that depend on its momentum, or you can measure the momentum with corresponding ignorance of the position. Does this remove your scruples??' (The word 'Impuls' in the margin of the original of Schrödinger's letter looks in fact like Bethe's hand.) Bethe evidently thought that even the product of the accuracies of the *separate* measurements should not violate the uncertainty relations, and thought that the suggested violation was an artefact of switching freely between momentum and velocity.

That Schrödinger at this time was in possession of all the elements of the EPR argument is not just confirmed but radically strengthened by the extraordinary findings of Jos Uffink and Christoph Lehner – who have unearthed Schrödinger's notes relating to the seminar, including notes in which Schrödinger actually works out a regularised version of the EPR state (and is thus being even more rigorous than EPR themselves). As Uffink (2017, 2018) remarks, it may be that the discussions mentioned in the letter of 7 June are in fact ones in which Schrödinger detailed *to* Einstein how the argument could be further developed formally.

What is also remarkable in this passage is that Schrödinger appears to shy away from fully endorsing the conclusion that the photon indeed '*possesses* at all times a fully definite position *and* a fully definite momentum'. He rather returns to the idea voiced in his Munich seminar of 1930 that the quantum mechanical description of the photon is in some appropriate sense a description of the *relation* between the probing subject and the object probed. While Schrödinger does not spell out his reservations here (but he seems to imply that the idea of simultaneous possession of position and momentum had been given up with good reason), we shall see in several places in his published and unpublished reactions to the EPR paper that Schrödinger finds (or will come to find) the conclusion of the EPR argument more truly *paradoxical* than Einstein – for whom it was a clear argument for the incompleteness of quantum mechanics.

The final form of Schrödinger's 'Einstein paradox' will crystallise only during the summer of 1935 (specifically, as we shall suggest in the next section, through his development of a remark by Pauli). But Schrödinger appears to express some doubts on the epistemic reading of uncertainty in the photon-box example also in two pages of notes of uncertain dating, but presumably no later than 1932.[43]

In these notes, Schrödinger first summarises the 'Bohr–Einstein example' in the exact version Einstein gave in Berlin (insofar as one can surmise from the report), including hints that Einstein in fact gave details of how each separate measurement on the box makes the result of the other uncertain: a short signal leads to an uncontrollable shift in weight, accurate determination of the energy makes the duration of the opening of the shutter indeterminate ('I am forced to assume that for observer B the *temporal duration* of the lid opening is indeterminate').

Then Schrödinger states (see p. 153):

[43] As regards dating: Schrödinger mentions a remark by Wigner told him by Becker over the phone (very many thanks to Jos Uffink and Christoph Lehner for pointing out it is Becker that Schrödinger is referring to, and not Bethe as we had originally misread). Wigner had been Becker's assistant in Berlin, and in the period 1930–32 he was half-time at Princeton and half-time at Berlin, losing the latter job with the advent of the Nazis in early 1933.

So Einstein said to me. I believe, because of a remark by Wigner that Becker related to me over the phone, that this temporal duration *is* indeterminate. An automatic shutter with the presupposed properties *has no sharp weight*.

It would appear that Schrödinger is just agreeing with the conclusion that Einstein has just stated (which he then seems to shore up with an explicit calculation). But then, what is the point Schrödinger is making? And why does he need Wigner's input for that?

Our suggestion is that here Schrödinger is worrying about Einstein's idea that the uncertainty (about the properties of the macroscopic box!) is to be understood *epistemically*. Indeed, his phrasing in the passage above is 'An automatic shutter with the presupposed properties *has no sharp weight*', and the final paragraph of the notes reads:

The thing to be weighed is coupled with the scales. It is by all means admissible to imagine that the energy of the total system box + scales is sharp. But it is evidently inadmissible to imagine that in addition also the reading of the scales is sharp.

The evidence may not be conclusive, but it seems that Schrödinger – who had been worrying about energy exchange leading to entanglement ever since 1927 – is seeing the same problem returning on a macroscopic level, and has thus reservations about concluding to an epistemic reading of the uncertainty involved.

New explicit reservations were to come in the summer of 1935. Furthermore, in the famous 'cat' paper macroscopic entanglement was itself to become an argument against the completeness, or rather against the direct literal interpretability of the wavefunction as a description of reality.

3.4 Schrödinger's Reactions to EPR

We turn finally to Schrödinger's reactions after the EPR paper came out in the 15 May issue of the *Physical Review*. During the summer of 1935, Schrödinger wrote both his first paper on entanglement for the *Proceedings of the Cambridge Philosophical Society* (Schrödinger 1935e, reprinted here as Chapter 10), to be followed by a second one in 1936 in which he generalises the notion of 'steering' further (Schrödinger 1936, final section reprinted here as Appendix to Chapter 10), and his famous 'Schrödinger cat' paper (1935f), reprinted here as Chapter 11.[44]

Especially the first of these papers relates directly to EPR, and develops the paradox further (we shall return to the last paper mainly in Section 3.4.3). Schrödinger names and introduces entanglement as '*the* characteristic trait of quantum mechanics': systems can be such that 'the best possible knowledge of a *whole* does not

[44] As Furry notes (1936b, p. 476; see p. 220), Furry and Schrödinger independently derived some of the same mathematical results.

necessarily include the best possible knowledge of all its *parts*, even though they may be entirely separated' (p. 555; see p. 164). Schrödinger then develops much of the now-standard machinery for discussing entanglement, in particular he develops the notion of steering a system into one of two generally disjoint sets of states by the choice of an appropriate 'measurement programme' on another system, as well as giving a proof of the biorthogonal decomposition theorem. He then shows that the EPR state (even though it is mathematically very singular) is what we now call maximally entangled. Indeed, he shows that for each observable of one of the two systems there is a corresponding observable on the other system that is perfectly correlated to the first, so that the EPR argument applies to *any* observable whatsoever. A schoolboy (or 'scholar' to use Schrödinger's Germanism) will know the answers to *all* possible questions the teacher will care to ask, if he knows already the answer to the two that EPR explicitly consider. But – and this is the new reservation in his version of EPR – Schrödinger also shows that the values for these observables inferrable from the EPR argument will fail to satisfy the *functional relations* between the corresponding operators, concluding that (p. 561/170):

> Our complete lack of insight into the relationship between the different answers in *one* system is all the more bewildering, since we have proved, on the other hand, that the one-to-one correspondence between the answers of the two systems necessarily extends to *all* pairs of observables whenever it holds for two of them.

He concludes the paper by extending the analysis to the time dependence of correlated pairs, and by speculating that the paradox might be eased if predictions did not refer to specific times (which, however, they do in quantum mechanics).

The two 1935 papers were written against the background of a dense correspondence between Schrödinger and a number of other physicists, mainly Einstein (with a correspondence stretching from June to October), but notably also Pauli and Bohr. As we shall see, the correspondence with Pauli is central to understanding the new twist that Schrödinger gave to the EPR paradox (and is illuminating also for Bohr's reply to EPR). In this section, we shall first briefly provide an overview of and comments on Schrödinger's correspondence, then focus on how Schrödinger settled on his final truly paradoxical reading of EPR as in the first Cambridge paper, and conclude with a short analysis of the cat paper. The aspects of the correspondence relevant to Bohr's reply shall be postponed to Chapter 5. Parts of this section follow Bacciagaluppi and Crull (2009) and also the previous study of Schrödinger's version of EPR by Fine (1994).[45]

[45] We had overlooked this excellent paper in our (2009), and are glad to be able to correct that oversight now.

3.4.1 Schrödinger's EPR Correspondence

On 7 June, Schrödinger opens the first letter of his EPR correspondence with Einstein with the words 'I was very pleased that in the paper just published in *Phys. Rev.* you have caught the dogmatic quantum mechanics also publicly by the scruff of the neck, about which we had talked so much already in Berlin' (see p. 281). As we have seen (and as has become even clearer through the work of Uffink and Lehner), all the main elements of the EPR argument (perhaps significantly with the exception of the 'criterion of reality') had indeed already been developed in Berlin. Schrödinger then makes a few clarificatory suggestions (which we shall presently discuss), and it is obvious that he has started working out the results of his first Cambridge paper: he points out that the EPR state is maximally entangled, and in a handwritten addendum he includes the proof of the uniqueness of biorthogonal decompositions.

By the end of June, Schrödinger has received a very substantive reply from Einstein (written 19 June, as well as a 17 June reply to a previous letter), where Einstein clarifies his own preferred version of the reasoning (*without* the criterion of reality). Schrödinger has also had exchanges of letters with Teller (11 and 14 June) and with Born (28 and 29 June). By 1 July he has further had some correspondence with Arnold Berliner about publishing an essay in *Naturwissenschaften* (after having considered publishing a shorter note in *Nature*). There follows an exchange of letters with Pauli (ca. 2 and 9 July), part of which we discuss in detail below (and part of which we discuss when we treat Bohr's reply to EPR in Chapter 5).

Then, on 13 July Schrödinger writes to Einstein that 'Now I am having fun and taking your note as an opportunity to prod a variety of clever people on this topic – London, Teller, Born, Pauli, Szilard, Weyl' (see p. 317). He includes a miscellany of remarks (variously connected with ideas we have encountered in the previous section), and informs Einstein that he is writing an extended piece. On 25 July he writes to Berliner (who replies on 29 July) that his manuscript for *Naturwissenschaften* is completed (and has changed considerably[46]). On the same day he also writes to Laue describing several of the results he will publish in the Cambridge paper.

On 8 August Einstein describes to Schrödinger his 'gunpowder' thought experiment. Schrödinger replies on 19 August, describing both his cat experiment and his own twists on the EPR paradox.

In the meantime, Schrödinger's two papers had been sent off and received (both on 14 August) by the respective journals. The Cambridge paper was communicated to the *Proceedings of the Cambridge Philosophical Society* by Born, but no accompanying letter from Schrödinger to Born seems to be extant. Instead we have

[46] The original plan for the paper must have been much more directly connected to EPR, since in this letter Schrödinger emphasises that he now mentions the EPR paper only three-quarters of the way in.

the letters between Schrödinger and Berliner (11 and 14 August), about whose dismissal from *Naturwissenschaften* for racial reasons Schrödinger also writes to Einstein on 18 August.

In his last letter to Schrödinger in this series (4 September) Einstein asks for further clarifications, and on 4 October Schrödinger provides a particularly clear explanation of his concerns (and sends his Cambridge paper). Finally, in October (13 and 26) Schrödinger has an exchange of letters with Bohr, concerning Bohr's own reply to EPR.

All of these letters are translated and included in Part V. Most of them were collected in a folder titled 'Korrespondenz betreffend das Einstein-Paradoxon' ['Correspondence concerning the Einstein paradox'], microfilmed as section 2 of AHQP 92, which in addition contains a carbon copy of a one-page typescript by Schrödinger on 'The Observation of Canonically Conjugates' (included herein as Chapter 20), and a letter of 14 June 1935 from the editor of *Nature* to Schrödinger (not reproduced here) about reviewing *Science and the Human Temperament* (Schrödinger 1935d), which included several of his foundational essays of the early 1930s discussed above[47], and expressing some interest in the possible piece that Schrödinger had proposed to the journal.[48] Nothing came of that, however. Enclosed with the letter from *Nature* were the proofs of Flint (1935, reprinted herein as Chapter 19).

Among this material, the correspondence with Einstein is of course best known, because it has been analysed by Howard, Fine, Held and others in the context of Einstein's views on EPR (as discussed in Chapter 1), as well as by Fine (1994) in the context of Schrödinger's views. Some aspects will be commented on in this section, others in Section 3.4.2. Among the rest of the correspondence, we shall comment further also on the exchanges with Teller and with Born (in this section), but especially on the ones with Pauli (in Section 3.4.2) and with Bohr (in Chapter 5).

Let us start with Schrödinger's correspondence with Einstein. His first letter presents what 'looks at first like objections, but they are just points I would like to have formulated more clearly' (see p. 281). As Schrödinger explains, taking the case of a measurement of a single quantity, one might be able to sometimes predict with probability 1 that the distant particle has a value A' for A, and sometimes also with probability 1 that the distant particle has a value A''. That does not mean that it must possess both contradictory values, so why should it follow with two different

[47] The book was later reprinted and expanded as *Science, Theory and Man* (Schrödinger 1957).

[48] 'I have been occupied today in outlining a note on "The completeness of quantum mechanical description" (of roughly 2000 words). It is immediately occasioned by a paper of Einstein, Podolsky and Rosen in the 15th May issue of Physical Review [*sic*]. To a large extent my paper may be called a free report of Einstein's views, divested from mathematical technique. Would you care for inspecting it, whether you think it suitable for *Nature*?' (Schrödinger to the Editor of *Nature*, 12 June 1935, AHQP 92, section 1-906, carbon copy, 1p., in English).

quantities? The reason is that (filling in some mathematical details) if one takes the biorthogonal decomposition of a state, one sees that there is a measurement on system 1 that *guarantees* that a quantity A will exhibit some value or other on system 2 if measured, while if one shifts to another decomposition, one sees that there are other measurements on system 1 such that at least sometimes, some other *incompatible* quantity B of system 2 will exhibit a corresponding value if measured (the conditions can be relaxed even further).[49] This is somewhat obscured in the EPR paper by the use of a maximally entangled example, where the biorthogonal decomposition is maximally degenerate. (The letter and the addendum include further details on biorthogonal decompositions.) Schrödinger concludes with some short remarks on how he thought that the current non-relativistic quantum mechanics was ill-suited to treat separated systems.

Einstein's letter of 17 June includes remarks about a possible call for Schrödinger to the Institute of Advanced Studies and other personal issues, then he comments on an argument in a previous letter by Schrödinger, in which Schrödinger suggested that the Maxwell equations are only a linear approximation to some set of non-linear equations – for example, those in the theory of Born and Infeld (1934a, b, 1935), to which Schrödinger also contributed (Born and Schrödinger 1935, Schrödinger 1935g). Schrödinger had noticed that the usual techniques used for deriving energy and momentum density from the (linear) Maxwell equations can be applied also to sound waves, giving the correct results – even though in the case of sound one believes that the correct derivation of these results requires considering non-linearities. This could have been an indication that the usual techniques were in fact a cheap trick, and non-linearities were required also in the electromagnetic case.[50] Einstein then reports on results in Einstein and Rosen (1935) – where wormholes (or 'Einstein–Rosen bridges') were first introduced, which in Einstein's view might have been relevant to his project of recovering quantum mechanics from a unified field theory – and he concludes with some colourful remarks about the sins of quantum mechanics.

Einstein's letter of 19 June is particularly illuminating regarding Einstein's views on completeness and his own preferred version of the incompleteness argument, and it has already been discussed in detail in Chapter 1. (We shall return to it briefly when we discuss Schrödinger's correspondence with Born below.) Einstein concludes it by declaring himself sceptical about the possibility that the problems with

[49] Schrödinger does not mention it here, but the two 'measurement programmes' can also steer the system into two disjoint sets of states when neither of the latter is orthogonal. See Uffink (2020) for a discussion of the possible further significance of Schrödinger's remarks.

[50] See the translation of the relevant passage in Meyenn (2011, vol. 2, p. 536, note 47), as well as Schrödinger's further brief comments on this point in his reply to Einstein on 13 July (see p. 320). Schrödinger's argument was published (in Spanish) in Schrödinger (1935i), and on the same subject there appears to be a draft letter by Schrödinger (to a journal) in AHQP 37, section 1-004.

quantum mechanics may be an artefact of the non-relativistic formulation (which in any case is the only consistent theory available).

From Schrödinger's subsequent letter of 13 July we hear further about the proto-EPR developments in Berlin – which appear to also have taken place independently of Einstein (see p. 317):

> From your letter I can see that I completely agree with you in terms of the attitude toward the *existing* theory. Indeed, after you pointed them out already years ago in Berlin, we discussed the issues in the seminars often and heatedly. Again and again the others pointed out to me that there is no magic in a crass sense, say, in which the 'American' system gives $q = 6$ when I do *nothing* or the *one* thing with the 'European' system (you see, we emphasised the separation!), instead $q = 5$ when I do the *other*; again and again I said: it need not be *that* bad in order to cause trouble. In fact, by abusing the European system I can steer the American one *at will* into a 'q-sharp' state or into one that definitely does *not* belong to this class, e.g. into a 'p-sharp' one. That is *also* magic!

Besides informing Einstein of his 'fun' and that he is writing down his considerations in essay form, Schrödinger comments briefly on some short published responses to EPR (Kemble 1935, Flint 1935, Bohr 1935a; all reprinted herein as Chapters 17, 19 and 21). Further remarks focus on how quantum mechanics is fixated on the idea that one is measuring classical quantities, albeit with fundamental uncertainties. For instance, quantisation of the values of energy or indeed of angular momentum with respect to an arbitrary axis suggests that classical quantities are the wrong quantities to be thinking about. These thoughts are given their fullest formulation in the opening sections of the Schrödinger cat paper, where Schrödinger explicitly characterises orthodox quantum mechanics as making (or pretending to make) probabilistic predictions *on the classical model* (see Section 3.4.3 and Chapter 11).

Even further, it is unclear whether the quantum mechanical representation of canonically conjugate pairs corresponds at all to anything we can in fact measure. Indeed, canonically conjugate pairs cannot be approximated by finite matrices, in the sense that the trace of the commutator of two finite matrices is always zero, so if the commutator is a multiple of the identity, it must be zero. This raises doubts as to how one is in fact supposed to measure canonically conjugate quantities. It also connects to Schrödinger's various pronouncements in the early 1930s about our observing capacities being finite. The point is also discussed in the one-page note on 'The Observation of Canonically Conjugates' (included herein as Chapter 20), but the letter is in some ways even more explicit than the note, and certainly more colourful. All of the more technical remarks lead back to the critique of the notion of measurement. Schrödinger concludes with some intriguing side-remarks on his acoustic analogy for the Maxwell equations.

Einstein's letter of 8 August (also discussed in Chapter 1) summarises Einstein's statistical ('Bornian') view, contrasting it with the allegedly 'Schrödingerian' view that the wavefunction is indeed a literal representation of physical reality. Einstein then presents his gunpowder example to show directly that this cannot be so, and concludes with a brief remark on Einstein–Rosen bridges. In his reply of 19 August, Schrödinger then tells of his own cat example (introduced by 'I am long past the phase when I thought one can consider the ψ-function somehow directly as a description of reality'; see p. 333).[51]

The 19 August letter is also relevant to Schrödinger's further thoughts about the paradoxical nature of the EPR result (on which more in the next subsection). Schrödinger also comments on the Einstein–Rosen bridges, on Berliner's dismissal, and on his likely move to Graz.[52]

The final two letters between Einstein and Schrödinger are mainly relevant to the discussion in the next subsection, and will be commented on there, but they also continue the discussions of the Einstein–Rosen paper and of what was happening to Berliner (Einstein in particular chiding Schrödinger for not retracting the paper from *Naturwissenschaften*).

The two short exchanges with Teller and with Born are both revealing of further aspects of Schrödinger's thinking. Teller sends to Schrödinger on 12 June a letter with some mildly uninteresting and other mildly confused arguments. These are dismissed in the first few lines of Schrödinger's reply two days later.[53] He then launches into a remarkably colourful characterisation of the real problem, in terms not tied just to the EPR example (as clear already from his 7 June letter, he realised very well that the argument could be run in the generic, not maximally entangled case). Indeed, he writes (see p. 289):

> According to quantum mechanics, the preparation of a system, whereby it is brought into a certain given state, does not merely consist in material treatment of the system with tools of all kinds, but, rather, what happens afterwards depends on what one does with the tools – whether one burns them, melts them down, tramples them or preserves them in a museum – but in particular whether and what kind of attention one pays to the signs of wear on the tools.

[51] Fine (1986, chapter 5), followed by others, sees the gunpowder example as the source of Schrödinger's cat. Specifically, Fine suggests that Schrödinger *added* material to his paper (which was sent off on 12 August). As a matter of fact, Schrödinger's letter of 25 July to Berliner includes a word count of the then finished manuscript, of '10230 words of text, without the §-titles, which are still missing' (see p. 321). The published version is some 12,170 words long, excluding footnotes, references, figure captions, section titles and the table of contents; thus Schrödinger *did* add material to it. However, we find the case for the specific additions suggested by Fine to be inconclusive. See Uffink (2020) for further discussion.

[52] This was an ill-timed move, just two years before the Anschluß, to be followed soon afterwards by Schrödinger's more permanent move to Dublin. Cf. e.g. Moore (1989, chapter 9).

[53] As Teller's son Paul remarked to us when shown the letters, Schrödinger is clearly oozing with sarcasm in setting this young upstart right.

In the standard Copenhagen view, knowledge procured about a system that was in a pure state must mean that other knowledge gets lost. But if the knowledge that gets lost depends on the choice of manipulation of the apparatus *after* it has ceased to interact with the system, then the view becomes untenable. The argument is not quite spelled out in the letter to Teller, but the missing step becomes clear when we compare to what Schrödinger writes to Pauli in very early July (see p. 311):[54]

> [O]ne can quite easily see that, and in what sense, a system *itself* changes if its psi-function changes. Indeed, each psi-function conveys the *same amount* of knowledge. When it changes, it is never only a gain but always also a loss of knowledge. But one can only acquire, not lose, knowledge (except by going soft in the head, which we shall ignore here). A loss of knowledge can only come about if the circumstances have changed. In this sense I say: different psi-functions definitely correspond to different circumstances – or states. I do not consider this an illegitimate dragging in of the reality dogma.

As Schrödinger puts it in the letter to Teller, the way quantum mechanical state preparations work 'stands on a level with the savage, who believes that he can harm his enemy by piercing his image with a needle' (see p. 289).

As regards Born, his position in his letter to Schrödinger of 28 June seems rather puzzling (as it clearly seemed also to Schrödinger). Born, too, focuses directly on the generic non-maximally entangled example that Einstein uses for his preferred version of the argument in his letter of 19 June, with dependence of the final state of system II on the measurement performed on system I. But Born calls it an 'age-old swindle' that he had already thought about for his paper on the adiabatic theorem (Born 1926c); he does not mention it in the paper, however. The problem according to Born seems to be that EPR neglect the fact that the two measurements on system I must be performed on *different* pairs of systems. He then sketches an analogous classical example in which one can make corresponding deductions about system II depending on the measurements one makes on system I, and in his reply of 29 June, Schrödinger seizes precisely on this classical example (see p. 303):

> [O]ur problem begins where your arguments leave off. The reference to the classical case, where the phases certainly have meaning and ignoring them makes the description incomplete, is surely not at all appropriate to explain that and why the quantum theoretical phases are meaningless and ignoring them does *not* affect the completeness of the description.

No further reply from Born is extant, however.

Back in 1926, when he first introduced the 'statistical interpretation', Born appears to have been espousing a *strictly* statistical view, in which any system is indeed always in a stationary state, and performs quantum jumps between the stationary states. The quantum probabilities are then probabilities for the actual occurrence of this or that stationary state ('Zustandswahrscheinlichkeiten',

[54] See also the detailed exposition of this argument in sections 9 and 10 of the cat paper (Chapter 11).

i.e 'state probabilities'). The statistical interpretation then went through further transformations, notably in Born and Heisenberg's report at the 1927 Solvay conference (Born and Heisenberg 1928), and von Neumann's approach to quantum mechanics as a 'statistical theory' (von Neumann 1927).[55]

While Born co-authored or at least endorsed these developments, his own position remains somewhat unclear. (As late as 1948 in his Waynflete Lectures (Born 1949, pp. 99–101) he appears to still endorse the view that quantum systems are in stationary states and performing quantum jumps when in interaction with other systems, even though the distinction between the system and the environment with which it interacts appears to rest for him on a *subjective* choice by the observer.)

In his letter to Born, Schrödinger also emphasises that the logic of the EPR argument is (supposedly) as follows: from the assumption of the completeness of quantum mechanics it follows *both* (1) that canonical conjugates are not determinable simultaneously, *and* (2) that canonical conjugates must be determinable simultaneously. This establishes that quantum mechanics is incomplete, but does *not* establish that canonical conjugates must be determinable simultaneously. None of the 'recalcitrant obstructionists' (Einstein, Laue and himself) should be accused of believing *that.*

Note that Schrödinger is saying something quite extraordinary here: not merely that the assumption of completeness implies (2′) that there exist elements of reality corresponding to canonical conjugates, but that it implies (2) that *canonical conjugates must be determinable simultaneously*. Can he really mean this, or is it a slip of the pen?

Note further that if Schrödinger does mean the less radical (2′), he is saying that Einstein (of all people) does *not* believe that there are elements of reality corresponding to both position and momentum in the EPR example (because that is derived only for the purpose of bringing out the contradiction).

Even if we leave aside for a moment this additional twist, Schrödinger is clearly endorsing a reading of the logic of EPR in which the argument is a *reductio* proceeding from the assumption of completeness, while most reconstructions of the argument proceed without this assumption.[56] We have discussed this puzzle about the logic of EPR in Chapter 1, suggesting that EPR may have seen completeness in the sense that the quantum state must be a complete description of a quantum system as also implying that any residual probabilistic elements in the description of a measurement can be explained by taking account of the 'insufficiently known factors foreign to the described system' (see p. 297) that are introduced by the observation (and suggesting that Podolsky failed to bring out this point explicitly).

[55] See the references in note 10 above, in particular Bacciagaluppi (2022) for discussion of von Neumann.

[56] That includes Bohr in his famous reply published in October 1935 (Bohr 1935b, reprinted here as Chapter 16); see Section 5.3.

This might explain why Schrödinger is saying that completeness implies the *simultaneous determinability* of position and momentum. Indeed, if we assume that the 'foreign factors' determine the outcomes of all measurements, then, in principle, knowing these foreign factors we could simultaneously predict the values of all physical quantities.[57]

The passage is equally intriguing if we believe that Schrödinger's pen slipped. Let us follow the suggestion that Einstein did not believe that position and momentum are real in the EPR example, and that establishing their reality is just a step in the *reductio* argument against completeness. That actually means that, according to Schrödinger, and contrary to popular belief, even Einstein's notion of *completing* quantum mechanics did not involve adding definite values for position and momentum to the quantum mechanical description. This may suggest that Einstein was very open-minded as to what kind of theory might provide a 'complete' description of reality (as suggested notably by Howard (1985)).

Finally, Born and Schrödinger discuss some remarks by Schrödinger (maybe from his previous letter to Born, which, however, does not seem to be extant[58]) on light quanta and the reduction of the wavepacket. Schrödinger appears to be considering the idea that in an experiment on a single photon going through a beam-splitter, the detection of the photon in one beam should *not* change the probability for detection in the other beam, but that the *expectation value* for the number of photons should be 1. (This may or may not have been related to Einstein's two-box thought experiment in the letter of 19 June.)

3.4.2 The 'Einstein Paradox'

We now zoom in to Pauli's letter to Schrödinger of 9 July, which we shall suggest is of central importance for understanding Schrödinger's final judgement of the EPR argument as truly paradoxical.[59]

After describing Bohr's reply (which Pauli had seen in draft[60]), and commenting as we shall see in Section 5.4 on the notion of quantum state, in the second half of his letter Pauli provides an argument of his own against the possibility of hidden variables in quantum mechanics.

Pauli states (see p. 315, entire passage emphasised in the original):

[57] Further support for this reading is perhaps provided by the cat paper, where Schrödinger talks about the belief that 'a *complete* model, agreeing with reality in *complete exactness* would permit predictive calculation of outcomes of all experiments with complete exactness' (Schrödinger 1935f, p. 324; see herein p. 176).

[58] Our thanks to Nancy Greenspan for correspondence on this point.

[59] This section builds mainly on Bacciagaluppi and Crull (2009) and on Fine (1994).

[60] As is clear from Schrödinger's letter to Bohr of 13 October, however, at least by that later date Schrödinger had also seen an early draft of Bohr's reply; see pp. 345–346.

[I]f one subdivides the quantum mechanical ensembles mentally into sub-ensembles and in the latter *simultaneously* assigns numerical values with greater accuracy than the uncertainty relation allows to quantities A, B belonging to non-commuting operators, then one comes into conflict with statistical claims of quantum mechanics about the total ensemble.

He then presents the following example. Take an ensemble of particles with definite momentum, and subdivide it in sub-ensembles with definite position such that the values of position are distributed uniformly in the total ensemble. By so doing, one can reproduce correctly both the distributions and expectation values of P and of functions $F_1(P)$, and the distributions and expectation values of Q and of functions $F_2(Q)$. Because expectation values are linear, one further reproduces correctly also the expectation values of functions

$$F(P, Q) = F_1(P) + F_2(Q), \tag{3}$$

but not the *distributions* of such functions.

Indeed, take the energy of the harmonic oscillator,

$$F = P^2 + \omega^2 Q^2, \tag{4}$$

or take angular momentum along the z-axis,[61]

$$F = P_x Q_y - P_y Q_x. \tag{5}$$

As Pauli points out, according to quantum mechanics both (4) and (5) take only quantised values, but in the ensemble so constructed they take continuously many values. Thus, assuming values for both P and Q gives in general the wrong distribution for functions $F(P, Q)$, therefore the wrong dispersion for $F(P, Q)$, and therefore in particular the wrong expectation value for $F(P, Q)^2$, coming into conflict with the statistical predictions of quantum mechanics.

Schrödinger took over a variant of this argument in the first paper on entanglement (1935e, pp. 560–561/170, notation adapted):[62]

One might presume [...] that the answer prepared for the variable $[F(Q, P)]$ is simply $[F(q, p)]$, if $[q]$ and $[p]$ are those prepared for $[Q]$ and $[P]$ respectively. But this is not so. For consider, e.g., the *series* of observables

[61] Note that angular momentum does not have the form $F_1(P) + F_2(Q)$, but it does have the form

$$F_1(P_x, Q_y) + F_2(P_y, Q_x),$$

where the arguments of F_1 commute and the arguments of F_2 commute. Thus, also in this case assigning values to P_x, Q_y, P_y and Q_x reproduces the distributions and expectation values of F_1 and F_2 and the expectation value of $F_1 + F_2$.

[62] Schrödinger presumably did take the idea from Pauli. Not only are the similarities between the arguments extremely close, but in the notebook in which Schrödinger is working out the details of the first Cambridge paper, it appears that all the rest of the material is in place already before this argument appears, so that it may have indeed originated through Pauli's external input. Cf. Schrödinger's notebook on 'Grundlagenfragen. Dichtematrix' ['Foundational Issues. Density Matrix'] (AHQP 46, section 007-001, in German and English).

$$[F(Q, P, b) = \frac{1}{b}P^2 + bQ^2],$$

where b is to be a positive c-number parameter. With every value for b we are confronted with a new observable, to which an answer must be pending. Moreover the answer must be, irrespective of b, an odd integral multiple of $\hbar$ (though not necessarily independent of b). This shows plainly that all these answers cannot conform to the results which would be obtained by inserting into the expression the same pair of c-numbers, $[p]$ and $[q]$.

In his variant, Schrödinger points out not only that there are observables $F(P, Q)$ and values p and q that assign a wrong value to F: he points out that *for all* values p and q there exists an observable F that is assigned a wrong value. Thus, not only is there a statistical prediction of quantum mechanics for the total ensemble that is violated, but even for each individual member of the ensemble there is some probability-unity prediction of quantum mechanics that is violated.[63]

This argument appears also in Schrödinger's letters to Laue of 25 July and to Einstein of 19 August. But on 4 September Einstein confesses he had not followed the argument, so on 4 October Schrödinger changes the example, this time taking angular momentum $F = P_xQ_y - P_yQ_x$ and varying continuously the choice of the z-axis.

Spelling this out, we have, for example,

$$F(\mathbf{P}, \mathbf{Q}, b) = P_xQ_y - P_y(Q_x - b). \tag{6}$$

For any values p_x, q_y, p_y, q_x, there are only discretely many positions of the z-axis for which the above expression is an integer multiple of h. So, again, for all values of $\mathbf{P}$ and $\mathbf{Q}$, there are observables that are assigned the wrong values. In Schrödinger's own words (see p. 343):

> Each element of the ensemble can have an integer momentum only with respect to some privileged reference points. For an arbitrarily chosen reference point, the momentum will turn out to be a non-integer. This is *terribly* trivial. [Das ist *furchtbar* trivial.]

This version of the argument appears in print in Schrödinger's cat paper (p. 811/182).

Already in Section 3.3.2 we suggested that Schrödinger, while long familiar with EPR-type arguments, had some reservations about fully embracing the conclusions that Einstein wanted to draw from such arguments. Now, presumably thanks to Pauli's argument, these reservations take their sharpest form. Schrödinger presents

[63] Note there is one more assumption that needs to be made in order to say that Schrödinger and Pauli obtain a contradiction with a prediction of quantum mechanics. This is the assumption that if in an individual case $F(P, Q)$ has the value $F(p, q)$ (because P and Q have the values p and q), then a *measurement* of $F(P, Q)$ would in fact also yield $F(p, q)$. This is again a standard assumption of non-contextuality: measurements reveal the 'hidden' values irrespective of the further details of how a quantity is measured.

a view of the EPR paradox as a *dilemma*: on the one hand, EPR-type considerations force us to infer that every observable has a value even when it is not measured, and on the other hand, these values fail to satisfy the functional relations between the corresponding quantum mechanical observables. As Schrödinger puts it in his letter to Laue of 25 July (see p. 325): 'My devilish system knows all its answers; whichever I ask it first, it will give. But I have not the foggiest idea by what mnemonic technique it masters all these answers.'

The strategy that Schrödinger is following in constructing this dilemma is an early example of a pattern of reasoning familiar from the more recent literature on nonlocality. As is well known, there is a long tradition of impossibility theorems for *non-contextual* hidden variables theories, that is, theories that assign values to (or probabilities for values of) quantum mechanical observables irrespective of the details of the measurement context. However, as Bell (1966) emphasised, actual examples of hidden variables theories (e.g. de Broglie–Bohm pilot-wave theory) are clearly contextual: the result of a measurement is determined not only by the hidden variables but also by the specific experimental context. Therefore, when considering EPR pairs, it is more helpful to impose locality constraints: in particular, the values or probabilities on Alice's side should not depend on the experimental context on Bob's side. Because of the perfect correlations for the case of maximal entanglement, however, this condition translates into a non-contextuality condition for the values or probabilities already on Bob's side. Thus, one can apply the standard impossibility theorems for non-contextual hidden variables theories.[64] And the dilemma proposed by Schrödinger can in fact be read as: either give up on locality, or run afoul of von Neumann's impossibility theorem against non-contextual hidden variables theories.

Indeed, the reasoning employed by Schrödinger (and Pauli) is closely related to von Neumann's theorem, as first pointed out by Fine (1994). Von Neumann's theorem (von Neumann 1927, 1932) is a corollary of his characterisation of quantum mechanics as a 'statistical theory', that is, his characterisation of all possible normalised positive continuous linear functionals $\langle A \rangle$ on the quantum mechanical observables as having the form $\mathrm{Tr}(\rho A)$ for some density operator ρ. But the extremal elements of the convex set of density operators are simply the quantum mechanical vector states, which (in dimension greater or equal 2) are not dispersion-free for all observables. Thus, as a corollary, it follows that (in dimension greater or equal 2) there are no dispersion-free normalised continuous linear functionals on the quantum mechanical observables.

[64] The nonlocality theorem by Stairs (1978) and Heywood and Redhead (1983) (which provides the basis for the Free Will Theorem by Conway and Kochen (2006)) is the classic example of this strategy.

Now, if we focus on Schrödinger's example of the series of observables

$$F(Q, P, b) = \frac{1}{b}P^2 + bQ^2, \tag{7}$$

we see that the only functional relations that he needs are $\langle A^2 \rangle = \langle A \rangle^2$ and $\langle \alpha A + \beta B \rangle = \alpha \langle A \rangle + \beta \langle B \rangle$, that is, just dispersion-freedom and linearity, as von Neumann requires. It is further easy to see that for functionals that are continuous and dispersion-free the assumptions of linearity and of respect of functional relations are in fact equivalent.[65]

Note that the form of the result as used by Pauli and by Schrödinger is slightly different from von Neumann's formulation, because the latter shows that dispersion-free linear functionals on the quantum observables do not exist in the first place, while Pauli and Schrödinger merely show that dispersion-free functionals respecting functional relations lead to contradictions with quantum mechanics (in Schrödinger's case, specifically to values of physical quantities that are not always eigenvalues of the corresponding operators). But this is a minor difference. In fact, it is not difficult to strengthen also Schrödinger's result to a proof of the non-existence of such functionals as in von Neumann's case.[66]

More importantly, however, there is presumably a difference in the intuition behind von Neumann's and Schrödinger's results. Von Neumann's starting point in 1927 are the algebraic relations between the quantum mechanical quantities, and he is investigating on a rigorous basis the form that a statistical theory *of these quantities* should take – as opposed to the *ad hoc* and piecemeal way he thought the 'statistical interpretation' had originally been introduced (cf. Bacciagaluppi 2022). Schrödinger instead is wondering whether one might interpret quantum mechanics quite literally as a theory that poses fundamental statistical limitations on what one can say about *classical quantities*. In this sense, even though each classical quantity has a quantum counterpart, when they are thinking of algebraic relations between physical quantities, von Neumann and Schrödinger are thinking of two different algebraic structures.

[65] One direction is trivial of course. For the other direction, note that Schrödinger defines functions of observables as symmetrised power series. It is easy to show that linearity and dispersion-freedom imply respect of functional relations for all symmetrised polynomials (by induction on the degree of the polynomial), and the result extends to power series by continuity (cf. Bacciagaluppi and Crull 2009).

[66] This follows because one can show that a dispersion-free functional respecting functional relations between the operators must in fact always assign to an operator one of its eigenvalues (if the operator has discrete spectrum). Thus the values must always be, but (by Schrödinger's argument) cannot always be, eigenvalues. The proof is trivial if one uses the modern functional calculus (thanks to Arthur Fine for emphasising this point in correspondence), but it can be shown also by elementary techniques that Schrödinger himself might have used.

Be that as it may, as regards the interpretation of the result, Fine cautions us not to automatically conclude that Schrödinger's proof of von Neumann's corollary implies some form of contextuality or nonlocality (the kind of conclusion one tends to draw from the more recent versions of these arguments). Indeed, one can still imagine (at least) two alternatives: (a) the values (x_1, p_1) and (x_2, p_2) determine the values of all observables for each of the two systems, but not in terms of the functional relations; (b) the measured values depend also on the experimental context. Only were Schrödinger to establish the latter would the dilemma posed for EPR be sharpened, because contextuality together with perfect correlations implies action at a distance.[67] Thus, choosing *either* horn of the dilemma would lead to a form of nonlocality.

But Schrödinger's own version of the 'Einstein paradox' explicitly only presents us with the choice between the violation of the separability principle and option (a). As Fine highlights, this second horn of the dilemma means not only that we cannot in general define values for observables in terms of functional relations, but that if we cannot do this, then according to Schrödinger it is maddeningly difficult to explain how matching pairs of observables always have matching values when measured.

Thinking of Schrödinger's characterisation of quantum mechanics as ostensibly making probabilistic predictions on the classical model, what Schrödinger is showing in his twist on EPR is that no conceivable assignment of values *in the classical model* is consistent with the predictions of quantum mechanics. For Schrödinger, thus, the 'Einstein paradox' shows that *both* alternative interpretations of uncertainty are untenable: the ontic interpretation because of the EPR argument and the epistemic interpretation (in this sense) because of the Pauli–Schrödinger argument.

3.4.3 'The Present Situation in Quantum Mechanics'

The *summa* of Schrödinger's views on quantum theory in 1935 is his paper in *Naturwissenschaften* (1935f and Chapter 11). As we have seen, this also started as a commentary to EPR, but it turned into a more extended essay containing broader reflections on the foundations of quantum theory. Before concluding, we shall thus briefly look at the role that the EPR argument plays within the 'cat paper'.

The framework that Schrödinger sets up summarises nicely various features we have seen above. The paper starts by describing the anschaulich ideal of the 'classical model' (derived, Schrödinger says, from kinetic gas theory), characterised by

[67] Again, assume the result of the measurement on Bob's side depends on the details of Bob's experimental context. Since his result fixes also the result on Alice's side, also the latter depends on the details on Bob's side.

a specific way of filling in the unobservable details in-between the phenomena. Specifying values (at one time) for appropriate subsets of variables (determining data) will fix the values of all variables, and indeed for all times. These, however (in particular also determinism), are features of the *model*.

According to Schrödinger, what is truly characteristic about quantum mechanics is that such classical models turn out to be inadequate. Remarkably, however, quantum mechanics *maintains the fiction of the classical model*, but declares it will make only probabilistic statements about it. Specifically, each classical variable is deemed measurable by itself (even though in the classical models the determining data were only inferred very indirectly – one should think again of classical gases), which in general fixes the *probabilities* for other variables. Thus the functional relations holding between the variables of the classical model no longer allow one to fix the values of all variables, let alone fix their future development. We also see again the idea that quantised values, most crassly in the case of angular momentum, do not fit this picture – suggesting one should instead consider different quantities altogether.

Against this background, Schrödinger proceeds to discuss in turn the epistemic and ontic readings of uncertainty. The epistemic (i.e. ensemble) reading is ruled out using the Pauli–Schrödinger example (specifically the angular momentum version), in the sense that no conceivable choice of values for the classical variables is consistent with all predictions of quantum mechanics.

Having thus dismissed the epistemic reading of uncertainty, Schrödinger considers its ontic reading: values of physical quantities are literally smeared out. He stresses that this picture is perfectly consistent, indeed it is a ('Schrödingerian') picture in which the quantum state is taken as a literal description of reality. Given that the quantum state also evolves deterministically according to the Schrödinger equation, this picture appears to be in close analogy to the ideal of the classical model. However, ontic indeterminacy is acceptable only at the *microscopic* level. As soon as one considers coupling to macroscopic systems (e.g. a hemispherical screen *à la* Einstein or a Wilson cloud chamber), this view is seen to be *empirically* inadequate. This is where Schrödinger brings in the example of the cat in the steel chamber, thus using it to argue specifically against the 'Schrödingerian' notion of completeness. But the orthodox notion of completeness is *not* the Schrödingerian notion, and the EPR argument was deployed against the orthodox notion.[68]

[68] This is not altogether surprising. As we have already seen in discussing Einstein's letters to Schrödinger in Section 1.5.2, the 'Schrödingerian' view could be disposed of simply through the gunpowder experiment, but in order to beat the 'Talmudist' and argue for a distinct 'Bornian' view, Einstein needs in addition to invoke the 'separability principle'.

If uncertainty is to be read neither epistemically nor ontically, is there an alternative? Orthodoxy does purportedly have an answer, even if it is an epistemological *fiat* (remember the Talmudist): there is *no distinction* between knowledge and state description. Indeed, the idea of a model is meaningless ('measurability' of single quantities does not in fact refer to any reality we are trying to model). And the theory becomes one that relates results of one measurement to results of another. But declaring the state to be 'knowledge', or a measurement not to be referring to an underlying reality, does not make the theory devoid of objectivity. 'Measurements' can be defined in operational terms through a repeatability criterion (repetition on the same system), while the theoretical predictions for such measurements can be checked by 'repeating' measurements in an ensemble of different but identically prepared systems.

In this orthodox context, the quantum state is just such a catalogue of theoretical predictions (a 'catalogue of expectations'), and since statements about models are meaningless, the catalogue of expectations provided by the quantum state (in a 'pure case', i.e. a wavefunction) is *ipso facto* complete. This is the orthodox sense of completeness that Schrödinger is eager to put pressure on.

The wavefunction evolves deterministically between measurements but *sui generis* in case a measurement is performed. According to Schrödinger, this aspect is what conflicts with a naively realist interpretation of the theory, because in a realist interpretation a 'measurement' should be analysable as a natural phenomenon and not be an exception to the rules.

On the other hand, while a change in the state through measurement could easily be interpreted in a purely epistemic way insofar as it concerns a *gain* in knowledge, the claim of completeness implies that a gain in predictive capacity in some respect must be compensated by a corresponding loss in predictive capacity in different respects. But such a loss cannot be interpreted purely in epistemic terms: we cannot *lose* knowledge in a measurement ('except by going soft in the head', as Schrödinger had written to Pauli), unless the 'measurement' actually affects how the object *itself* will behave under further measurements.

Thus, even behind the epistemological smokescreen of orthodoxy, the conception of the quantum state as an expectation catalogue for measurement results implies a conception of the quantum state as a description of the state of the system (one obtains 'a kind of new reality – in entirely legitimate fashion', p. 825/190), and orthodoxy can no longer brush aside the request to understand what happens physically in the process of measurement.

Schrödinger now introduces his analysis of entanglement and brings it to bear on the process of measurement. In terms of expectation catalogues, in a measurement interaction the maximal knowledge about the system and the apparatus gets translated into maximal knowledge of the *total* system, which now essentially involves *conditional* expectation catalogues, and the individual systems in general no longer have their own maximal catalogues of expectations. '*The insufficiency of the ψ-function as model replacement rests solely on the fact that one doesn't always have it*' (p. 827/193, original emphasis). Only once the observer somehow gains knowledge about the apparatus does the system also acquire again its own wavefunction. But since there is no longer any interaction with the measured system, this acquisition of a state cannot in fact correspond to a change in the quantum system.

This is the background against which Schrödinger deploys the EPR argument in the cat paper. He runs the argument in a now familiar way, largely as in his first Cambridge paper (but less technically). He discusses different 'measurement programmes', emphasising the possibility (in both the maximal and non-maximal case) of having one type of measurement guarantee that a state from one given set will arise, and another type of measurement guarantee that a state from another disjoint set of states will arise. He then runs the EPR argument along the lines of Einstein's preferred version.[69] Then he analyses specifically the EPR example. He emphasises that the 'schoolboy's' answers to the 'teacher's' questions must be there in advance. He points out in the EPR case that for every single observable on one system there is a perfectly correlated one on the other. And he points out the bewildering lack of functional relations between their values. He finally extends the results to include time dependence of the correlated pairs, and speculates whether a different treatment of time, perhaps in a relativistic setting, might be needed to ease the paradox.

A postlude to Schrödinger's worries about entanglement is his second paper in the *Proceedings of the Cambridge Philosophical Society* (1936, received 21 April, read 26 October). The bulk of the paper deals with generalising further the results about steering discussed in the first paper.[70] The concluding section then picks up the speculations from the first paper about whether the paradox might get resolved in the proper relativistic setting. As Schrödinger puts it fairly strongly in the paper's summary (1936, p. 452):

> It is suggested that these conclusions, unavoidable within the present theory but repugnant to some physicists including the author, are caused by applying non-relativistic quantum mechanics beyond its legitimate range.

[69] '[S]ince after all I neither have actually touched the system A, nor in the imagined other case would have touched it, the statements of the other catalog, whatever it might be, must all *also* be correct. They must therefore be entirely contained within the first, since the first is maximal. But so is the second. So it must be identical with the first' (p. 844/196).

[70] Schrödinger reviews some results about mixed states, then shows that for entangled systems, every convex decomposition of the mixed state of one system into states that are linearly independent can be realised by an appropriate measurement on the other. In the generic case, a system can thus be steered into any state whatsoever with non-zero probability.

The following speculation, found at the end of the last section of Schrödinger's second Cambridge paper[71] and envisaging 'an alternative possibility' for resolving the paradox, was eventually to be proven wrong, but it makes for a fitting end to our story of Schrödinger's long struggles with entanglement (Schrödinger 1936, pp. 451–452/173):

It seems worth noticing that the paradox could be avoided by a very simple assumption, namely if the situation after separating were described by the [biorthogonal] expansion

$$[\Psi(x, y) = \sum_k a_k g_k(x) f_k(y)],$$

but with the additional statement that the knowledge of the *phase relations* between the complex constants a_k has been entirely lost in consequence of the process of separation. This would mean that not only the parts, but the whole system, would be in the situation of a mixture, not of a pure state. It would not preclude the possibility of determining the state of the first system by *suitable* measurements in the second one or *vice versa*. But it would utterly eliminate the experimenter's influence on the state of the system which he does not touch.

This is a very incomplete description and I would not stand for its adequateness. But I would call it a possible one, until I am told, either why it is devoid of meaning or with which experiments it disagrees. My point is, that in a domain which the present theory does not cover, there is room for new assumptions without necessarily contradicting the theory in that region where it is backed by experiment.

[71] The full concluding section of the paper is reproduced as the Appendix to Chapter 10.

4

Heisenberg on EPR*

4.1 Introduction

Immediately following the publication of the EPR paper in the spring of 1935, Pauli wrote to Heisenberg suggesting that Heisenberg develop his own response independently of Bohr (see p. 291). Heisenberg reacted positively to Pauli's suggestion, and set to work on a draft entitled: 'Ist eine deterministische Ergänzung der Quantenmechanik möglich?' ['Is a Deterministic Completion of Quantum Mechanics Possible?']. At the end of August, Heisenberg sent a carbon copy to Bohr, asking for comments, and mentioning he was thinking of sending it to the *Naturwissenschaften* for publication. Bohr was puzzled by certain aspects of the argumentation. Although Heisenberg clarified these aspects in a subsequent letter, he postponed the decision to publish until further discussion with Bohr in October in Copenhagen. As it happens, the paper was never published during Heisenberg's lifetime. A German transcription was published in volume 2 of Pauli's scientific correspondence (Pauli 1985, pp. 409–418), and we include our translation as Chapter 14. We have given an analysis of Heisenberg's draft in a previous paper (Bacciagaluppi and Crull 2009); what follows also serves to situate this manuscript more thoroughly within the broader context of debates in 1927–35 regarding completeness and nonlocality.

More than a direct reply to the EPR paper, Heisenberg's draft is a presentation of Heisenberg's own views about the completeness of quantum mechanics. In particular, the draft contains what appears to be Heisenberg's fullest presentation

* The material in this chapter relates to talks given to the Australasian Association of Philosophy annual conference, Melbourne (July 2008), New Directions in the Foundations of Physics in Washington, DC (April 2010), HQ-3, Berlin (July 2010), the annual History of Science Society meeting, Boston (November 2013), the Boston Colloquium for Philosophy of Science, Boston University (October 2015), as well as to Bacciagaluppi and Crull (2009) and Crull and Bacciagaluppi (2017a).

of his argument about the movable 'cut' between what is counted as part of the system to be observed and what is counted as part of the means of observation (this constitutes the bulk of Heisenberg's §1). It also contains (at the beginning of §2) one of the earliest distinctions between what are now called contextual and non-contextual hidden variables. The crux of Heisenberg's argument is presented in the remainder of his §2, wherein he argues that the assumed supplementation of the quantum mechanical description by way of additional variables (say, position) is incompatible with the quantum mechanical predictions for the measurement of the complementary variables.

The latter point is illustrated with an example used by Heisenberg in 1927, and indeed other key facets of Heisenberg's position appear already in the late 1920s (e.g. the relation of uncertainty to causality and the question of completeness). This is evident in both his correspondence with Einstein in the early summer of 1927 and his contributions to the Solvay conference in the fall of 1927. Our Section 4.2 describes Heisenberg's pre-EPR views on pertinent concepts such as these.

Heisenberg's final §3 summarises the preceding argument by reference to Grete Hermann's regrettably little-known essay 'Die naturphilosophischen Grundlagen der Quantenmechanik' ['The Natural-Philosophical Foundations of Quantum Mechanics'] (Hermann 1935a, c). Indeed, one can recognise the similarity with some of Hermann's main views in most of the section, and as we shall argue in our Section 4.4, the influence of Hermann on Heisenberg's thinking is evident at several additional points in the draft. Indeed, no treatment of Heisenberg's views at this time can afford to neglect mention of Hermann's role.

To begin to appreciate this connection, we describe aspects of Hermann's recently discovered 1933 manuscript on determinism in quantum mechanics (which was read by Bohr and Heisenberg, among others) and draw attention to several ideas common to both Heisenberg's and Hermann's essays of 1935. By way of an *amuse-bouche*, we have included in this volume our translation of Hermann's 1935 article in *Naturwissenschaften*, which is a substantially abridged version of her essay from the spring of that year (see Chapter 15). Detailed analyses of Hermann's work and full translations of both her 1933 and 1935 essays can be found in Crull and Bacciagaluppi (2017a).

Finally, a historiographical point: due to difficulties obtaining certain primary sources from the 1930s, Heisenberg scholarship in this time period has often been forced to lean heavily on his writings from the 1940s and 1950s, which were no doubt greatly influenced by the War and Heisenberg's role therein. In this chapter we shall provide details of Heisenberg's philosophical leanings specifically through examination of his unpublished response to EPR. This in tandem with critical investigation of new primary sources from the 1930s will, we hope, open up a richer perspective on Heisenberg's work and thought in these earlier years.

4.2 Heisenberg's Pre-EPR Views

In this section we highlight aspects of Heisenberg's thinking that we consider (in)formative regarding his response to EPR, namely: his views on indeterminism as set out in particular in correspondence with Einstein in 1927 (Section 4.2.1) and his views on interference as set out in particular in his joint report with Born at the 1927 Solvay conference (Section 4.2.2).[1] The important interactions between Heisenberg and Grete Hermann from 1933–35 shall receive separate treatment in Section 4.4.

Since the details of Heisenberg's thinking in this time period have been examined elsewhere, this section is intended to function primarily as a reminder of the manner and extent to which the relevant concepts (e.g. completeness and locality) were already forming in Heisenberg's thought prior to EPR.

4.2.1 The 1927 Heisenberg–Einstein Correspondence

Bacciagaluppi and Valentini (2009) and Bacciagaluppi and Crull (2009) both note that already in early 1927 Heisenberg considers the indeterminism arising in quantum mechanics as a necessary feature of the theory rather than a provisional one. Traces of this thinking can be seen in his uncertainty paper in March of that year (see Jähnert and Lehner 2012, pp. 21–22), but he is explicit on this point in his correspondence with Einstein a few months later.

As discussed in Chapter 1, in May 1927 Einstein proposed, but never published, an alternative version of pilot-wave theory. Heisenberg had heard of the paper and wrote to Einstein about it (it is unlikely that he ever saw a copy, though). Of this correspondence two letters survive, both from Heisenberg to Einstein.[2] The second letter gives in detail one of (if not the) earliest of Heisenberg's arguments for the view that in quantum mechanics indeterminism is '*necessary*, not just consistently possible' (emphasis original).[3]

In the second letter Heisenberg imagines free electrons impinging on a grating with lattice spacing comparable to the electrons' de Broglie wavelength. If electrons indeed have trajectories as Einstein suggests, yet also behave like particles (i.e. engage in short-range interactions), then it would be possible for a small obstacle, cleverly placed on the grating, to deflect the electrons in an arbitrary direction. However, as both Heisenberg and Einstein well knew from various experimental

[1] For more details on Heisenberg's thinking in this period, see Bacciagaluppi (2008), Bacciagaluppi and Valentini (2009) and Camilleri (2009).

[2] Heisenberg to Einstein, 19 May and 10 June 1927 (EA 12-173 and 12-174, handwritten, 2pp. and 3pp., in German).

[3] Einstein's 1927 unpublished draft and Heisenberg's response are discussed in connection with de Broglie's talk on pilot-wave theory at the 1927 Solvay conference in section 11.3 of Bacciagaluppi and Valentini (2009).

results, electrons diffracted in such a way are diffracted only in certain preferential directions, suggesting that each electron's 'trajectory' depends on global properties of the grating and not solely on its local properties, as one would expect in particle interactions.

One can recover diffraction phenomena by letting the trajectory depend on global properties of the wave, but Heisenberg insists that such a move is tantamount to giving up the 'particle' notion altogether. The essence of his argument can be understood as follows: the existence of hidden variables describing the quanta (here in the sense Einstein describes in his manuscript – of deterministic electron trajectories) would be inconsistent with measured interference effects.

4.2.2 Born and Heisenberg's Solvay Report

Born and Heisenberg's report at the 1927 Solvay conference (Born and Heisenberg 1928) contained arguably the most detailed presentation of the statistical interpretation of quantum mechanics to date. Until this report, the most explicit presentation of the statistical interpretation was in Born's paper on the adiabatic theorem (1926c), which lacks, however, a separate discussion of interference; and this is the crucial point where the report by Born and Heisenberg goes further, with a treatment of interference that is quite remarkable.

The details of this report are discussed elsewhere (Bacciagaluppi 2008, Bacciagaluppi and Valentini 2009, Bacciagaluppi and Crull 2009); for present purposes we note that Born and Heisenberg's treatment of the law of total probability in the Solvay report is analogous to Heisenberg's famous discussion of the 'law of causality' in his uncertainty paper: the law holds as a conditional statement although the state of the system is defined only to within the accuracy given by the uncertainty principle.

Born and Heisenberg's transition probabilities are well-defined even when there is no value that could make a transition. But if such values have a well-defined distribution, the law of total probability automatically holds. We see again the idea that the existence of hidden values would destroy interference, but the mechanism for this is made more explicit than in Heisenberg's 1927 letter to Einstein. Born and Heisenberg here explain interference *precisely* through the absence of any 'hidden' values, specifically for the energy, stating that – necessarily – if the distribution of energy values were well-defined, the interference formula would cease to hold.

4.3 Heisenberg's Unpublished Response to EPR

As mentioned, following the publication of EPR, Pauli wrote to Heisenberg and suggested that Heisenberg develop his own response, despite the fact that Bohr was already composing one. Heisenberg, as Pauli well knew, would provide a rather different argument in defence of quantum mechanics (or the impossibility of ever rendering it 'complete' as Einstein, Podolsky and Rosen desired).[4] Heisenberg took up Pauli's suggestion. On 2 July he wrote back to Pauli, saying 'Since this [Bohr's] reply is very different from the thoughts that one might for instance express in connection with the "cut", I actually still want to write something about the cut in line with your suggestion' (see p. 308). This is the draft we translate as Chapter 14 herein.[5]

As promised, Heisenberg's response centrally relies on the cut argument. Although he had spoken of the 'cut' several times in the years prior to 1935 and would continue to use this specific argument in writings and lectures afterwards, it is in his response to EPR that we find its most detailed instantiation. The general idea is to demonstrate not just the incompleteness but the 'incompletability' of quantum mechanics through appeal to the necessary conceptual divide between observer and observed – this divide being referred to as the 'Schnitt', or cut, between those parts of a system to be treated classically and those to be treated quantum mechanically. It is within the wider context of Heisenberg's cut argument as presented in his 1935 draft that one encounters his fullest argument against the possibility of hidden variables.

The strategy of Heisenberg's response to EPR is as follows: he attempts to demonstrate that unless one wishes to give up on quantum mechanics entirely (including its empirically verified predictions), one must accept that a deterministic completion of quantum mechanics is impossible. Heisenberg first presents a brief review of the assumptions of quantum mechanics and its basic formalism, after which he introduces the cut argument, whose two main components we discuss below. After the cut argument is explained, Heisenberg moves to an evaluation of the possibility of hidden variables, and ends the essay by discussing the unique role of observational context in quantum mechanics.

4.3.1 The Cut Argument

In the Chicago lectures delivered in the spring of 1929, Heisenberg presents aspects of the cut argument in conjunction with a discussion of interference that parallels

[4] Pauli to Heisenberg, 15 June 1935 (CERN-ARCH-PLC heisenberg_0215, and Pauli 1985, pp. 402–405; for translation see p. 291).

[5] The original manuscript is in Heisenberg's Nachlaß in the Werner-Heisenberg-Archiv in Berlin, and is reproduced in the Archive for the History of Quantum Physics (AHQP 45, section 11-001). We have also used a typescript in the Albert Einstein Archives (EA 5-207). The differences are negligible, but in the typescript there are a couple of small annotations in Einstein's hand (some of which are discussed below). Both sources are taken into account in our translation in this volume.

his discussion with Born in their 1927 Solvay report.[6] In 1934 Heisenberg gave a lecture in Hannover for a gathering of the Gesellschaft deutscher Naturforscher und Ärzte [Society of German Scientists and Physicians] titled 'Wandlungen in den Grundlagen der exakten Naturwissenschaft in jüngster Zeit' ['Changes in the Foundations of the Exact Natural Sciences in Most Recent Times'] (Heisenberg 1934). In this lecture Heisenberg mentions the impossibility of completing quantum mechanics based on considerations related to the cut. While he does present the claims that would become central to the cut argument in his 1935 manuscript, he does little in the way of arguing for them. Likewise, in a lecture at the University of Vienna in November of 1935, 'Prinzipielle Fragen der modernen Physik' ['Questions of Principle in Modern Physics'] (Heisenberg 1935b), he again mentions the basic elements of the cut argument, this time including the same thought experiment described in his 1927 letter to Einstein involving particle diffraction. As mentioned, Heisenberg's 1927 argument for the impossibility of hidden variables functions as part of his wider arguments involving the cut. Before addressing the question of hidden variables, then, let us first see how Heisenberg presents the cut argument in his 1935 draft.

The cut argument in the 1935 paper is initially set up as addressing the issue of where the description of the world according to wavefunctions (i.e. the quantum formalism) is to be cut off from the 'classical-anschaulich description' given to macroscopic phenomena, like a line on a photographic plate. Discussion of the cut involves two main components: the claim that the placement of the cut is arbitrary, and the claim that the statistical element can only appear at the place of the cut. Each claim is argued for separately; while the first is essential for the entire argument and possibly uncontentious for EPR, it also serves to emphasise Heisenberg's conviction that the 'classical-anschaulich' aspect must remain a part of quantum mechanics if objectivity is to be maintained.

Heisenberg argues for the arbitrary placement of the cut by imagining the different components of an experimental apparatus: the atomic system *A* to be measured, a chain of measuring devices *B*, *C*, etc., and the observer. He then claims that the probability for registering a specific reaction in the process of measurement is independent of the placement of the cut (up to the sensitivity of the measuring devices).

Three different scenarios are considered in demonstration of this point: (i) the cut is placed between atomic system *A* and the first measuring device *B*; (ii) the cut is placed beyond *B* yet before *C* (in which case *C* is considered the classical

[6] See chapter 4 in Heisenberg (1930a, b). Although in this treatment Heisenberg evidently thinks these two issues are connected, he does not make the connection explicit.

measuring device); and (iii) the cut is placed at any other point after device C yet not beyond the observer.

In case (i) the atomic system A is of course treated quantum mechanically while measuring device B is treated classically. Heisenberg thus assigns the time-dependent wavefunction $\psi_A(q_A, t)$ to the processes of system A. It follows that the probability that the coordinates q_A of A will assume the particular value q'_A at the time of interaction between the atomic system and the measuring device t' is given by $|\psi_A(q'_A, t')|^2$.

In case (ii), the atomic system and the device B are treated quantum mechanically as a composite system $A + B$, and the other side of the cut contains the new measuring device C treated classically. Here Heisenberg must demonstrate that the probability for B showing the specific value q'_A upon interaction with system A corresponds to the probability given in case (i). Prior to any interaction between A and B, the wavefunction of the composite system is given by the product of the undisturbed individual wavefunctions: $\psi_A(q_A, t) \cdot \psi_B(q_B, t)$. Now let H_A be the Hamiltonian of system A, H_B that of system B, and H_{AB} the interaction Hamiltonian. Schrödinger's equation for this system is then:

$$\left(\frac{\hbar}{i}\frac{d}{dt} + H_A + H_B\right)\psi(q_A, q_B, t) = -H_{AB}\psi(q_A, q_B, t), \tag{1}$$

where for $t = t'$ the state $\psi(q_A, q_B, t)$ is given by the product of the undisturbed wavefunctions, $\psi_A(q_A, t') \cdot \psi_B(q_B, t')$. Furthermore, since H_{AB} is zero other than for $q_A = q'_A$, the right-hand side of (1) becomes $-H_{AB}\psi_A(q'_A, t')\psi_B(q_B, t')$. Heisenberg then states without proof that the solution to the Schrödinger equation has the form

$$\psi(q_A, q_B, t) = \psi_A(q_A, t)\psi_B(q_B, t) + \psi_A(q'_A, t')\varphi(q_A, q_B, t, t'), \tag{2}$$

where $\varphi(q_A, q_B, t, t')$ is independent of the state of system A before $t = t'$.

Heisenberg now writes the probability of B obtaining the specific result q'_A in the measurement of coordinate q_A of A at time t' as

$$\int dq_A dq_B |\psi_A(q'_A, t')\varphi(q_A, q_B, t, t')|^2$$
$$= |\psi_A(q'_A, t')|^2 \int dq_A dq_B |\varphi(q_A, q_B, t, t')|^2, \tag{3}$$

which is indeed proportional to the probability of the outcome for case (i).

In case (iii), when the cut is removed beyond any number of measuring devices, the calculation is essentially the same as in (ii). Given these three cases, which cover in general all placements possible for the cut, Heisenberg concludes that the predictions of quantum mechanics regarding a specific measurement outcome are the same irrespective of the placement of the cut (again, up to a factor depending on the sensitivity of the measuring instruments).

An important qualification is made here: the placement of the cut is not completely arbitrary, for the measuring device B could be a microscopic system that always needs to be considered on the quantum mechanical side of the cut. Thus while the cut can be removed arbitrarily far in the direction of the observer, it cannot be moved arbitrarily far in the direction of the atomic system.

There are problems with Heisenberg's proof. For instance, without explicitly invoking the sensitivity of the measurement in case (i), Heisenberg cannot show that the probabilities for registering the outcome are the same in both cases (i) and (ii). Furthermore, (3) is presumably meant as an instance of the standard Born rule, which assumes that the two terms in (2) are orthogonal. But that is inconsistent with the normalisation of the wavefunction because the first term is already normalised. For our purposes, however, these problems are not essential. In fact, a rigorous version of the proof for the movability of the cut is given by von Neumann in chapter VI of his book.[7]

The second component of the cut argument is Heisenberg's claim that the statistical element can only appear at the location of the cut. His reasoning is quite simple: the quantum-mechanical side of the cut is described by a wavefunction evolving deterministically (given well-defined initial values), and the classical side of the cut is described by fully deterministic equations of motion. Thus the statistical element must be introduced precisely at the place of the cut, for the evolution is deterministic everywhere else.

It is at times unclear in what sense Heisenberg understands the cut, and we notice that it plays a slightly different role in the two components just discussed. *Qua* demarcation between the system to be observed and the system performing the observation, the cut is an imaginary divide necessarily imposed by the experimentalist on any system to be investigated, classical or otherwise. *Qua* element of stochasticity introduced by the switch from quantum to classical mechanics the cut takes on a slightly different flavour, as here it is unique to the study of quantum systems. The cut understood in this way is a wholly new aspect required of physical descriptions by quantum mechanics.

Heisenberg does not provide an explicit conclusion based on these two elements of the cut argument in his 1935 draft, nor indeed in any of his presentations of this argument. Instead, both in his Vienna lecture and in his 1935 draft Heisenberg reiterates, with minimal variation, the central argument made in correspondence with

[7] Note that when introducing his proof, von Neumann states that he owes 'essential elements' of the discussion to conversations with Szilard (presumably while they were together in Göttingen), and writes '[c]f. also the similar considerations of Heisenberg', followed by a reference to the Chicago lectures (von Neumann 1932, p. 262, endnote 208; 1955, p. 421, footnote 208).

Einstein in 1927. We interpret this argument as follows: in order to be rid of the statistical element and provide a deterministic completion of quantum mechanics, one must postulate the existence of as-yet unknown (hidden) variables that describe the system's evolution *through the cut.* Since the position of the cut is (largely) arbitrary, one can imagine a scenario in which the cut is removed far beyond the measuring apparatus. But then the state of affairs at the initial place of the cut is overdetermined, since the quantum mechanical description was already determined at that place. Hence additional predictions to be made on the basis of these hidden variables will, in general, contradict the predictions made using quantum mechanics. Thus, no such completion of the theory is possible. We shall now see in detail how Heisenberg develops this argument.

4.3.2 Hidden Variables

In the second section of his EPR response, Heisenberg segues from the cut argument to a specific consideration of hidden variables theories. He opens with the following statements, which deserve special attention (see p. 228):

> We shall thus investigate the assumption that the physical systems about which quantum mechanics makes statistical statements are bearers of hitherto unknown physical properties that determine univocally the behaviour hitherto known only statistically. For example, let the nucleus of the radium atom possess, apart from the properties that are fixed by the knowledge of its stationary state, still other hitherto unknown properties whose knowledge makes possible an exact prediction of when the atomic nucleus will emit an α-particle. And we assume specifically: this statement should hold independently of the means of observation that are used to detect the α-particle.
>
> One might at first be tempted to drop this last special assumption, and consider that the radioactive emission may also depend on the properties of the means of observation.

Clearly Heisenberg recognises as a separate assumption the independence of hidden variables from the means of observation. Here we must pause to correct a claim made in our 2009 paper: we wrote that with these remarks it is Heisenberg – not Bell, as is usually thought – who first clearly delineates contextual from non-contextual hidden variables. However, in the meantime we discovered a 1933 manuscript written by Grete Hermann in which *she* makes this distinction. It is an essay titled 'Determinismus und Quantenmechanik' ['Determinism and Quantum Mechanics'],[8] and the relevant text appears in a section examining the scope of the Heisenberg uncertainty relations. She argues that the uncertainty principle does not

[8] Dirac Archives, DRAC 3/11, translated in full in Crull and Bacciagaluppi (2017a, chapter 14), and now published in the original German (together with other fascinating material) in a volume edited by Kay Herrmann (2019, pp. 185–203) (with two 'r's).

rule out a deterministic completion of quantum mechanics (Crull and Bacciagaluppi 2017a, pp. 226–227):

> [T]he possibility remains, however, of discovering other, as yet unknown physical quantities, such that the knowledge of their values together with the position and momentum of the electron are sufficient to predict the result of the position measurement. These quantities need not necessarily be determinants just of the observed electron itself; they could also pertain to the measurement apparatus used for the position measurement, since this indeed exerts a demonstrable influence on the electron.

We know that Heisenberg read this manuscript with interest (Crull and Bacciagaluppi 2017b), thus the idea of contextual hidden variables as distinct from non-contextual ones originated with Hermann. Heisenberg remains, however, the first to provide separate arguments against both classes of hidden variables.

Indeed, after Heisenberg makes the distinction in his own manuscript of 1935, he quickly dismisses the possibility of contextual hidden variables from grounds of the movability of the cut. If measurement results were dependent upon the observational context, in a scenario wherein the cut is placed well within the domain describable by classical mechanics, this would imply that the manner in which a macroscopic event occurs depends upon some later means of observing it. As this is totally unreasonable, so too are hidden variables of this kind. Using the familiar example of the decay of a radium atom, he writes (see p. 229):

> [O]ne could reckon as part of the system to be observed the counter device and the photographic plate that registers its counts. Then the question of the blackening of the plate replaces the question of the emission of the α-particle. And here, one is surely forced to assume that whether or not the blackening of the plate occurs is entirely independent of how the observer later takes a look at the plate. For our description of nature is based in the end on the assumption that one can speak of objective events in space and time. It would be altogether impossible to speak of the correctness of the predictions of any theory if one did not assume that the occurrence of a particular event is an objective fact that does not depend on our observation of this event.

That Heisenberg errs in ruling out contextual hidden variables is obvious when one considers Bohm's theory, in which a measurement is modelled precisely as the coupling of an arbitrary quantum mechanical observable to an appropriate position observable, which instead can be observed independent of context.[9]

Despite the argument's invalidity, it raises a few points worth considering. For one, here Heisenberg provides a clearer understanding of the nature of the cut, specifying that it does not physically divide the quantum-mechanical world from the classical world. If it did, this would imply a discontinuity 'at some place between

[9] Of course this comes at the cost of rendering the other, arbitrary observable *dependent* upon context. It is this sense in which Bohm's theory is a contextual hidden variables theory: position is not context-dependent, but rather all other variables depend on context so that position need not.

the macroscopic and the microscopic events [...] for which not the slightest indications are present either in experience or in the quantum mechanical formalism' (see p. 229). Imagining the cut to be physically located somewhere is therefore inappropriate. For another, while Heisenberg does not explicitly mention spatial nonlocality as the trouble with contextual hidden variables, he is remarking about the strangeness such a theory would entail in terms of *temporal* nonlocality or even retrocausality (the observer later taking a look at the plate).

Let us now examine Heisenberg's more substantial argument against context-*independent* hidden variables. The method employed is proof by contradiction, and the details nicely parallel those of his letter to Einstein in 1927: assuming the existence of (context-independent) hidden variables will lead to contradiction with the empirically verified predictions of quantum mechanics, ergo such hidden variables cannot exist.

Recall case (i) in which the cut is placed between the atomic system A and the first measuring device B. Assume there exist non-contextual hidden variables at the place of the cut – specifically, variables that assign a definite value to q_A of system A. Under this arrangement, however, one has the freedom to choose at any time after the system's interaction with B whether to measure q_A or p_A. If one chooses to measure q_A at B, then there is no problem. But one could just as well have chosen to measure p_A at B, and then either the hidden variables' precise determination of a value for q_A must lead to a determination of p_A that generally contradicts the predictions of quantum mechanics, or the hidden variables 'learn' which variable to make determinate after B is suitably arranged, and Heisenberg believes he has already demonstrated the unacceptable violation of objectivity this latter option entails.

So it would not be possible for hidden variables to exist were the cut placed between A and B. Since the placement of the cut does not matter for one's calculations (Heisenberg's first component of the cut argument), the impossibility of hidden variables between the system and the first measuring device should be sufficient to demonstrate their impossibility *tout court*. To show this, Heisenberg returns to the same case as in his proof against the possibility of contextual hidden variables, now moving the cut downstream from B yet upstream of an observer. Using the radioactive decay example, he lets A be an α-particle interacting with a diffraction grating B. In such cases it is experimentally known that the particle will scatter in a well-defined direction determined by its de Broglie wavelength. If non-contextual hidden variables exist that determine the angle of diffraction as well as the time of radioactive decay, then one should infer that the particle interacted in a localised way with the diffraction grating – that is, the interaction depended only on a subregion of the grating's surface. However, it is also experimentally known, for example from the analogous case of sodium emission, that the radiation's interaction with a diffraction grating gives rise to interference fringes. These wave phenomena indicate, contrary to the assumption above, that the interaction with the grating is *not* localised. And indeed a wave-like interaction is what one assumes when applying Schrödinger's wave mechanics.

The typescript of Heisenberg's essay belonging to Einstein includes several telling annotations throughout this section. For one, Einstein underlined the sentence where Heisenberg claims that the statistical character in quantum mechanics comes about 'only because at the location separating the observer with his devices from the system to be observed, a fundamentally uncontrollable disturbance of the system by means of observation prevents us from following the causal connections' (see pp. 229–230). Another instance of Einstein marginalia appears where Heisenberg claims that the existence of hidden variables would entail that the angle of diffraction depends on local properties of the grating. Next to this, Einstein simply writes 'no', followed by 'way of speaking as in the usual mechanics' (see p. 231). The existence of hidden variables applied to a quantum system need not entail strict dependence upon short-range interactions with the diffraction grating. Einstein is accusing Heisenberg of reverting to classical notions and thinking of a particle too *classically*.

4.3.3 The Context of Observation

Before describing the role of mutually exclusive observational contexts in the final section of his paper, Heisenberg summarises the argument against hidden variables just given and notes that its contents are essentially captured by the idea of the causal completeness of quantum mechanics as discussed in Grete Hermann's essay, 'Die naturphilosophischen Grundlagen der Quantenmechanik' (Hermann 1935a).

Already in the 2 July 1935 letter to Pauli in which Heisenberg agrees to write his own response to EPR, he made mention of Hermann's essay, even providing Pauli with its publication details (see p. 308):

> In particular I would like to treat the question of the incompleteness of quantum theory. By the way, a half-philosophical essay on this point by Miss Grete Hermann has appeared recently, which has grown out of discussions in Leipzig, and which I find reasonable on the whole (but for the philosophically-rationalist tendency).

Prima facie this does not exactly read as a ringing endorsement of Hermann's work. But one should take away something slightly different given the timbre of other relevant source materials (e.g. the enthusiastic preface Heisenberg composes for the separate publication of Hermann's 1935 essay (Hermann 1935c)). One should read his comments to Pauli more positively: while Heisenberg is understandably less interested in the consequences of quantum mechanics for the specific philosophical programme to which Hermann ascribed, he does make use of central aspects of Hermann's work, including her views on causal completeness (more on which below) as well as her perspicuous and lengthy analysis of Bohrian complementarity.

In this section we take for granted the overlap between Hermann and Heisenberg on the validity of the cut argument and the importance of the context of observation, and focus instead on the interestingly different roles these two features of quantum mechanics play within Heisenberg's arguments for incompletability as compared to Hermann's.[10] We argue that compared to Hermann's 1935 essay – wherein the cut argument *is undergirded* by the relative context of observation – it is less clear in Heisenberg's 1935 essay what role his final section on the context of observation plays within the overarching argument.

In the first chapter of Hermann's 1935 essay she argues that the relative context of observation is the solution to certain puzzles arising in quantum mechanics, namely (i) the conflicting dual descriptions of light (and matter) as both wave and particle as evidenced by various duality experiments; and (ii) the insurmountable limits placed upon one's ability to predict the outcomes of quantum mechanical experiments due to the uncertainty relations. It is out of these puzzles that one is motivated to search for further causal explanations that might complete the theory, and indeed one sees that a complete causal explanation *can* be given in terms of either the wave or particle description *if* one takes into account the context of observation. She explains (Hermann 1935a, as translated in Crull and Bacciagaluppi 2017a, p. 256):

> [T]he quantum mechanical description by which, on the basis of some observation, a physicist determines his system, does not characterise this system completely and absolutely, but (so to speak) reveals only one aspect of it – precisely the aspect that presents itself to the researcher on the basis of the observation made here. From the point of view of this observation – relative to it – the system has no sharp values with respect to certain physical quantities; hence – relative to it there are therefore also no features of the system from which the result of a sharp measurement of these quantities can be read off. However, if one makes such a measurement – which necessarily disturbs the system and brings it into a different state – then one obtains for this new state not only a quantum mechanical description that assigns this quantity a sharp value, but moreover in the context of this mode of description one can also find causes for precisely this unpredictable value having had to result. Nevertheless, those causes were not utilisable for a prediction of this result, since they also determine the system only relatively – exactly like the description given before the measurement – and, indeed, relative to the observation that was made in the first place with the measurement itself.

[10] Though for a more thorough treatment of the role of the context of observation in Hermann's 1935 essay, see Crull (2017).

It is only after Hermann thus argues for the resolution of apparent causal incompleteness via the relative context of observation that she invokes the freedom to measure different quantities after the interaction has ceased, specifically in the context of Heisenberg's γ-ray microscope thought experiment – examined in more detail in our next section. It is only after that, at the tail end of section 13 of the essay, that Heisenberg's cut enters into her considerations. Even then Hermann uses the cut quite differently from Heisenberg in *his* 1935 essay – she invokes it as a means to fully elucidate Bohr's twin principles of complementarity and correspondence. Here is the context for Hermann's introduction of the cut (Crull and Bacciagaluppi 2017a, p. 267):

> From the point of view of *theory*, the classical and quantum mechanical modes of description are joined at the '*cut*' that is inevitable in the theoretical treatment of any measurement event. The calculation that tracks such a process [...] must limit the quantum mechanical approach, i.e. the use of wave functions, to the state of the observed system. For the measuring instrument, for the state of one's own body – which likewise interacts with the observed system during measurement – the physicist is dependent upon the intuitive classical approach that makes intelligible to him the meaning of the registered measurement result. What in this division he considers to be the observed system and what the measuring apparatus, thus how far he follows in the quantum mechanical formalism the strictly causal development of the change in state, is largely left to his fancy. [...] And at the place where he makes [the cut], he necessarily ceases to keep strict causal track of the quantum mechanical states in favour of the classical-intuitive – but fragmented and accordingly unsharp with respect to certain variables – interpretation of the interaction between measuring apparatus and system.

Thus far her use of the cut closely parallels Heisenberg's, both in his earlier lectures and in his draft reply to EPR. But Hermann then goes on to use the cut to motivate considerations that ultimately anticipate – and respond to – the question at the heart of Einstein's worries. To wit: a few pages after her introduction of the cut quoted above, she writes (p. 270):

> The quantum mechanical formalism prescribes for this [the object–measuring device interaction] the multiplication of the two wave functions assigned to the object and the measuring apparatus prior to interaction, then the determination of the operator that is assigned to this interaction, and the application of this operator to the product of the wave functions. The result will generally be a linear combination whose individual terms each connect one state of the object to one of the measuring apparatus. These various possibilities, among which observation decides, are combined on an equal footing by a symbolic addition into a combined wave function that develops in a phase space determined by the degrees of freedom of both systems. But this means that one must forgo *within this observational context* the intuitive tracking of each of the two individual systems.

If the first system of an EPR pair is considered a measuring device for the second system, then one can see that in a way Hermann not only anticipates an EPR-like scenario from the above quote, but furthermore provides a correct explanation based on the failure of separability (entanglement!) for *why* such an interaction necessarily results in the loss of one's ability to describe each *individual* system in an intuitive (read: classical) way.

By way of comparison, in the final section of Heisenberg's draft he writes (see p. 233):

> The experimental results accumulated in quantum mechanics have further shown that the observation of a system in general leads discontinuously from one observational context to another. The causal flow can be followed only within a given observational context; in the discontinuous transition from one observational context to another (in fact a 'complementary' one in the Bohrian sense), only statistical predictions are possible. The possibility of different complementary observational contexts, unknown to the classical theory, is thus responsible for the occurrence of statistical laws.

This statement and the discussion of the context of observation Heisenberg provides after do not fit seamlessly with Heisenberg's larger strategy in his response to EPR; one wonders if his citing the relative context of observation was mainly an attempt to include the nice results of his protégé Hermann, or perhaps a nod in the direction of the strategy preferred by his mentor, Bohr, in defending quantum mechanics. Whatever his reasons for including this particular discussion, it will occasion some complaints of ambiguity from Bohr in their correspondence later that year, as we shall see below. We now turn to a discussion of Hermann and her significance for Heisenberg's draft reply to EPR.

4.4 The Role of Grete Hermann

Hermann, a doctoral student of Emmy Noether's in the mid-1920s, then an enthusiastic member of Leonard Nelson's neo-Kantian school thriving at that time in Göttingen, began investigating the import of quantum mechanics for transcendental idealism soon after the publication of Dirac's 1930 *Principles of Quantum Mechanics*.[11] Drawing upon her unique dual training in mathematics and critical philosophy, in 1933 she attempted to demonstrate that determinism remained an open possibility within quantum mechanics.[12] She sent a copy of this essay not only to Dirac but also to Copenhagen, where in December 1933 it was read with interest by Bohr, Heisenberg and Weizsäcker. As a consequence, Heisenberg invited

[11] See Hansen-Schaberg (2017) for further biographical details on Hermann.

[12] See note 8 above.

Hermann to join his Leipzigerkreis, sit in on his quantum mechanics seminar and learn still more about the physics.[13]

And join she did, not more than a few weeks later.[14] While attending Heisenberg's lectures, she began composing a new essay for the purpose of reconciling the neo-Kantian law of causality with quantum mechanics. She also had many discussions with Heisenberg and Weizsäcker at this time about the interpretation of quantum mechanics, and these three figures certainly influenced each others' thinking in that regard.

Detailed analyses of Hermann's two essays on the philosophy of quantum mechanics from 1933 and 1935 can be found in Crull and Bacciagaluppi (2017a);[15] in this section our aim is to investigate specifically the potential influence of Hermann's thinking on Heisenberg, as evident in his draft response to EPR. Hermann had composed and published her second essay by March of 1935, and thus even her later arguments concerning the completeness (ergo incompletability) of quantum mechanics antedated the publication of EPR by two months and Heisenberg's reply to EPR by four.

4.4.1 Hermann on Completeness

In her first essay of 1933, Hermann's goal is to argue that the indeterministic aspect of quantum mechanics can be excised from the formalism without affecting its explanatory power or logical cohesion. One crucial aspect of her argument is to demonstrate that the possibility of hidden variables remains open, and this she does in part by arguing for the failure of von Neumann's famous proof against such a possibility. Indeed, in both her 1933 and 1935 essays Hermann devotes an entire section to a critique of von Neumann's impossibility proof for hidden variables – the relevant section in her 1935 essay is titled 'Der Zirkel in Neumanns Beweis' ['The Circularity in Neumann's Proof']. In both versions Hermann argues that von Neumann's linearity assumption in the case of non-commuting observables is unjustified, and that it in fact characterises quantum mechanical states; thus von Neumann's proof becomes circular. Hermann's critique is roughly similar to that of Bell (1966),[16] and in this way she anticipates a pivotal insight usually attributed to this latter-day saint of quantum theory.

Since Heisenberg knew Hermann's work well, it is curious that in his treatment of hidden variables he mentions neither this aspect of Hermann's essays nor

[13] For an analysis of Hermann's 1933 essay and comparison to her essay of 1935, see Crull and Bacciagaluppi (2017b).

[14] We know from a letter to Weizsäcker that Hermann was still in Berlin on 14 January 1934, but a letter to her mother on 5 March 1934 reveals that she was in Leipzig by February. See Herrmann (2019, p. 463).

[15] On Hermann's unique interpretation of quantum mechanics, see Crull (2022).

[16] Though see Seevinck (2017) for a detailed comparison of Hermann's and Bell's criticisms of no-go proofs.

von Neumann's proof itself. (He does cite von Neumann's book at the beginning of his draft, but not specifically in connection to hidden variables.[17]) Perhaps we gain insight from a letter Weizsäcker writes to Hermann on 17 December 1933 – shortly after she had sent her 1933 manuscript off to Copenhagen (Herrmann 2019, pp. 439–444). In it, Weizsäcker writes that he agrees with Hermann's main argument that (as yet) there exists no strictly logical deduction leading from the quantum mechanical formalism to indeterminism. Weizsäcker notes that he is responding to Hermann also on behalf of Bohr, and this invites the following historical speculation: if Weizsäcker and Bohr (and, one assumes, Heisenberg) together not only read Hermann's arguments but agreed with her that indeterminism was not strictly entailed by the formalism, then they must also have agreed with her that von Neumann's proof against hidden variables was inconclusive.

Hermann would not long maintain the hope of a deterministic completion of quantum mechanics, however. Writing to her mother on 16 June 1934, Hermann reports a complete turnabout on the question, confessing that it was Heisenberg who convinced her to abandon the train of thought established in her 1933 essay and accept indeterminism as a permanent feature of quantum mechanics (Herrmann 2019, pp. 467–468).

The task for her 1935 essay, then, was to find a way to argue for the causal completeness of quantum mechanics on behalf of the Kantians while simultaneously doing justice to her new-found conviction that the theory was inescapably indeterministic. This she does by appeal to retroduction: although the ability to predict the outcomes of measurements on quantum systems is certainly lost, this does not entail the nonexistence of the requisite chain of cause–effect relations. Such a chain, stretching continuously from initial conditions to pointer reading, can indeed be identified – but only after measurement – by inferring backwards the evolution of the system from the measurement outcome to the initial state.

It is in this sense, according to Hermann, that quantum theory is already causally complete, hence in no need of emendation via hidden variables. Although Heisenberg tackles the completeness question differently by using the movability of the cut, the final analysis is the same for both: the existence of hidden variables would over-determine the measurement results.

4.4.2 Hermann on Contextuality and Separability

Heisenberg's thinking may also have been sharpened through interaction with Hermann regarding contextuality and separability. The first instance of such is

[17] The question of why Heisenberg neglected to mention both von Neumann and Hermann in his discussion of hidden variables has also been raised by Soler (2017) and Seevinck (2017).

in Hermann's 1935 discussion of the context of observation. There she regards Heisenberg's well-known treatment of retrodiction in the Chicago lectures as mistaken (Crull and Bacciagaluppi 2017a, pp. 263–264):

> Since every physical description and explanation of processes is valid only relative to its respective observational context, so the calculation of, say, a corpuscle trajectory combining the variables of different observational contexts into one representation remains physically vacuous precisely to the extent that it exceeds the uncertainty relations: it is uncheckable and provides no grounds for future predictions. The controversial representation Heisenberg has given of this situation – that it is purely a question of taste whether one should assign physical reality to such a calculation of trajectories – is therefore to be amended in the sense that this calculation disregards the quantum mechanically crucial relation each specification of a physical variable [physikalische Bestimmung] bears to the respective observational context, and in this sense becomes physically meaningless. Conversely, the possibility of checking even only indirectly the causal claims that arise in the interpretation of a measurement process depends on the fact that this interpretation does justice to duality, and thus implicitly to the relative character of quantum mechanics.

As we remark in Section 1.4.3, there are certain analogies between Hermann's reasoning here and that of the ETP paper. However, unlike Hermann (and Bohr), in their analysis ETP do not discuss observational context and presuppose a purely epistemic reading of the uncertainty relations. Bohr and Hermann take observational context to be central to their reasoning, and arguably read the uncertainty relations as ontic.

Another instance of Hermann's keen insight concerns separability. This topic arises for Hermann in her unique treatment of the γ-ray microscope thought experiment. In its original form, the thought experiment was deployed both by Heisenberg (1927a) and Weizsäcker (1931) for the specific purpose of illustrating the uncertainty relations. Weizsäcker's treatment considers two different measurement possibilities: if one examines an electron moving in a plane using a microscope with suitably high resolving power, one can either precisely determine the location of the electron (by placing a photographic plate at the eyepiece of the microscope) or precisely determine the momentum of the electron (by placing the photographic plate in the focus of the lens so that it gathers light entering the microscope through the full aperture angle), but the precise measurement of one variable makes impossible a simultaneous measurement of the other, complementary one.[18]

In her 1935 essay Hermann introduces a *third* case for the thought experiment, wherein one removes the photographic plate to an infinite distance. In this *de facto* 'no-measurement' context, the quantum mechanical formalism yields a description

[18] As Jammer has noticed, the Heisenberg microscope in this instance is equivalent to Bohr's single-slit thought experiment in his reply to EPR (Jammer 1974, pp. 96–97).

of the post-collision state which is 'not in the same way intuitive' as is the post-collision state in the first two cases. She writes: 'Through this linear combination the light quantum and the electron are thus not described each by itself, but only in their relation to each other. Each state of the one is associated with one of the other' (Crull and Bacciagaluppi 2017a, p. 258). And so it is from the perspective of relative observational context that Hermann encounters and articulates – two months prior to the publication of EPR – the very notion of nonseparability undergirding Einstein's troubles.

Furthermore, in her concluding statements on the microscope thought experiment she anticipates the situation that would rankle EPR. She writes (Crull and Bacciagaluppi 2017a, pp. 258–259):

> The coexistence of these different [measurement] possibilities now evidently means that, depending on how one procures one's knowledge of the observed system, or, as we can say for this, depending on the relevant observational context, one can obtain different wave functions for the same system and for the same instant – namely for the electron at the time immediately after the collision with the light quantum. Thus the quantum-mechanical characterisation, unlike the classical one, does not pertain to the physical system still somehow 'in itself', and this means here: independently of which observation one uses to procure one's knowledge of it.

EPR, recall, wish to infer the existence of individual elements of reality independently of the means of observation; Einstein wishes for the existence of bijective relations between wavefunctions and measurement outcomes. Hermann points to the futility of both desires by stressing the particular *un*intuitive state description necessitated by quantum mechanics in certain cases – namely, those now known as entangled.

Surely Heisenberg noticed Hermann's new application of his own γ-ray microscope thought experiment here; one wonders what he thought about it, specifically regarding the larger issue of completeness. Hermann seems to be the first to connect these dots; that Heisenberg would also appreciate this connection becomes evident only later, when, in correspondence with Bohr in the autumn of 1935, Heisenberg attempted to elucidate the reasoning behind his arguments in the draft reply to EPR. To wit: in his letter to Bohr on 19 September, Heisenberg appeals to the γ-ray microscope in order to justify the precise way in which the cut renders impossible the existence of hidden variables. And so Heisenberg, like Hermann, comes to see the microscope thought experiment as more than a mere illustration of the position–momentum uncertainty relation, but as a means of showing the incompletability of the entire theory.[19]

[19] One may legitimately ask whether Hermann's analysis of the γ-ray microscope and her argument for the completeness of quantum mechanics had an influence also on *Bohr*'s reply to EPR. Certain is that Hermann visited Bohr's institute and gave a seminar on her 1935 work just 10 days after the EPR paper was published,

4.5 Discussions with Bohr

Aspects of the Bohr–Heisenberg correspondence directly concerning EPR have been mentioned above; in this section we provide fuller treatment of letters in this exchange that have received little or no attention before now.

On the same day Heisenberg writes to Pauli agreeing to compose a response to EPR based on the cut, Bohr writes to Heisenberg reporting that *he* has written a response to EPR, and that he looks forward to Heisenberg's comments on it. Maintaining an optimistic view of the whole affair, Bohr writes that he believes Einstein will eventually come around.[20] Apparently Heisenberg already knew of the existence of Bohr's response – and even its contents, to some degree – for in his letter to Pauli he sketches the main contours of Bohr's argument. On 14 July he writes to Bohr in a congratulatory mood, claiming that Bohr has articulated EPR's argument more clearly than EPR themselves.[21]

In a letter to Heisenberg on 10 August 1935, Bohr writes that he has heard from Pauli of Heisenberg's own reply to EPR. Presumably soon thereafter Heisenberg sends his manuscript to Bohr for comment. Bohr, unsurprisingly, takes Heisenberg's essay as an occasion to wrangle over subtle differences in their respective interpretations. In a short letter on 10 September, Bohr promises to send to Heisenberg in a few days not only a revised version of his own response to EPR, but also comments on Heisenberg's manuscript.[22]

True to his word, five days after this Bohr delivers another letter with the promised content: a revision of Bohr's paper to be submitted to *Physical Review* as well as questions for Heisenberg regarding the cut argument. In the interim period, Heisenberg had written to Margrethe Bohr, saying that there was no need for Bohr, who was at that time still recovering from illness, to rush to return comments on his manuscript; Heisenberg promises Mrs Bohr he will not send it off for publication 'until there is agreement over the content' (see p. 338).

so just about the time when Bohr had famously dropped any other work to concentrate on the reply to EPR's challenge. Hermann herself commented explicitly on the EPR paper only once, in a remarkable letter of 8 July 1936 to logicist Heinrich Scholz and physicist Adolf Kratzer (Herrmann 2019, pp. 517–523). Hermann had invited Scholz to a meeting she was organising in Heidelberg in September, and Scholz had sent her a reply to EPR by Kratzer (who presented it at the Heidelberg meeting). In it, Kratzer appears to suggest that the interaction between a system and a measuring apparatus produces a random and discontinuous transition, which establishes correlations between the two systems, and that reading off the result is accordingly just an extension of the experimenter's knowledge. Hermann objects that if the read-out operation is just extending the experimenter's knowledge, EPR are correct in claiming that the quantum mechanical description is incomplete, because the quantum mechanical description includes unknown elements only if the total state is mixed. On the other hand, if one assumes that a discontinuous transition in the state takes place at the moment of reading the result, then one is assuming a 'mystical interaction' (p. 522) between the system and the apparatus. Hermann's understanding of the EPR paper is characteristically sure-footed, although she states that she knows the paper only from Bohr's reply!

20 Bohr to Heisenberg, 2 July 1935 (AHQP-BSC 20, section 2, carbon copy, 1p., in Danish).

21 Heisenberg to Bohr, 14 July 1935 (AHQP-BSC 20, section 2, handwritten, 2pp., in German).

22 This and other letters mentioned below are translated in Part V, pp. 337ff.

Unfortunately, the copy of the typescript Heisenberg sent to Bohr is not in the Niels Bohr Archive. But one learns from Bohr's subsequent letter to Heisenberg that Bohr has had considerable trouble making sense of the relationship Heisenberg appears to assume between the cut and a given context of observation. Bohr writes, 'Any well-defined quantum-mechanical problem must be concerned with a certain classically described experimental setting; and if one changes the kind or use of the measuring instruments, and thus of the [experimental] setting, the phenomenon will usually change completely' (see p. 339). To illustrate, he cites Heisenberg's example of moving the cut from between a system A and measuring device B to a point beyond B but prior to C, in which case one evaluates the combined system $A + B$ entirely quantum mechanically. In the manuscript Heisenberg stated that system B might yet be used to measure the variable complementary to the one originally considered (i.e. when the cut was between A and B), but Bohr thinks this claim worrisome.

Certainly Bohr is correct when considering Heisenberg's example of system B reacting to variable q_A having the specific value q'_A (Bohr cites his own example of the fixed diaphragm with a slit in a well-defined position). If that is the form of the interaction, then B cannot be used to measure anything else. But Heisenberg was aware of this point already. This much is clear – if not from his (admittedly ambiguous) manuscript, then from his letter to Bohr of 19 September. There, Heisenberg insists that they are in agreement about one's inability to use B to measure also the conjugate variable in the specific case where the cut is removed beyond B, yet B is used to measure q_A.

In fact, Heisenberg considers it unnecessary even that B represent the same system in both cases of measuring conjugate variables on A. He explains this to Bohr as follows. In the case where B *is* the same system both times, it is helpful to think of the γ-ray microscope thought experiment: the photon (B) interacts with the electron (A) then enters the microscope (C). One can examine B in such a way so as to obtain a sharp value for either the position or momentum of A; *which* property B reveals depends on the physical location of the photographic plate with respect to the microscope. Naturally one considers the photographic plate to be part of the measuring apparatus, but then it is the arrangement of C *alone* which determines what is sharply measured after interaction between A and B.

Heisenberg ends his defence against Bohr's criticism by pointing out that although they agree on essential points, Heisenberg does not think that Laue and others who are convinced of the completability of quantum mechanics will be satisfied with Bohr's blunt assertions that (i) there is no room for such a causal completion, and (ii) the issue is primarily an epistemological one. Instead, Heisenberg has tried to adopt a tack emphasising the *formal* inconsistencies arising from attempts at determinism of the kind hoped for by Laue and others (presumably

Schrödinger, as well as, of course, EPR). And this strategy, he believes, is better realised through discussion of the cut than through appeals to complementarity and the relative context of observation.

In a brief letter Bohr sends to Heisenberg shortly thereafter (1 October 1935), he affirms that their residual disagreements are over substantial matters.[23] Heisenberg visited Bohr later that month, and although the exchange is not documented, we guess that complete agreement was *not* reached between them. Or perhaps Bohr managed to convince Heisenberg that publishing a reply to EPR in addition to his own was unnecessary. All that is certain is that Heisenberg's draft was not in the end submitted for publication to *Naturwissenschaften*, or elsewhere.

Writing to Hermann some months after this visit to Copenhagen, Heisenberg insists that while he may not have always been absolutely precise in formulating the cut argument, he was so in his response to EPR. He also reiterates the importance of the cut argument for establishing the incompletability of quantum mechanics. He writes (Heisenberg to Hermann 9 July 1936, translated from Herrmann 2019, pp. 525–526):

> If one places value on the movability [of the cut], then one emphasises a definite and important point of quantum theory, namely this wonderful 'fitting together' of quantum theory and classical theory, which is crucial for non-predictability. But there is also a point in saying that the place of the cut is prescribed. For already through the formulation of the question that one wants to put to nature with the measurement apparatus, one has essentially made the decision about the placement of the cut. One can for instance ask 'How probable is it that a specific point in the Wilson chamber is hit by an electron', or one can ask 'How probable is it that at this point in the Wilson chamber a water droplet will form'. In the first case the cut lies between the electron and the atoms in the Wilson chamber; in the second case between the vapour in the Wilson chamber and the light with which I view the water droplets. With the Bohrian emphasis one recognises very well how classical physics in a sense has already been absorbed into the language [in die Sprache übergegangen ist].

In short, it appears that the heart of Heisenberg's and Bohr's disagreement over responding to EPR concerns the best strategy of proof, not interpretation of the physics itself.

4.6 Concluding Remarks

Although the interpretation of the physics is indeed quite close among the three primary players of this chapter – Heisenberg, Hermann and Bohr – a few subtler aspects remain which put Heisenberg's draft response to EPR in a slightly different light from the others, as well as from other responses examined in this book.

[23] Bohr to Heisenberg, 1 October 1935 (AHQP-BSC 20, section 2, carbon copy, 1p., in Danish).

On the first point, Hermann and Bohr approach the question of completeness in a manner similar to Heisenberg in that all three think in terms of which actions or experimental arrangements will allow for which kinds of measurement results. However, while Bohr and Hermann place emphasis on the whole context of observation, Heisenberg prefers to couch things in terms of specific contradictions: postulating the existence of hidden variables at one time will destroy interference effects at a later time, and in this way engender a conflict with quantum mechanical predictions. Bohr does not even mention hidden variables, while in as much as Hermann is interested in the notion of causality (hence concerned with the possibility of explicating causal chains throughout a measurement process), her analysis of quantum-mechanical completeness likewise focuses directly on hidden variables.

Along these lines, Heisenberg – following Hermann – is the only author engaged in the EPR debate who considers that hidden variables theories might be contextual, in the sense of a joint dependence of measurement results on both hidden variables and measurement context. These two aspects may well be connected: dependence on the measurement context is a dynamical feature of hidden variables theories, and its possibility may have been realised by Heisenberg precisely *because* he thought of hidden variables in a dynamical way – in terms of how they might produce (or fail to produce) the results of particular measurement scenarios.

5
Bohr on EPR*

5.1 Introduction

Bohr's reply to EPR is the most famous among the reactions elicited by their paper and the most discussed (Bohr 1935b, reprinted here as Chapter 16). While conscious of previous debates, our own discussion aims to be both self-contained and to provide some new angles on this endlessly intriguing subject.[1]

It is well known (as we have seen in Chapter 1) that the EPR argument was a development of Einstein's 1930 version of the 'photon-box' thought experiment. While traditionally – arguably following Bohr (1949) – the photon box has been seen as an attempt at a critique of the empirical validity of the uncertainty relations, it appears rather to have been a critique of the *ontic reading* of uncertainty, i.e. of the idea that (in the case of the photon box) time and energy cannot in fact be simultaneously definite. In the terminology of the EPR paper, it was not a critique of the correctness of quantum mechanics but of its completeness. In this way it is presented by Ehrenfest directly to Bohr in his letter of 9 July 1931 (see p. 15):

> [I]t is interesting to be clear about the fact that the projectile [the photon], which is already flying about in isolation "on its own", must be ready to satisfy very different ""non-commutative"" prophecies, "without even knowing" which one of these prophecies one will make (and check).

* The material in this chapter relates to talks given at the XVIIth Seven Pines Symposium, Stillwater, Minnesota (May 2013), at SPHERE, Paris (November 2014) and at the Max Planck Institute for Quantum Optics, Munich (April 2015), as well as to Bacciagaluppi (2015, 2016a, 2017).

[1] For a lucid overview of the main aspects of Bohr's understanding of quantum mechanics, see Scheibe (1973, chapter 1). Thanks to Harvey Brown for pointing us to this excellent book.

However, as Einstein himself remarked to Ehrenfest on 5 April 1932, in this version of the argument the incompleteness of the quantum mechanical description does not yet logically follow (see p. 20).

The version of the argument given in the EPR paper can be seen as including a missing premise that turns the argument into a logically valid one, namely EPR's famous 'sufficient criterion of physical reality' (p. 777/158, emphasis in the original):

> *If, without in any way disturbing a system, we can predict with certainty (i.e., with probability equal to unity) the value of a physical quantity, then there exists an element of physical reality corresponding to this physical quantity.*

We saw in Chapter 1 that the exact role played by this criterion may be less than obvious, and it is well known that the published version of the argument was not Einstein's preferred one. But Bohr reacted to the published version, and in his reply of October 1935 he sought to undermine precisely the criterion of reality as 'ambiguous' and not applicable to the case at hand, so that EPR's claim would not follow.

In a nutshell, as Bohr explains, the ambiguity in the criterion of reality is that while there is no 'mechanical disturbance' of the second particle when the first one is measured, there is '*an influence on the very conditions which define the possible types of predictions regarding the future behavior of the system*' (p. 700/250, emphasis in the original). And it is this kind of disturbance that EPR would need to rule out according to Bohr in order to be able to apply their criterion of reality.

Bohr's argument relies on an analysis of a particle diffracted through a single slit, which in fact is one of his most detailed expositions of complementarity. For Bohr – and these formulations will require some unpacking – 'the impossibility of controlling the reaction of the object on the measuring instruments if these are to serve their purpose' means that (p. 697/245, emphasis in the original)

> the *finite interaction between object and measuring agencies* conditioned by the very existence of the quantum of action entails [...] the necessity of a final renunciation of the classical ideal of causality and a radical revision of our attitude towards the problem of physical reality.

His final conclusion from the single-slit example is (p. 699/249):

> Indeed we have in each experimental arrangement suited for the study of proper quantum phenomena not merely to do with an ignorance of the value of certain physical quantities, but with the impossibility of defining these quantities in an unambiguous way.

Immediately afterwards, he states:

> The last remarks apply equally well to the special problem treated by Einstein, Podolsky and Rosen, which has been referred to above, and which does not actually involve any greater intricacies than the simple examples discussed above.

He then describes a thought experiment with a double slit that realises the EPR state, adapts very briefly his previous points, and carries over the conclusion.

This chapter will try to cover a number of issues, starting from purely textual ones (Section 5.2) and stretching further from the overall logic of Bohr's reply (Section 5.3) to the question of whether Bohr in the end provides a successful reply to EPR (Section 5.7). We shall focus on two main points: (A) the respective role of the two examples used by Bohr (Section 5.4), and – after a detour through Pauli's arguably Bohrian theory of measurement (Section 5.5) – (B) Bohr's notion of non-mechanical disturbance (Section 5.6)

As regards (A), we take it there are two *prima facie* options for the respective role of the examples: (i) The single slit is a familiar example whose analysis is a straightforward issue of disturbance. The double slit is intended as the actual EPR example, and the single slit is merely providing an analogy: also in the double slit case we have disturbance, but of a more subtle kind. (ii) The single and double slits are *strictly* analogous, and the discussion of complementarity in the first one is meant to be illustrative also of the second (and already an example of *non*-mechanical disturbance). Our main claim is going to be that option (ii) is correct.

As regards (B), we shall explore two ways of trying to make sense of Bohr's notion of non-mechanical disturbance (which, if our analysis of (A) is correct, is going to apply equally in the single-slit and the EPR cases): (α) the idea that the *frame of reference* (as defined by the heavy laboratory bench) may play a crucial role, and (β) Don Howard's (1994) analysis of the role of classical concepts in Bohr's thinking. The former we are going to criticise using an argument in fact already suggested by Schrödinger in correspondence with Bohr in October 1935 (in one of the letters we translate below, p. 345). The latter instead seems to us to show the right way for understanding non-mechanical disturbance – perhaps supplemented by Grete Hermann's (1935a, b) ideas about the relative character of quantum mechanical descriptions (cf. Chapters 4 and 15).[2]

We shall eventually point out two possible problems with Bohr's account in Section 5.7. The first relates to the fact that in the EPR example one can consider that the measurements on the two particles could be performed by two different observers at *spacelike* separation, rather than merely involving predictions on different systems that could be checked later by the same observer. The second problem instead relates to the special choice of a maximally entangled example – as opposed to Einstein's preferred versions of his incompleteness argument – and also to which kinds of measurements can be treated satisfactorily in the Copenhagen interpretation.

[2] For a more detailed discussion of Hermann's views on quantum mechanics, see Crull and Bacciagaluppi (2017a).

5.2 The Text

The definitive version of Bohr's reply is the published one in the 15 October issue of the *Physical Review* (Bohr 1935b), and reproduced here as Chapter 16. We know from Rosenfeld's often-quoted recollections (1967, pp. 128–129) that the gestation of the paper was quite frantic:

> This onslaught came down upon us like a bolt from the blue. Its effect on Bohr was remarkable. We were then in the midst of groping attempts at exploring the implications of the fluctuations of charge and current distributions, which presented us with riddles of a kind we had not met in electrodynamics. A new worry could not come at a less propitious time. Yet, as soon as Bohr had heard my report of Einstein's argument, everything else was abandoned: we had to clear up such a misunderstanding at once. We should reply by taking up the same example and showing the right way to speak about it. In great excitement, Bohr immediately started dictating to me the outline of such a reply. Very soon, however, he became hesitant: 'No, this won't do, we must try all over again ... we must make it quite clear...' So it went on for a while, with growing wonder at the unexpected subtlety of the argument.

The manuscript was eventually complete by the beginning of July,[3] and the paper was received by the journal on 13 July. Around that time, the manuscript was circulated to various colleagues.[4] On 10 August Bohr writes to Heisenberg that he has heard from Pauli,[5] and that he will try to make some small changes in proof. Unfortunately no correspondence between Bohr and Pauli on EPR appears to have survived.[6] After a period of illness, on 10 September Bohr informs Heisenberg that he has made the small changes 'to put forward the principled point of view a bit more clearly' (see p. 337), and five days later he sends him the 'improved manuscript' with the changes in proof, which he characterises as being 'of a formal nature only' (see p. 339). We also know from the two letters between Schrödinger and Bohr of that October (see pp. 345ff.) that Schrödinger by then had also seen the manuscript of the paper (from which in fact he quotes). Unfortunately, this manuscript has not been preserved either in the Niels Bohr Archive[7] or in the Schrödinger *Nachlaß* in Alpbach.[8,9]

[3] See Bohr to Heisenberg, 2 July 1935, AHQP BSC-20, section 2 (typed, 1p., in Danish) and Heisenberg to Pauli, 2 July 1935, p. 308 herein.

[4] Pauli reports on Bohr's reply to Schrödinger on 9 July (see p. 312), and both Klein and Kramers comment on it in letters to Bohr on 20 July and 9 August, respectively (Bohr 1996, pp. 459–462).

[5] AHQP BSC-20, section 2 (typed, 1p., in Danish)

[6] As noted by Kalckar (1996, p. 31, note 23), Bohr burned a lot of letters of this period from German correspondents just before his escape to Sweden from Nazi-occupied Denmark in 1943.

[7] Our thanks to Felicity Pors for confirming this.

[8] As consulted in the copies at the MPIWG in Berlin.

[9] The work that was dropped in haste followed on from the famous paper on the measurability of fields by Bohr and Rosenfeld (1933). As late as 22 May Bohr was expecting to send a manuscript to Heisenberg within days (Kalckar 1996, p. 46). But in reply to a query by Schrödinger on a passage in the version he had of Bohr's reply to EPR ('The removal of any incompleteness in the present methods of atomic physics ... might indeed

The Niels Bohr Archive contains two typescripts of the paper, one in English and one in German.[10] The English version includes abstract and footnotes, while the German one has none, but there are no other significant differences between them.[11] We assume that the English version is the original, and the German a rather accurate if a bit ponderous translation. Schrödinger's quotations from the earlier draft are in English, and the manuscript corrections in the German version are not in Bohr's hand.[12]

We can subdivide Bohr's reply as follows. An introductory part sets the stage for the argument to come (pp. 696–697/244–246). Then comes the analysis of the single-slit experiment in terms of complementarity (pp. 697–699/246–249), followed by the realisation and analogous analysis of the EPR example (pp. 699–700/249–250). This leads to a crucial criticism of the EPR criterion of reality (p. 700/pp. 250–251). There follow an extension of previous considerations to the case of time measurements, and some general consequences for complementarity (pp. 700–701/251–252). An analogy with relativity brings the paper to a close (pp. 701–702/252–253).

only be effected by a still more radical departure from the method of description of classical physics, involving the consideration of the atomic constitution of all measuring instruments, which it has hitherto been possible to disregard in quantum mechanics'; see p. 346), Bohr on 26 October states that he has omitted that passage because he is (still) 'just finishing a paper together with Rosenfeld on the measurement problems of the electron theory, which illuminates this question somewhat further' (see p. 349). A typescript on 'Field and Charge Measurements in Quantum Theory' from AHQP-BMSS 19, section 2 (17pp., in English) is reproduced in Bohr (1996, pp. 195–209), and is discussed in the introduction by Kalckar (1996, pp. 37–48), who dates it to the mid-1930s (p. 39, note 31). There is close similarity in language between the passage quoted by Schrödinger and the published typescript, which closes with the words: 'As regards the unsolved difficulties of the theory of electrons, the preceding analysis stresses only the consistency of the formalism as a correspondence theory, but leaves entirely open such questions as the possible necessity in a future theory to take into account the atomic constitution of the measuring instruments, which would entirely modify the outlook of the problem of measurements in atomic theory' (Bohr 1996, p. 209). We conclude it is highly likely that this manuscript is indeed the one Bohr and Rosenfeld were working on at the time of EPR's 'bolt from the blue', when Bohr switched to working on the reply to EPR, and that they resumed work around September/October and completed it presumably in late 1935. Instead, another paper on space-time concepts in atomic physics, which Bohr announces in a footnote in his reply to EPR (pp. 701–702/253) and also promises in his letter of 26 October to send to Schrödinger, does not appear to have ever been finished. Bohr had given a related talk already on 21 March 1935, 'Rum og Tid' ['Space and Time'] (AHQP-BMSS 14, section 1, typescript, 14pp., in Danish), and there are 58 pages of handwritten notes, partial drafts and figures, several bearing the title 'Rum og Tid i Atomfysikken' ['Space and Time in Atomic Physics'] from 1934–35 also in AHQP-BMSS 14, section 1 (in Danish). See also the brief descriptions in Bohr (1996, p. 521).

[10] AHQP BMSS-14, section 1 (English 15pp., German 16pp.).

[11] The only possible point of note is the German rendering of the phrase 'influence on the very conditions which define the possible types of predictions regarding the future behavior of the system', which is originally given as 'Einflussnahme auf die zugrundelegenden Bedingungen, die die möglichen Typen von Voraussagen, betreffend das spätere Verhalten des Systems, definieren'. This is a very literal translation (except for the use of 'zugrundelegend' [underlying] instead of 'very'), but it is then modified in ink so that 'zugrundelegenden Bedingungen' becomes 'Verhältnisse' [circumstances], and 'definieren' becomes 'bestimmen' [determine] – which suggests a more *practical* than *logical* reading of 'defining conditions'.

[12] The repeated (but corrected in ink) misspelling of 'Diaphragma' as 'Diafragma' might possibly suggest a Dutch speaker as translator. Rosenfeld was Belgian, but there also seem to be some differences with Rosenfeld's handwriting.

5.3 Structure of the Argument

As mentioned above, the fulcrum of Bohr's argument is his attempt to undermine EPR's criterion of reality. In his introductory sketch, after describing the EPR criterion of reality, Bohr writes (p. 696/245):

> By means of an interesting example, to which we shall return below, [EPR] next proceed to show that in quantum mechanics, just as in classical mechanics, it is possible under suitable conditions to predict the value of any given variable pertaining to the description of a mechanical system from measurements performed entirely on other systems which previously have been in interaction with the system under investigation.

Here, we take it that Bohr is presenting what he takes to be common ground between EPR and himself. But while he considers the criterion of reality as formulated to be perfectly well applicable in classical mechanics, it 'contains [...] an essential ambiguity when it is applied to the actual problems with which we are here concerned' (p. 697/246).

How is such an ambiguity to block the application of the criterion of reality? This of course depends on how one understands the structure of the EPR argument itself. Different authors disagree on how exactly the criterion of reality is applied, or what it leads to.[13] So for instance, in his discussion of Bohr's reply, Dickson (2002a, b) proposes a modal reading of the conclusion of the EPR argument, as the *possibility* of the simultaneous reality of position and momentum. Instead, Fine (2007, p. 18) in our opinion convincingly interprets EPR as concluding to the *actual* and not merely possible simultaneous reality of position and momentum. Similarly, there is disagreement about whether the argument requires merely the *possibility* of performing different measurements, or whether for the argument to go through the different measurements would need to be *actually* performed simultaneously. As Whitaker remarks (2004, p. 1320, emphases original):

> Let us spell out three distinct positions. The first, ours here, is that there is no requirement even to discuss measurement of more than one quantity. The second, which we tentatively assign to EPR, is that measurement of either of two quantities must be considered, but there is no necessity in any part of the argument to discuss two actual measurements being performed. (The full required results may be obtained *without* any suggestion that two measurements be performed.) The third, tentatively assigned here to Bohr and to Dickson [2002b], is that, to achieve what EPR required, two measurements would [have to] be performed. (To put things a slightly different way, this third position is that it is because two measurements are *not* performed, and cannot be performed without disturbance in the EPR scenario, that EPR's results do not hold.)

[13] For a detailed discussion of the logic of the EPR paper, see Fine (2017). See also Section 1.5.1 for a more unusual take on the question.

We agree with Whitaker about the fact that EPR consider measurements of the two observables without requiring that they be simultaneously performable. Indeed, if we consider the classical case – as Bohr enjoins us to and where he has no objections to invoking the criterion of reality – we see that the argument does not require performing two simultaneous measurements (even though one could).[14] Instead, we disagree with Whitaker's tentative interpretation of Bohr, and take also Bohr as understanding the EPR argument to involve only the *possibility* of separate measurements.

Indeed, we see Bohr as interpreting the EPR criterion as a straightforward inference (involving no modal logic) from the possibility of certain measurements to the actuality of certain values. Schematically (letting 'predict a value' be short for 'predict a value with certainty without disturbing the system', and 'reality' be short for 'a corresponding element of reality'), the premises of the argument then read:

(1) $\Diamond$(predict a value for Q_2) $\rightarrow \exists$(reality for Q_2)
(2) $\Diamond$(predict a value for P_2) $\rightarrow \exists$(reality for P_2)
(3) $\Diamond$(predict a value for Q_2)
(4) $\Diamond$(predict a value for P_2)

Premises (1) and (2) follow from the EPR criterion, and (3) and (4) from the EPR example. Therefore:

$$\exists(\text{reality for } Q_2) \wedge \exists(\text{reality for } P_2).$$

Bohr explicitly disambiguates between the following two senses of 'without disturbing the system':

- 'without mechanically disturbing the system',
- 'without disturbing the conditions affecting the possibility of predictions on the system'.

Thus, we propose to interpret the strategy of Bohr's reply as follows:

- If 'without disturbing the system' means 'without mechanically disturbing the system', one can accept premises (3) and (4), but one should reject (1) and (2).
- If it means 'without disturbing the conditions affecting the possibility of predictions on the system', then one can presumably accept (1) and (2), as in the classical case, but one should reject (3) and (4).

In either case, two out of four of the premises in the EPR argument (as we think Bohr understands it) are blocked.

[14] We also agree with both Whitaker and Fine (2017, section 1.3) that an argument can indeed be constructed using only one observable, although that is not what is done in the EPR paper.

It thus must be the case that Bohr's analysis of his two examples (the particle going through a single slit, and his realisation of the EPR example) serves to establish these two points. To this analysis we now turn.

5.4 Bohr's Examples

The two examples of the single slit and of the EPR case form the core of the paper – especially the former; and it may appear odd that Bohr spends such a large part of his reply on the single-slit case (essentially the same example he had discussed with Einstein at the 1927 Solvay conference), because the crucial and novel point of the EPR paper is surely the separation between the two particles in the EPR example.

Indeed, at least on a first reading, Bohr's arguments in the case of the single particle appear to be grounded in immediate physical intuitions, for example, the idea that in a measurement of the position of the particle an uncontrollable amount of momentum passes from the diaphragm used for the measurement into the rigid support defining the laboratory frame. In the EPR case, instead, with its threat of 'spooky action at a distance', the physical grounding of Bohr's arguments seems rather less immediate. However, Bohr appears to minimise the difference between the single slit and the EPR example.

We shall argue that the two examples are indeed closely analogous in the relevant sense, and that Bohr is correct in taking EPR to involve no 'greater intricacies', so that the complementarity analysis performed on the first example can be transferred without further ado to the second. (There is a separate question, of course, of whether this analysis provides a fully satisfactory answer to Einstein's worries, in particular in the case in which measurements on the two particles can be performed at spacelike separation, but we shall discuss that later.)

Bohr's first example is that of a particle passing through a single slit in a screen. Even if the momentum of the particle is known before the passage (as Bohr explicitly mentions), the aperture of the slit narrows down the uncertainty in the particle's position, giving rise to a corresponding uncertainty Δp in its momentum. Bohr adds (p. 697/246):

> Obviously the uncertainty Δp is inseparably connected with the possibility of an exchange of momentum between the particle and the diaphragm; and the question of principal interest for our discussion is now to what extent the momentum thus exchanged can be taken into account in the description of the phenomenon to be studied by the experimental arrangement concerned, of which the passing of the particle through the slit may be considered as the initial stage.

Bohr then specifies the scenario further, in *two alternative* ways: by considering that the screen is throughout 'rigidly fixed to a support which defines the space frame of reference'; or that it is freely movable in the direction perpendicular to the passage of the particle, and its momentum in this direction is initially known.

In the first case we have obviously gained the ability of predicting the position of the particle immediately after its passage through the slit (within the uncertainty defined by the width of the slit). But, as Bohr explains, 'the momentum exchanged between the particle and the diaphragm will [...] pass into this common support' (p. 697/246), thus we are unable to reconstruct the exchange of momentum between the particle and the diaphragm, and we have lost the ability to make predictions regarding the particle's momentum (and thus also to make predictions about its later position when it hits a subsequent screen).[15]

In the second case, conversely, by making a measurement of momentum on the diaphragm after the passage of the particle, we are able precisely to reconstruct the exchange of momentum between the particle and the diaphragm, and thus make accurate predictions about the particle's momentum. But then the diaphragm, as Bohr states (p. 698/247),

> can no longer be used as a measuring instrument for the same purpose as in the previous case, but must, as regards its position relative to the rest of the apparatus, be treated, like the particle traversing the slit, as an object of investigation, in the sense that the quantum-mechanical uncertainty relations regarding its position and momentum must be taken explicitly into account.

That is, the position of the diaphragm will now be completely uncertain, and we have lost the ability to predict the position of the particle immediately after its passage through the slit.

In either case, what is crucial for Bohr is the possibility of reconstructing some relevant aspect of the original interaction between the object and the measuring instruments, which he calls 'controlling the reaction of the object on the measuring instruments if these are to serve their purpose' (p. 697/245). Note that Bohr is talking about how in the interaction with the object it is the measuring instruments that are 'uncontrollably' affected (whatever that means), not the other way round (a point well emphasised already by Dickson (2002b)).

At this stage, an analogy with the EPR case is not at all apparent. But now Bohr makes a crucial observation, namely that in the second case – with the movable diaphragm – the experimental arrangement is *not yet fixed* after the particle has passed through the slit. Indeed, the passage through the slit is only the 'initial stage' of the measurement, and we still have 'a *free choice* whether we wish to know the

[15] The fact that the prediction regarding position can be made only immediately after the passage might be seen as a limitation of Bohr's example, but neither in the EPR paper are there considerations of dynamics: depending on how the EPR state evolves further, the observable whose values we can predict with certainty on system II, and the measurement we need to perform on system I, will in general change in time, as discussed explicitly by Schrödinger (1935e, section 4; reprinted here as Chapter 10).

momentum of the particle or its initial position relative to the rest of the apparatus' (p. 698/248, emphasis in the original). If we now measure the momentum of the diaphragm, we have realised the experimental arrangement for the measurement of the particle's momentum. But we can alternatively measure the diaphragm's position immediately after the passage of the particle and thus use the diaphragm as in the first case to measure the particle's position. And this, as in the first case, entails the impossibility of reconstructing the exchange of momentum between the particle and the diaphragm (indeed, Bohr treats this case as reconducible to that of rigidly fixing the diaphragm to the support). Thus, we have now a situation in which immediately *after* the particle and the diaphragm have exchanged momentum, we still have the ability to make manipulations *of the diaphragm* that allow us to predict *either* the position of the particle immediately after its passage through the slit *or* its momentum after the passage.

Note that in completing the measurement we manipulate only the diaphragm, so that we *no longer mechanically disturb* the particle (Pauli, in his letter to Schrödinger of 9 July 1935, in which he describes the draft of Bohr's reply, emphasises 'in both cases *without* disturbing the particle mechanically'; see p. 312). Note also that the original interaction between particle and diaphragm does 'mechanically disturb' the particle, but that is beside the point: the same happens also classically, and we are not interested in the initial state of the particle, which is in fact known. This is in perfect analogy to EPR if we understand one of the particles as taking over the role of the diaphragm.

And this is in fact what happens in Bohr's version of the EPR case (p. 699/249):

> The particular quantum-mechanical state of two free particles, for which [EPR] give an explicit mathematical expression, may be reproduced, at least in principle, by a simple experimental arrangement, comprising a rigid diaphragm with two parallel slits, which are very narrow compared with their separation, and through each of which one particle with given initial momentum passes independently of the other. If the momentum of this diaphragm is measured accurately before as well as after the passing of the particles, we shall in fact know the sum of the components perpendicular to the slits of the momenta of the two escaping particles, as well as the difference of their initial positional coordinates in the same direction; while of course the conjugate quantities, i.e., the difference of the components of their momenta, and the sum of their positional coordinates, are entirely unknown.[16]

[16] The state in Bohr's thought experiment is only approximately equal to the EPR state (which is anyway improper, i.e., not mathematically representable by an element of the Hilbert space) – due to the finite width of the two slits in the diaphragm. Note also that both of Bohr's examples do not merely require interaction of two systems to create an entangled state. They also require a measurement: a measurement of the momentum of the screen in his version of the EPR state, but also (implicitly) a measurement confirming the passage through the slit(s) (rather than the particle(s) having been reflected back or absorbed by the diaphragm). We believe this is an inessential complication, and that the arguments are independent of the procedure that in fact prepares the initial entangled state.

This example is entirely analogous to the single-particle one, with the role of the (single-slit) diaphragm played now by the second particle. Indeed, the only difference between this form of the EPR case and the single-slit case is that the interaction between the two particles is indirect, since it is mediated by the double-slit diaphragm. As a consequence, when Bohr describes, for example, the exchange of momentum being 'taken into account', he needs to take into account that the momentum was exchanged by the intermediary of the double-slit diaphragm. The latter plays no further role, and this difference is therefore inessential. Should this point need emphasising, think again of the classical case, imagining two small balls colliding with a screen with two slits. If we know all initial momenta and measure the momentum of the screen after the balls have passed through the slits, we have a classical analogue of the EPR state (which of course raises no conceptual problems). If we measure the momentum of one ball, we can reconstruct the (mediated) exchange of momentum between the two balls. Similarly, if we measure the position of one ball immediately after passage, we know the position of the screen and thereby that of the other ball.

Bohr hints at one further example, involving time measurements, but that is then fairly straightforward: presumably Bohr is now thinking of the photon box as a variant of the EPR argument based on energy and time, but he takes it to be completely analogous to the examples he has already discussed. There is a very close analogy between Bohr's version of the EPR example and the ETP example (cf. Section 1.4.3). In Bohr's version, the total momentum of diaphragm and particles is known initially, and the momentum of the diaphragm is measured again after the passage of the particles so that the total momentum of the particles – as well as the difference in their positions – is known (while the position of the diaphragm and particles is indeterminate). In the ETP example, the total energy of box and particles is known initially, and the energy of the box is measured again after the emission of the particles so the total energy of the particles – as well as the difference in their times of emission (zero) – is known (while the time when the box emitted the particles is indeterminate, as noted in the conclusion of the ETP paper). And in both the ETP paper and Bohr's example, the predictions on the second particle are made via retrodictions on the box or diaphragm, respectively.

To further support the view that Bohr's single- and double-slit examples are strictly analogous, and thus that the analysis performed on the first can be transferred to the second, let us look at Pauli's just-mentioned letter to Schrödinger of 9 July 1935. After he has summarised Bohr's argument, Pauli comments on the following important query Schrödinger had raised on the concept of 'state' (Schrödinger to Pauli, early July 1935; see p. 311):

Now, while 'state' is a word that everybody uses, even the holy PAM [i.e. Dirac], this does not make it more meaningful. But one can quite easily see that, and in what sense, a system *itself* changes if its psi-function changes. Indeed, each psi-function conveys the *same amount* of knowledge. When it changes, it is never only a gain but always also a loss of knowledge. But one can only acquire, not lose, knowledge (except by going soft in the head, which we shall ignore here). A loss of knowledge can only come about if the circumstances have changed. In this sense I say: different psi-functions definitely correspond to different circumstances – or states. I do not consider this an illegitimate dragging in of the reality dogma. But I would very much like to know what you think about this. And whether you really think that the Einstein case – let us call it like this – gives rise to no further worry, but is totally clear and simple and self-evident.

Pauli replies (see p. 313):

A pure case of A is a global situation in which the results of particular measurements on A (a maximal set) are predictable with certainty. I have nothing against calling this the 'state' – but even then it *is* the case that changing the state of A – i.e. that which is predictable of A – lies within the ***free choice*** of the experimenter even without directly disturbing A itself – i.e. even *after* isolating A.

This circumstance required by quantum mechanics appears to me *inescapable* in order to avoid contradictions, if the conservation laws as well as the superposition principle are valid. (Of course, this is precisely a ***difference*** as compared to the *classical* concept of state.) In my opinion ***there is indeed no problem here*** – and one knows the fact in question even without the Einstein example.

But Schrödinger did not have to be told by Pauli. As we have seen in Chapter 3, Schrödinger himself characterised this kind of situation in quantum mechanics using arbitrary quantum mechanical measurements already in his letter to Teller on 14 June. Unlike Pauli, this is precisely where Schrödinger thought there was a problem (see p. 289: 'This assumption stands on a level with the savage, who believes that he can harm his enemy by piercing his image with a needle'). Schrödinger agrees with Bohr that the single-slit example is essentially the same as the EPR example, but he believes that precisely for this reason, *both* examples are equally problematic.

5.5 Interlude: Pauli's Measurement Theory

The examples discussed by Einstein and by Bohr are clearly not the most general examples of quantum measurements, because in both cases the (post-interaction) state of the two systems considered is maximally entangled, or approximately so.

It may be appropriate at this stage to briefly interrupt our discussion of Bohr's argument in order to summarise Pauli's general treatment of measurement in quantum mechanics from his 1933 handbook article (Pauli 1933, 1990, section A.9).[17] This is presumably the closest one can get to an 'official' account of measurements along Bohr's lines, since Pauli was a keen spokesman for Bohr. It is also an account that is particularly relevant to the present discussion, because Pauli had long been sensitive to issues of systems and subsystems and to how these issues related to treating measurements quantum mechanically.[18]

The first obvious generalisation to consider concerns arbitrary measurements of the 'first kind' (Pauli's terminology). Measurements of the first kind are those in which the system and 'apparatus' (which may also be another microscopic degree of freedom, as in the EPR case) interact as

$$|u_n\rangle|U_0\rangle \mapsto |u_n\rangle|U_n\rangle, \tag{1}$$

where the $|u_n\rangle$ are the eigenstates of the measured observable, $|U_0\rangle$ is the initial state of the apparatus, and the $|U_n\rangle$ correspond to the different measurement readings.[19] Thus we have

$$\sum_n c_n|u_n\rangle|U_0\rangle \mapsto \sum_n c_n|u_n\rangle|U_n\rangle, \tag{2}$$

and when we read off the value n, the system is left in the state $|u_n\rangle$. A measurement of the first kind is a preparation procedure (the system is left in an eigenstate of the measured observable, and if the measurement is repeated, the same result is obtained with probability 1), but it is also a 'measurement' in the sense that one can draw 'an unambiguous conclusion back from the measured value to the quantity of the system under consideration before the measurement' (Pauli 1990, p. 97).

This last statement can be read in two ways. The unremarkable way is that if the system is initially assumed to be in an eigenstate $|u_n\rangle$ as in (1), then the measurement reliably establishes which one. But a way that better fits Bohr's analysis is that *even* if the system is in a superposition of eigenstates $|u_n\rangle$ as in (2), one is in a position to reconstruct the interaction between the system and apparatus as coupling the values of the measured observable to the ones of the relevant apparatus observable (as is the case for both momentum and position in the special case of the single-slit experiment).

[17] Pauli's handbook article is available in modern edition as Pauli (1990). This is a reprint of the second edition, but apart from the omission of the last three sections on the relativistic theory, the differences between the two editions are very minor.

[18] For instance, Pauli had been instrumental in suggesting to Heisenberg how Born's strategy in deriving transition probabilities in his treatment of collisions could be mimicked within matrix mechanics using systems and subsystems. He had also urged Born to discuss measurements from this point of view at the 1927 Solvay conference (Bacciagaluppi and Valentini 2009, sections 3.4.4 and 6.2.1). Cf. also Pauli's comments to Heisenberg in his letter of 15 June 1935 (see p. 291).

[19] Pauli's treatment is meant to cover both the discrete and (unrigorously) the continuous case.

A measurement in this sense consists of two stages: an initial stage of interaction between the system and either a macroscopic apparatus or an ancillary microscopic system, and then a manipulation of the latter resulting in an observation of a read-out and allowing an 'unambiguous conclusion' about the measured system.

Besides measurements of the first kind, Pauli discusses what he calls measurements of the 'second kind', namely the following. He considers first a *completely general* unitary interaction between system and apparatus, which he writes as

$$|u_n\rangle|U_0\rangle \mapsto \sum_{m,k} c_{mk}^{(n)}|u_m\rangle|U_k\rangle =: \sum_k |v_k^{(n)}\rangle|U_k\rangle, \tag{3}$$

so that

$$\sum_n c_n|u_n\rangle|U_0\rangle \mapsto \sum_{n,k} c_n|v_k^{(n)}\rangle|U_k\rangle =: \sum_k |\psi_k\rangle|U_k\rangle. \tag{4}$$

He then *explicitly imposes* as a condition that one may unambiguously retrodict the value of the measured quantity from the reading of the apparatus, in the sense that for each k there is at most one n such that $|v_k^{(n)}\rangle \neq 0$. In other words, the set K over which the index k ranges decomposes into disjoint subsets K_n such that

$$|u_n\rangle|U_0\rangle \mapsto \sum_{k\in K_n} |v_k^{(n)}\rangle|U_k\rangle. \tag{5}$$

For simplicity, we now specialise Pauli's assumption further and assume that each K_n contains exactly one element, which we relabel by n. Thus,

$$|u_n\rangle|U_0\rangle \mapsto |v_n^{(n)}\rangle|U_n\rangle, \tag{6}$$

and $|\psi_n\rangle = c_n|v_n^{(n)}\rangle$. In this case, although such a measurement does not leave the system in an eigenstate of the measured observable, we can still reconstruct the interaction between system and apparatus as one in which an initial state $|u_n\rangle$ couples to an apparatus state $|U_n\rangle$, but the backreaction of the apparatus on the system changes $|u_n\rangle$ to $|v_n^{(n)}\rangle$ (incidentally, the states $|v_n^{(n)}\rangle$ need not be mutually orthogonal).

Dropping our further requirement (6) might look puzzling: we then have a one-to-many correspondence between the states $|u_n\rangle$ of the system and the states $|U_k\rangle$ of the apparatus. But this is also perfectly intuitive if we imagine measuring, say, a discrete quantity such as the energy of the harmonic oscillator using a continuous pointer. The energy of the system determines the position of the pointer to within a certain interval, but when we look at the pointer we may well determine its position to a greater degree of precision than needed to 'unambiguously conclude' back to the value of the energy.

If instead we drop Pauli's own condition (5), then *even* if we know the quantum mechanical description of the interaction, it is impossible to 'reconstruct' it in a way that we could interpret as a 'measurement' of an observable of the system. In this sense, Pauli's notion of 'unambiguous conclusion' is the formal analogue of Bohr's statement about *needing to control* the (relevant aspects of the) interaction between the system and the measuring apparatus in order for the latter to 'serve its purpose'.[20]

From a modern perspective, this most general case is also taken to be a 'measurement'.[21] In Pauli's treatment, instead, even though he clearly conceives of general unitary interactions violating (5), the general case falls outside of his classification of measurements. Indeed, the cases that Pauli is excluding are cases in which one may well perform a measurement on the 'pointer observable', but performing such an operation fails to 'serve a purpose' as a measurement on the *system*. What is clear from (4), however, is that in such a case the system is still left in some state $|\psi_k\rangle$. That is, even though we have no measurement, we still have a preparation.

Now note that the choice of what observable to measure on the 'apparatus' (which, recall, could just as well be another microscopic system) is in fact arbitrary. Thus, indeed, even in the most general case, what state (or more precisely what not necessarily orthogonal family of states) a system is left in 'lies within the *free choice* of the experimenter even without directly disturbing [the system] itself'.

As we shall discuss more fully in Section 5.7, it is rather unsatisfactory that in these more general cases no picture is available in which to apply 'the idea of space-time coordination', or 'the conservation of momentum' or any other classical idea. Maybe Pauli and Bohr can bite the bullet here and not be expected to provide such a classical picture when in fact (in their view) *no* measurement is being performed (on the system). But it is rather ironic that precisely such a general case of preparation is actually described by EPR in their equations (7)–(8), just before the special example of the EPR state (p. 779/160). This is also the example used by Einstein to explain his preferred version of the incompleteness argument to Schrödinger in his letter of 19 June 1935 (see p. 298), in which Einstein complained that the main point of the paper had been 'buried in erudition'. It seems that, unwittingly, Podolsky phrased the incompleteness argument in a way that turned out to be particularly apt for a complementarity analysis on the part of Bohr!

[20] One might also wonder whether Pauli's otherwise somewhat cryptic remarks to Schrödinger above about combining the requirements of the conservation laws with the superposition principle might be related to his characterisation of measurements (a superposition of terms for each of which separately the conservation laws hold?).

[21] Modern measurement theory has generalised the notion of an observable to include so-called 'positive-operator-valued' (POV) observables. A measurement works by letting the system interact unitarily with (in modern jargon) an 'ancilla', then performing a traditional measurement on the ancilla (phenomenologically described by a collapse of the state in the pointer basis). This induces a so-called completely positive map on the states of the system, and any such completely positive map can be implemented in this way (Naimark dilation theorem). See e.g. Busch et al. (1991, 1997, 2016) for overviews.

5.6 Non-mechanical Disturbance

We now return to the details of Bohr's argument, specifically his notion of non-mechanical disturbance. As we have seen, an essential point according to Bohr is that if we wish to predict the position or the momentum of the particle by manipulating the diaphragm, we need to be able to reconstruct the relevant aspect of how the particle affected the diaphragm when they originally interacted. In particular: if we are to use the diaphragm for a position measurement on the particle, we must be able to reconstruct the 'space-time coordination' of particle and diaphragm; and if we are to use the diaphragm for a momentum measurement on the particle, we must be able to apply the conservation of momentum to the system consisting of particle and diaphragm.

So far, however – and Bohr effectively says so in his introduction – the analysis also applies classically. The crucial novelty in the quantum case is that complementarity considerations come into play which provide *fundamental limitations* to the applicability of such classical pictures. Namely: if we use the diaphragm for a position measurement, its momentum becomes completely uncertain, and we have *freely cut ourselves off* from the possibility of applying the conservation of momentum to the particle and diaphragm and reconstructing the exchange of momentum; similarly, if we use the diaphragm for a momentum measurement, its position becomes completely uncertain, and we have *freely cut ourselves off* from the possibility of reconstructing the space-time coordination of particle and diaphragm.

This is where Bohr's concept of 'non-mechanical disturbance' comes in. There is no mechanical disturbance of the particle when we choose the manipulation we make on the diaphragm, but the choice that we make determines which aspects of the *earlier* 'mechanical' interaction between the particle and the diaphragm we can reconstruct.[22] But what exactly is 'cutting oneself off' supposed to mean? And how does it translate into the kind of 'disturbance' that would support Bohr's sceptical claims about physical reality?

As far as Bohr's quantum mechanical description of the particle and diaphragm goes, EPR would presumably agree with him. After all, they accept the correctness of quantum mechanics, and they even explicitly use the fact that the choice of measurements performed on one system determines which quantum state we assign to another system. They would presumably also accept at least some form of 'cutting oneself off': if quantum mechanics is empirically correct, then it is the case

[22] Note that perhaps there is just a hint of turning the tables on Einstein here, in the sense that – as reported by Heisenberg in describing the inspiration for his own analysis of uncertainty – it is Einstein who insisted that '[i]t is the theory which decides what we can observe' (Heisenberg 1969, 1971, chapter 5).

that when we determine the position of the diaphragm, its momentum will become totally uncertain – in the *empirical* sense of becoming unpredictable.

This, however, is still compatible with the idea that by choosing to perform, say, a measurement of position on the diaphragm, we cut ourselves off merely *epistemically* from gaining access to the momentum of the particle. As compared to a naive disturbance theory of measurement, what we have is that instead of a position measurement on the particle uncontrollably disturbing the momentum of the particle (implicitly conceding that it exists), a position measurement uncontrollably disturbs our possibility of knowing that momentum (thus equally conceding it exists!).

For Bohr, however, 'cutting oneself off' is not meant epistemically. He states explicitly (p. 699/249):

> [W]e have in each experimental arrangement suited for the study of proper quantum phenomena not merely to do with an ignorance of the value of certain physical quantities, but with the impossibility of defining these quantities in an unambiguous way.

We shall now analyse two ways of reconstructing how cutting oneself off might be understood in an ontic rather than epistemic sense.[23] One, for which at first sight there is good textual evidence in Bohr's reply, is that the frame of reference (as defined by the macroscopic laboratory bench) plays a crucial role. We shall argue, however (using evidence from Schrödinger's correspondence with Bohr), that this is a red herring. The other is based on Don Howard's (1994) analysis of Bohr's notion of classical concepts, and we argue it is much more credible as a reading of Bohr – even though ultimately maybe not as unproblematic an answer to EPR as Bohr might have thought.

5.6.1 The Role of the Reference Frame

When discussing the single-slit experiment with the diaphragm rigidly bolted to the support, Bohr states (p. 697/246):

> Then the momentum exchanged between the particle and the diaphragm will, together with the reaction of the particle on the other bodies, pass into this common support, and we have thus voluntarily cut ourselves off from any possibility of taking these reactions separately into account in predictions regarding the final result of the experiment.

This phrasing may indeed indicate that the reference frame plays an important role in Bohr's considerations. It is particularly suggestive because it seems to provide a strong *physical* explanation for why a measurement on the screen would not

[23] The same ontic–epistemic distinction can be made with regard to the notion of a 'maximal' set of observations (as in Pauli's quote above), as was indeed made by Grete Hermann with reference to Dirac's use of the term (cf. Crull and Bacciagaluppi 2017b, sections 8.2 and 8.3).

disturb the particle mechanically, but rather *affect the conditions under which its momentum is well-defined.* Among recent commentators, Dickson (2002a, b, 2004) has taken up this point, and elaborated it in the language of quantum reference frames as a justification for the uncertainty between position and momentum. As Dickson (2004) shows explicitly, when one performs a position measurement inside the lab, the momentum of the lab is in fact disturbed in a way that fits nicely with the uncertainty relations for the particle. As we shall presently see, we do not think this is helpful in understanding Bohr's arguments. (To be fair to Dickson, he himself questions whether such a use of the reference frame is too tied to the specific example of the EPR state, and fails to generalise, in particular to the case of the spin singlet state.)

Let us see explicitly how this interpretation of the role of the reference frame in Bohr's argument is supposed to work in the case of the two-slit arrangement, in which we have the two particles prepared in the EPR state and then perform a measurement of the position of one particle by letting it pass through a further diaphragm bolted to the lab frame. The suggestion for how to read Bohr's argument is this: the second diaphragm transfers a certain amount of momentum to the frame of reference, so that now the momentum of the particle with respect to the frame of reference has changed, and we have lost track of this change because the frame of reference itself is what defines our measurements (so that once and for all we have renounced the possibility of taking this momentum transfer into account). This would also seem to explain why the momentum of the *distant* particle has become ill-defined. Thus, by affecting the definition of the reference frame, a measurement of the nearby particle would not disturb the distant particle mechanically but would affect the conditions under which its momentum is well-defined.

Support for this reading could be claimed from the fact that this seems to be Schrödinger's own reading of the argument in his letter to Bohr of 13 October 1935 (see p. 345):

> [P]erhaps making a very precise coordinate or momentum measurement of particle No. 2 *from the coordinate frame* (standing on it, so to speak) does not after all have an arbitrarily small effect on the coordinate frame, no matter how heavy and massive and solid the coordinate frame might be.

On the other hand, Schrödinger immediately points out that such an argument would in fact explain *too much*:[24]

> But this [...] in fact does not apply. For had we [...] first established the situation S' of particle No. 1 through a *direct* measurement, we would not believe that this S' can be modified through further measurements that we make from the same iron platform on other

[24] One might note at least a superficial similarity with Teller's discussion herein, on p. 287.

light particles, e.g. also on particle No. 2. The physical influence of such a measurement on the platform is therefore considered to be negligible.

What Schrödinger is saying is that, if one grants that a measurement on particle 2 affects the coordinate frame, the argument would apply to particle 1 *whether or not* it was entangled with particle 2. But if we first measure particle 1 (the distant particle), we disentangle the state of the two particles, and quantum mechanics definitely predicts *no* further effect of the measurement of particle 2 on the state of particle 1.[25]

Has Schrödinger exposed the essential weakness in Bohr's reply to EPR, or is there an alternative reading of Bohr's use of the frame of reference? (Note that in his letter to Schrödinger of 9 July, Pauli makes no mention of the frame of reference when summarising Bohr's argument.) The only plausible alternative, in our view, is to interpret Bohr's use of the reference frame as meaning that we must necessarily accept that the momentum of the *diaphragm* (not of the frame of reference!) is now indefinite. What then follows (in Bohr's terminology) is that we have lost the possibility of taking into account the exchange of momentum between the diaphragm and the nearby particle, and therefore we have also cut ourselves off from the possibility of tracking the earlier exchange of momentum between the nearby particle and the distant particle.

The explicit inclusion of the reference frame may thus not help us understand what Bohr means exactly by cutting oneself off. Nevertheless, this second interpretation of the role of the reference frame is the more charitable to Bohr, when we consider how he replied to Schödinger on this point (Bohr to Schrödinger, 26 October 1935; see p. 349):

Regarding my answer to Einstein's article, I believe it has already come out, and as you will see, I have tried in several places to formulate the thoughts somewhat more clearly, and I hope that your first objection regarding the measurement setup is thereby also answered.

Otherwise, not only would we conclude that Bohr made a rather crass mistake in writing the paper, but that he failed to see this mistake even after Schrödinger had made the point perfectly clear. Bohr may have been sometimes obscure in his writing, but he was not thick!

5.6.2 Bohr's Doctrine of Classical Concepts

Instead of focusing on the role of the reference frame, it may be helpful to recall the essentials of Howard's (1994) analysis of Bohr's doctrine of classical concepts. Two elements are important according to Howard.

[25] Schrödinger was commenting on a previous draft of the paper, which appears to be lost. Nevertheless, the tone of Bohr's answer we shall presently quote seems to indicate that on this point the final version was not essentially different from the draft.

First, that Bohr (*like* Einstein) took it that the separability of system and apparatus is a necessary condition for the objectivity of a measurement; but that Bohr recognised that the 'uncontrollable exchange of quanta of action' in a measurement interaction destroys this separability.[26]

Second, that according to Bohr separability and objectivity are regained in each observational context. Crucially, however, Howard argues that in defining a context Bohr does not consider the apparatus as classical and the system as quantum mechanical, but he treats *certain aspects* of the apparatus and *certain aspects* of the system classically, and renounces a classical description of other aspects.[27]

As Howard puts it, these two points can be formally 'reconstructed' (an aid to our own intuition rather than Bohr's way of putting it) as, respectively, the entanglement of system and apparatus, and substituting for the entangled state an 'appropriate mixture'.

Let us see how this applies to the EPR case. When we have the EPR state,

$$|\Psi\rangle = \int_{-\infty}^{\infty} e^{ix_0/\hbar} |-p\rangle |p\rangle \, dp = \int_{-\infty}^{\infty} |x + x_0\rangle |x\rangle \, dx, \tag{7}$$

we can choose to apply to both systems the idea of spatial coordination by substituting the mixture

$$\rho_Q = \int_{-\infty}^{\infty} |x + x_0\rangle\langle x + x_0| \otimes |x\rangle\langle x| \, dx, \tag{8}$$

which allows us to selectively exploit the position correlations in the EPR state. Or we can choose to apply to both systems the conservation of momentum by substituting the mixture

$$\rho_P = \int_{-\infty}^{\infty} |-p\rangle\langle -p| \otimes |p\rangle\langle p| \, dp, \tag{9}$$

which allows us to selectively exploit the momentum correlations in the EPR state.

Choosing to measure the position of particle 2 – whatever else it might mean – requires letting a third system interact (locally) with particle 2, such that the total state of the three systems evolves to a state

$$|\Psi\rangle|R\rangle = \int_{-\infty}^{\infty} |x + x_0\rangle |x\rangle |R\rangle \, dx \mapsto \int_{-\infty}^{\infty} |x + x_0\rangle |x\rangle |R_x\rangle \, dx. \tag{10}$$

[26] As Bohr (1928b, p. 580) wrote in the Como lecture, 'the quantum postulate implies that any observation of atomic phenomena will involve an interaction with the agency of observation not to be neglected. Accordingly, an independent reality in the ordinary physical sense can neither be ascribed to the phenomena nor to the agencies of observation'.

[27] As Howard notes, this may be a significant difference between Bohr's view and Heisenberg's treatment of the 'cut' (cf. Section 4.5).

Choosing to measure the momentum of particle 2 instead requires letting a third system interact (locally) with particle 2 such that the total state of the three systems evolves to a state

$$|\Psi\rangle|R\rangle = \int_{-\infty}^{\infty} e^{ix_0/\hbar}|-p\rangle|p\rangle|R\rangle\, dp \mapsto \int_{-\infty}^{\infty} e^{ix_0/\hbar}|-p\rangle|p\rangle|R_p\rangle\, dp. \qquad (11)$$

But now, when the state of the total system (particles 1 and 2 and apparatus) is (10), resp. (11), the *reduced state* of the particle pair becomes ρ_Q, resp. ρ_P.

Note that as long as the state of the two systems remains the pure state $|\Psi_0\rangle\langle\Psi_0|$, we still have the *free choice* to interact with the second system to produce either the state ρ_Q or the state ρ_P. Once we have implemented this interaction, however, and (say) turned $|\Psi_0\rangle\langle\Psi_0|$ into ρ_Q, this state *no longer contains* the correlations in momentum of the original state. That is, *without interacting with particle 1*, we have changed the state of the two particles to one that no longer contains any exploitable correlations in momentum. Unless, that is, we undo the (local) interaction with the second system. This is of course always possible in principle, but if we do so we are not using the second system as a measuring device for the position of the first. So, *if* we are to, say, use the second particle as a measuring device for the position of the first, we have to choose to give up on our ability to instead exploit the correlations in momentum. This gives us a fairly straightforward reading of Bohr's point that performing a measurement of either position or momentum on the nearby particle in the EPR case – while allowing us to predict the position of the distant particle – influences the 'conditions affecting the possibility of predictions' on the distant particle.

One might even go further and say (borrowing Everett's terminology of relative states) that once we have made a choice, then relatively to any of the pointer states both particles are in well-defined states, but relatively to any other states of the apparatus, the particles' state will be entangled and neither particle will be in a well-defined state.[28] In this sense, cutting oneself off seems indeed to have more than a merely epistemic significance (there are no 'Everett worlds' in which the apparatus has some well-defined property and the particles have well-defined momenta). Of course, this too is a 'reconstruction' in the same sense as employed by Howard, but if one considers that one of Everett's aims was precisely to re-derive the 'usual'

[28] The EPR state is of course highly singular, but for regular maximally entangled states this point can be made rigorous in the following sense. Maximally entangled states of bipartite systems have degenerate biorthogonal decompositions, but if we take a tripartite system, and expand its state in terms of linearly independent states of systems $(1+2)$ and 3, then there is at most one such decomposition for which the states of $(1+2)$ are product states (Elby and Bub 1994).

Copenhagen interpretation from his own, it may be an illuminating point of view on Bohr's claim that 'cutting oneself off' has implications for 'physical reality'.[29]

If a quasi-Everettian 'reconstruction' feels too anachronistic, we can alternatively borrow from Grete Hermann. As we saw in Chapter 4, in her essay of March 1935 Hermann was explicitly characterising Bohr's doctrine of complementarity, and discussed the example of the Heisenberg microscope – which is again perfectly analogous to the single slit (as already pointed out by Jammer 1974, pp. 96–97).[30]

For Hermann, quantum mechanical descriptions are always relative to a context of observation. Specifically – translating to (Bohr's version of) the EPR example – choosing to enter the context of observation determined, say, by a momentum measurement on one particle defines a description in which (much like Bohr's reconstructing the relevant aspects of the earlier interaction between the two particles) one can tell a causal story about the interaction between the two particles that explains the result of the measurement in terms of the earlier exchange of momentum between the particles (and the diaphragm). This causal story can be checked indirectly by performing a measurement on the other particle, which confirms that indeed the other particle has the exact momentum that it should have if the reconstructed exchange has taken place. In this sense, relative to the context of observation one has entered, we could say that also the momentum of the other particle has 'physical reality'. But this reality is only relative to the context of observation. Hermann is very clear that one cannot assume that both position and momentum are real irrespective of the context of observation, because she takes her argument to establish that quantum mechanics is already causally complete, and therefore *cannot be further completed* by any 'hidden variables'.

While Bohr might have resisted a reading in terms of causality,[31] Hermann's 'reconstruction' fits extremely well all the aspects of Bohr's reasoning we have discussed (indeed, it appears to be closer to Bohr's treatment of the single-slit thought experiment than to Heisenberg's own treatment of the microscope), and illuminates further how we might think of the relation between 'non-mechanical disturbance' and 'physical reality'.[32]

[29] For his part, Everett was perfectly convinced that he had subsumed complementarity in his own theory (see e.g. Everett 2012, pp. 18, 153 and 175).

[30] Hermann's (1935a) essay was published in an abridged version in October 1935 in the *Naturwissenschaften* (Hermann 1935b), which is translated here as Chapter 15.

[31] In his early formulations, Bohr associated causality specifically with the conservation laws, while in later years he associated it with the possibility of combining spatiotemporal coordination and conservation laws, and took complementarity to have replaced the classical ideal of causality (Scheibe 1973, pp. 13–16). He also connected the loss of ability to apply the law of causality with the emergence of the statistical element in quantum theory, e.g. in the Como lecture (Bohr 1928a, b).

[32] Even though Hermann's longer paper (1935a) was published before EPR, one of its main claims is that quantum mechanics is complete, so that in fact Hermann's, Heisenberg's and Bohr's papers, even though with some significant differences, all crucially put the same example (the Heisenberg microscope or single slit) to

5.7 Final Assessment

We have argued that Bohr's discussion of the single slit is intended as a presentation of the arguments he is using also in the case of the EPR state. Summarising, Bohr takes EPR's reasoning as arguing in both the case of position and that of momentum from the possibility of a measurement on one particle (and thus the possibility of predicting with certainty the result of a measurement on the other particle) to the physical reality of position and momentum, respectively. He wishes to argue that a measurement on one particle does in fact disturb the other particle, albeit non-mechanically, in such a way as to affect not only the possibility of our knowing the values of certain quantities on that particle, but even whether or not these quantities are well-defined. We have reconstructed Bohr's idea of non-mechanical disturbance in terms of the quantum mechanical fact that interacting with one particle (which acts as a measuring apparatus, or in modern parlance an ancilla) changes the state of the particle *pair* from the initial entangled state to a mixture. Depending on the interaction chosen, this mixture encodes correlations between either the particles' positions or the particles' momenta. If we choose once and for all how we interact with the nearby particle (no quantum erasure of measurement results, so that the apparatus indeed 'serves its purpose'), Bohr appears to suggest that these correlations give us clues as to the 'physical reality' pertaining to the other particle to the *exclusion* of other putative elements of reality. We have further tried to reconstruct this intuition borrowing either from Everett's later work or from Hermann's work of 1935.

Despite being one of the crucial and most influential texts on Bohr's views on complementarity, Bohr's reply to EPR is not a straightforward text to read. We hope, however, that our analysis makes clear that the strength and elegance of what Bohr achieves in his reply lie in the fact that he relies exclusively on the effects of the experimenter's *local* choices. We can in fact quote from Everett here, since the physics he and Bohr describe is the same (Everett 2012, p. 157):

> In the case of observation of one system of a pair of spatially separated, correlated systems, nothing happens to the remote system to make any of its states more 'real' than the rest. It had no independent states to begin with, but a number of states occurring in a superposition with corresponding states for the other (near) system. Observation of the near system simply correlates the observer to this system, a purely local process – but a process which also entails automatic correlation with the remote system. [...] The mouse does not affect the universe – only the mouse is affected.

the same use of arguing for the completeness (or incompletability) of the quantum mechanical description of physical reality.

But what does in fact Bohr's reply achieve? How effective is it in blocking the EPR argument, and – in its focus on the criterion of reality – is it effective at all in addressing Einstein's wider worries about incompleteness?

Let us focus first on the argument as published in the EPR paper. For this case, Bohr's treatment is indeed proposing an alternative explanation (not involving elements of reality) of why we should be able to make predictions with certainty on the distant particle: we can indeed reconstruct the relevant aspects of the interaction between the two particles and can thus, according to Bohr, legitimately infer from our measurement result to the future behaviour of the measured system. At the same time, this explanation precludes that we may reconstruct other aspects of the interaction. As emphasised for instance in the Everett quote above, Bohr's explanation is based on the local establishment of correlations between the observer and the nearby particle. The latter is a particularly attractive element of Bohr's reasoning, because it proposes to explain the apparent 'spooky action at a distance' as an effect of a purely self-imposed limitation.

But there may be a sting in the tail. Suppose we have entanglement between an 'American' and a 'European' particle (borrowing Schrödinger's terminology; see p. 317). And suppose now Bohr performs a measurement, say of position, on the European particle, cutting himself off from the possibility of applying the conservation of momentum to the particle pair. Let us grant that his account convincingly explains why he would be confident that if he went to the effort of crossing the ocean to perform a position measurement on the other particle, he would find it with probability 1 in the exact spot he predicts (say, 112 Mercer Street, Princeton, New Jersey). Does this tell us anything about what *Albert* would find if he performed a measurement on the American particle at spacelike separation from *Bohr's* measurement?

Let us spell this out. The physical story Bohr is telling us is that if he performs a position measurement, and we apply the picture of spatiotemporal coordination, then we can attribute reality to the position of both particles – in Howard's reconstruction describing them by a mixture of their joint position eigenstates. If we do so, then any account of further measurements on either particle (whatever the measurements) can proceed locally: we assume that a particle is in some eigenstate of position, and tell the appropriate story about how we interact locally with such a particle, either observing which value of position it has, or establishing a new context of observation for that particular particle.

As long as both Albert and Bohr perform the same measurement, or as long as Albert's measurement is in the future light cone of Bohr's, this is a straightforward extension of the physical story that Bohr tells us about his own measurements. But if Albert and Bohr are performing incompatible measurements at *spacelike* separation, say Bohr measures position on the European system and Albert measures momentum on the American system, then we have no justification for telling the physical story in terms of *spatiotemporal coordination*, as Bohr does, as opposed to telling a physical story in terms of *momentum conservation*, as Albert would (if we imagine him applying Bohr's reasoning!). Indeed, choosing one story over the other (say, the one in terms of spatiotemporal coordination) privileges Bohr's measurement over Albert's, thus introducing after all a preferred frame into the physical description (the absolutely earlier measurement decides), or maybe a (solipsistic?) notion of a preferred observer (Bohr's measurement decides, because of course Bohr is God[33]). And we clearly cannot describe the interaction between the two particles *both* in terms of their spatiotemporal coordination *and* in terms of conservation of momentum.

What Bohr could say, at a pinch, is that the two pictures are *complementary* to each other, in the sense that either picture is separately applicable, and provides a physical story that explains what happens in the measurements on the two sides, but the two pictures cannot be combined into one. In this sense, there appears to be a natural Bohrian answer to the question of spacelike separated measurements, namely that our notion of 'physical reality', or the kind of physical stories that we can tell about quantum systems, will depend holistically on the entire observational context, consisting of *both* Albert's and Bohr's local choices. This answer fits with Bohr's well-known entreaty that 'the unambiguous account of proper quantum phenomena must, in principle, include a description of all relevant features of the experimental arrangement' (Bohr 1958, p. 311).

Quite remarkably (in hindsight), all of Einstein's proto-EPR arguments involve *timelike* scenarios, in that there is always a *mirror* that reflects the distant particle back to some observation point. The version in Ehrenfest's letter of July 1931 to Bohr includes such a mirror, the ETP scenario includes such a mirror, the version in the Berlin seminar includes such a mirror, and the version in Einstein's letter to Ehrenfest includes such a mirror. It is only in the EPR paper (and in Schrödinger's letter to Sommerfeld of 11 December 1931) that no reflective device is included explicitly in the experimental setup.

Maybe even more remarkably (always in hindsight, given the emphasis on spacelike separation in modern discussions), not even EPR in their paper explicitly discuss the spatial (let alone spacelike) separation of the two particles in their example. They merely note that the particles have ceased to interact. And even though (as just sketched) the spacelike case might be given a Bohrian treatment, Bohr

[33] As in the parody of Goethe's *Faust* performed at Bohr's institute in 1932 (AHQP-66, section 1-001), translated as appendix to Gamow (1966).

himself does not go beyond a single-user perspective in his reply to EPR. As presented in his *Physical Review* article, his reply to EPR is local, but measurements performed by a single user are of course always at timelike separation from each other.

Finally, we need to ask whether Bohr's argument is relevant to Einstein's *wider* worries about incompleteness, irrespective of the criterion of reality and the specific form of the argument in the EPR paper. Here we leave aside Einstein's queries about the macroscopic adequacy of quantum mechanics, as in his gunpowder example and in Schrödinger's cat example: these rest on arguments quite different from any of the versions of the EPR argument, and Bohr's reply is clearly not relevant in addressing them. We should, however, include Einstein's own preferred version of the EPR argument; while it is spelled out in his letter to Schrödinger of 19 June (see p. 297), it is left more or less implicit in the EPR paper itself.

As discussed in Chapter 1, EPR consider two no longer interacting systems in a generic entangled state, which the authors note can be written in two alternative ways:

$$|\Psi\rangle = \sum_{n=1}^{\infty} |\psi_n\rangle|u_n\rangle = \sum_{s=1}^{\infty} |\varphi_s\rangle|v_s\rangle \tag{12}$$

(their eqs. (7) and (8), here rewritten in Dirac notation). The states $|u_n\rangle$ and $|v_s\rangle$ are, respectively, (normalised) eigenstates of observables A and B of one system, and the states $|\psi_n\rangle$ and $|\varphi_s\rangle$ are (not necessarily normalised or orthogonal) states of the second system. Depending on whether one measures A or B on the first system, the state of the second system becomes one of the (non-zero) $|\psi_n\rangle$ or one of the (non-zero) $|\varphi_s\rangle$, respectively. EPR (p. 779/161) remark that

> since at the time of measurement the two systems no longer interact, no real change can take place in the second system in consequence of anything that may be done to the first system.

They then introduce the EPR state along with the familiar version of their argument. But they could have argued, as Einstein does in his letter to Schrödinger, that incompleteness follows immediately, because if no real change can take place in the second system, then the complete description of the real state of the system, whatever it might be, must be independent of the choice of measurement of A or B. Therefore the final quantum state of the second system is not a candidate for providing such a complete physical description. This version of the argument does not depend on the criterion of reality, but it still crucially depends on invoking a lack of disturbance of one system when a measurement is performed on the other system. Presumably here, too (and recalling Pauli's comments to Schrödinger), Bohr can point out that while no 'mechanical' disturbance is present, there will be a kind of 'non-mechanical' disturbance that affects the types of predictions one can make on the second system (depending on whether one measures A or B, one prepares one of the states $|\psi_n\rangle$ or one of the states $|\varphi_s\rangle$), and this should lead us to revise our ideas about 'physical reality'.[34]

As hinted already in Section 5.5, however, there is an important difference between the EPR case (or the single slit) and the generic case of (12). In the EPR case, Bohr can do more than give a negative account of how either choice of measurement cuts the experimenter off from the ability to reconstruct some aspects of the interaction between the two particles: as just discussed, Bohr also gives a *positive* account of how either choice does give the experimenter the ability of reconstructing the *chosen* aspects of the interaction.

This positive (if – compared to the classical case – fundamentally limited) account of the interaction can be extended to the case of arbitrary measurements of the first and even of the second kind (the latter including an aspect of back-reaction onto the system). It provides a physical *explanation* for why one can make a certain type of prediction (with certainty) for the system, by providing a term-by-term reconstruction of the relevant aspects of the interaction between the system and the apparatus (even though the state of the combined system becomes entangled).[35] In the general case, however, an arbitrary choice of measurement by the experimenter, while steering the system into some states $|\varphi_s\rangle$, will *not* be analysable term-by-term via application of even some partial classical picture.

And this is bad news for Bohr. The difficulty in treating general measurements is a weakness of Bohr's Copenhagen interpretation, which as mentioned in Section 5.5 he could perhaps try to brush off,[36] but it becomes a particular weakness in the context of Einstein's version of the EPR argument. Bohr might still insist that the assumption of a separate physical reality for one system – independent of manipulations performed on another system – should be questioned. But now he *no longer has an alternative explanation* for why we can make certain predictions with probability 1 on the 'distant' system.

[34] Note that Born in his letter to Schrödinger of 28 June 1935 focuses precisely on this example, calling it an 'age-old swindle', even though he does not explain why (see p. 300). Also, Pauli's description of the EPR argument in his letter to Heisenberg of 15 June (see p. 291) makes no explicit mention and minimal use of the criterion of reality. Pauli largely appears to have seen through the 'erudition' of the published version down to the core of Einstein's argument – and have thought that Bohr's reply still applied.

[35] This is precisely Hermann's (1935a) point about the causal stories one can tell in quantum mechanics – which thereby show it is causally complete.

[36] Similar points can be made in the context of Hermann's views, as discussed in detail in Bacciagaluppi (2017). It is also noteworthy to mention that one of the advantages that Everett lists for his approach to quantum theory (as compared to the standard von Neumann formulation) is that Everett can describe 'approximate measurements'. This is emphasised both in his 'long thesis', which remained unpublished until 1973, and in the published article of 1957 (cf. Everett 2012, pp. 145–149, 176–177 and 182, note 6).

Our (tentative) final word on Bohr's reply to EPR is thus: Bohr provides a rather elegant and fairly effective critique of their criterion of reality, but he falls short of adequately addressing the wider issues in the debate on nonlocality and incompleteness.

Part II

Selected Pre-EPR Papers

6

Knowledge of Past and Future in Quantum Mechanics*

ALBERT EINSTEIN, RICHARD C. TOLMAN AND BORIS PODOLSKY

It is well known that the principles of quantum mechanics limit the possibilities of exact prediction as to the future path of a particle. It has sometimes been supposed, nevertheless, that the quantum mechanics would permit an exact description of the past path of a particle.

The purpose of the present note is to discuss a simple ideal experiment which shows that the possibility of describing the past path of one particle would lead to predictions as to the future behaviour of a second particle of a kind not allowed in the quantum mechanics. It will hence be concluded that the principles of quantum mechanics actually involve an uncertainty in the description of past events which is analogous to the uncertainty in the prediction of future events. And it will be shown for the case in hand, that this uncertainty in the description of the past arises from a limitation of the knowledge that can be obtained by measurement of momentum.

Consider a small box B, as shown in the figure, containing a number of identical particles in thermal agitation, and provided with two small openings which are closed by the shutter S. The shutter is arranged to open automatically for a short time and then close again, and the number of particles in the box is so chosen that cases arise in which one particle leaves the box and travels over the direct path SO to an observer at O, and a second particle travels over the longer path SRO through elastic reflection at the ellipsoidal reflector R.

The box is accurately weighed before and after the shutter has opened in order to determine the total energy of the particles which have left, and the observer at O is provided with means for observing the arrival of particles, a clock for measuring their time of arrival, and some apparatus for measuring momentum. Furthermore the distances SO and SRO are accurately measured beforehand, – the distance SO being sufficient so that the rate of the clock at O is not disturbed by the gravitational

* [The text follows the original publication in the 'Letters to the Editor' section of *Physical Review* (Einstein, Tolman and Podolsky 1931). Reprinted with permission from: A. Einstein, R. C. Tolman and B. Podolsky, *Physical Review* **37**, 780–781 (1931). Copyright 1931 by the American Physical Society (*eds.*).]

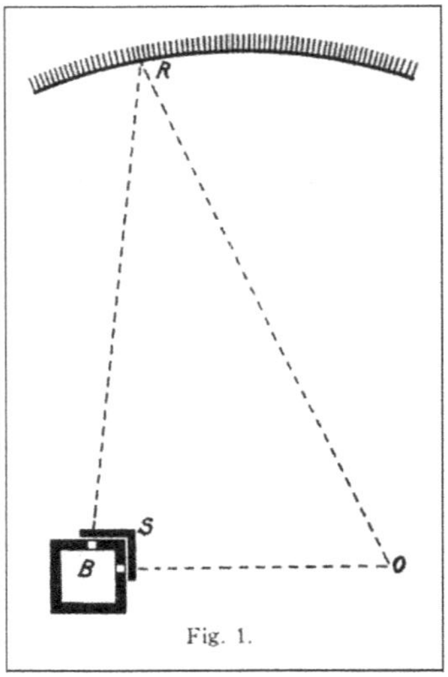

Fig. 6.1 [ETP's Fig. 1 (*eds.*).]

effects involved in weighing the box, and the distance *SRO* being very long in order to permit an accurate reweighing of the box before the arrival of the second particle.

Let us now suppose that the observer at *O* measures the momentum of the first particle as it approaches along the path *SO*, and then measures its time of arrival. Of course the latter observation, made for example with the help of gamma-ray illumination, will change the momentum in an unknown manner. Nevertheless, knowing the momentum of the particle in the past, and hence also its past velocity and energy, it would seem possible to calculate the time when the shutter must have been open from the known time of arrival of the first particle, and to calculate the energy and velocity of the second particle from the known loss in the energy content of the box when the shutter opened. It would then seem possible to predict beforehand both the energy and the time of arrival of the second particle, a paradoxical result since energy and time are quantities which do not commute in quantum mechanics.

The explanation of the apparent paradox must lie in the circumstance that the past motion of the first particle cannot be accurately determined as was assumed. Indeed, we are forced to conclude that there can be no method for measuring the momentum of a particle without changing its value. For example, an analysis of the method of observing the Doppler effect in the reflected infrared light from an approaching particle shows that, although it permits a determination of the momentum of the particle both before and after collision with the light quantum used, it leaves an uncertainty as to the time at which the collision with the quantum takes place. Thus in our example, although the velocity of the first particle could be determined both before and after interaction with the infrared light, it would not be possible to determine the exact position along the path *SO* at which the change in velocity occurred as would be necessary to obtain the exact time at which the shutter was open.

It is hence to be concluded that the principles of the quantum mechanics must involve an uncertainty in the description of past events which is analogous to the uncertainty in the prediction of future events. It is also to be noted that although it is possible to measure the momentum of a particle and follow this with a measurement of position, this will not give sufficient information for a complete reconstruction of its past path, since it has been shown that there can be no method for measuring the momentum of a particle without changing its value. Finally, it is of special interest to emphasize the remarkable conclusion that the principles of quantum mechanics would actually impose limitations on the localization in time of a macroscopic phenomenon such as the opening and closing of a shutter.

ALBERT EINSTEIN
RICHARD C. TOLMAN
BORIS PODOLSKY

California Institute of Technology,
February 26, 1931.

7

On the Indeterminacy Relation*

ALBERT EINSTEIN

Physics Colloquium.
Berlin, 4 November 1931.
In the chair: Prof. M. VON LAUE.

The indeterminacy relation set up by HEISENBERG[1] states that it is not possible to determine position and velocity of a corpuscle with the same accuracy, instead that it is only possible to determine with certainty one of the two components, position or velocity. The more accurate the determination of the one quantity is, the more unsharp that of the other one turns out. Imagine a box with a shutter that opens and closes automatically, which is further equipped with a clock; upon opening of the shutter, let a monochromatic light ray be emitted of about 100 wave trains, and let it be reflected by a mirror placed at a known distance (many light years) and return to an observation point. The energy (colour) of the emitted light ray can be determined through weighing before and after the emission of the ray, the time of emission through the clock. EINSTEIN shows in this very ingenious thought experiment that it is not possible to predict with the aid of measurements both the colour as well as the time of arrival of a light ray at the observation point. Only *one* measurement – of the time *or* the colour – can be carried out precisely, and in fact according to EINSTEIN one can still decide after the departure of the light ray which of the two predictions one wants to choose. – The American physicist TOLMAN has further extended this thought experiment in such a way that one can additionally show that also for the past one can make only one of the two statements precisely.

* [Originally published in *Angewandte Chemie* **45**(1) (1932), 23 (*eds.*).]

[1] Cf. this journ[al] **43**, 853 (1930) [Heisenberg 1930c].

8

Bohr–Einstein Example*

ERWIN SCHRÖDINGER

I have a box with an automatic shutter, which tomorrow at midday shall briefly open a lid. I use the meantime to weigh the box very exactly.

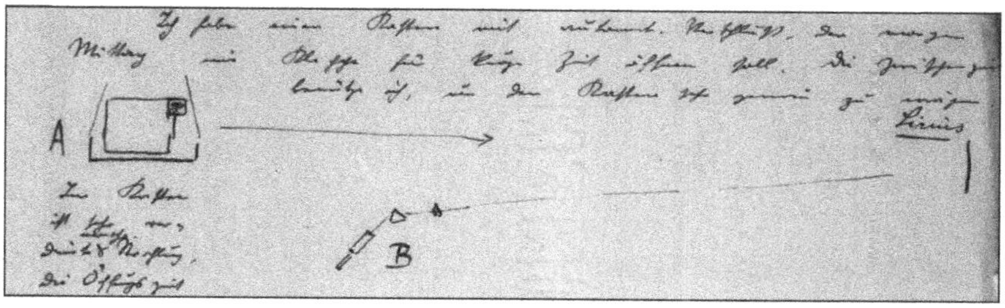

Fig. 8.1 [Beginning of Schrödinger's text, including the illustration of a weighable box (labelled 'A'), mirror ('<u>Sirius</u>') and observation point ('B') (*eds.*).]

In the box there is very diluted monochromatic γ-radiation, the time during which the box is open [is] determined in *such* a way that it is probable that one quantum comes out. After the light quantum is gone, I can still choose whether I wish to transmit to observer B exactly the time of departure, so that he knows exactly the time of arrival, or whether I wish to weigh my box exactly and let him know the colour.

Both together does not work, because through a *short* light signal to B, I change the weight uncontrollably. But already the possibility of an arbitrarily accurate communication of the colour is paradoxical, because the wave train *is* in fact short. I am forced to assume that for observer B the *temporal duration* of the lid opening is indeterminate.

So Einstein said to me. I believe, because of a remark by Wigner that Becker related to me over the phone, that this temporal duration *is* indeterminate. An automatic shutter with the presupposed properties *has no sharp weight.*

* [From AHQP 43, section 1-004, and ZBP-Wien W33-781/4 (handwritten, 2pp., in German), accessible via http://phaidra.univie.ac.at/o:167034 (*eds.*).]

The expectation values in a system with *completely* sharp energy are time-independent. To expect of such a system that at 12:00 it does something definite, [and] will function again at 12:00.0000001 in a[nother] quite particular way, is daft. If an event in a system is to be temporally localisable in advance with a sharpness Δt, somewhat different frequencies must be present in its ψ-function, and specifically one has a basic [elementar] unsharpness relation:

$$\Delta t \Delta \nu \sim 1$$

This gives a weighing error $\frac{h\Delta\nu}{c^2} = \frac{h}{c^2\Delta t}$, i.e. an uncertainty $\Delta E = \frac{h}{\Delta t}$ in the energy of the emitted light quantum, i.e. at least of the right order of magnitude.

One can also say it *like this*:

The thing to be weighed is coupled with the scales. It is by all means admissible to imagine that the energy of the total system box + scales is sharp. But it is evidently inadmissible to imagine that in addition also the reading [Einstellung] of the scales is sharp.

Part III

Core EPR Papers

9

Can Quantum-Mechanical Description of Physical Reality be Considered Complete?*

ALBERT EINSTEIN, BORIS PODOLSKY AND NATHAN ROSEN

In a complete theory there is an element corresponding to each element of reality. A sufficient condition for the reality of a physical quantity is the possibility of predicting it with certainty, without disturbing the system. In quantum mechanics in the case of two physical quantities described by non-commuting operators, the knowledge of one precludes the knowledge of the other. Then either (1) the description of reality given by the wave function in quantum mechanics is not complete or (2) these two quantities cannot have simultaneous reality. Consideration of the problem of making predictions concerning a system on the basis of measurements made on another system that had previously interacted with it leads to the result that if (1) is false then (2) is also false. One is thus led to conclude that the description of reality as given by a wave function is not complete.

1.

Any serious consideration of a physical theory must take into account the distinction between the objective reality, which is independent of any theory, and the physical concepts with which the theory operates. These concepts are intended to correspond with the objective reality, and by means of these concepts we picture this reality to ourselves.

In attempting to judge the success of a physical theory, we may ask ourselves two questions: (1) 'Is the theory correct?' and (2) 'Is the description given by the theory complete?' It is only in the case in which positive answers may be given to both of these questions, that the concepts of the theory may be said to be satisfactory. The correctness of the theory is judged by the degree of agreement between the conclusions of the theory and human experience. This experience, which alone enables us to make inferences about reality, in physics takes the form of experiment and measurement. It is the second question that we wish to consider here, as applied to quantum mechanics.

* [The text follows the original publication in *Physical Review* (Einstein, Podolsky and Rosen 1935), which was received 25 March 1935 and published 15 May. Reprinted with permission from: A. Einstein, B. Podolsky and N. Rosen, *Physical Review* **47**, 777–780. Copyright 1935 by the American Physical Society (*eds.*).]

Whatever the meaning assigned to the term *complete*, the following requirement for a complete theory seems to be a necessary one: *every element of the physical reality must have a counterpart in the physical theory*. We shall call this the condition of completeness. The second question is thus easily answered, as soon as we are able to decide what are the elements of the physical reality.

The elements of the physical reality cannot be determined by *a priori* philosophical considerations, but must be found by an appeal to results of experiments and measurements. A comprehensive definition of reality is, however, unnecessary for our purpose. We shall be satisfied with the following criterion, which we regard as reasonable. *If, without in any way disturbing a system, we can predict with certainty (i.e., with probability equal to unity) the value of a physical quantity, then there exists an element of physical reality corresponding to this physical quantity*. It seems to us that this criterion, while far from exhausting all possible ways of recognizing a physical reality, at least provides us with one such way, whenever the conditions set down in it occur. Regarded not as a necessary, but merely as a sufficient, condition of reality, this criterion is in agreement with classical as well as quantum-mechanical ideas of reality.

To illustrate the ideas involved let us consider the quantum-mechanical description of the behavior of a particle having a single degree of freedom. The fundamental concept of the theory is the concept of *state*, which is supposed to be completely characterized by the wave function ψ, which is a function of the variables chosen to describe the particle's behavior. Corresponding to each physically observable quantity A there is an operator, which may be designated by the same letter.

If ψ is an eigenfunction of the operator A, that is, if

$$\psi' \equiv A\psi = a\psi, \tag{1}$$

where a is a number, then the physical quantity A has with certainty the value a whenever the particle is in the state given by ψ. In accordance with our criterion of reality, for a particle in the state given by ψ for which Eq. (1) holds, there is an element of physical reality corresponding to the physical quantity A. Let, for example,

$$\psi = e^{(2\pi i/h)p_0 x}, \tag{2}$$

where h is Planck's constant, p_0 is some constant number, and x the independent variable. Since the operator corresponding to the momentum of the particle is

$$p = (h/2\pi i)\partial/\partial x, \tag{3}$$

we obtain

$$\psi' = p\psi = (h/2\pi i)\partial\psi/\partial x = p_0\psi. \tag{4}$$

Thus, in the state given by Eq. (2), the momentum has certainly the value p_0. It thus has meaning to say that the momentum of the particle in the state given by Eq. (2) is real.

On the other hand if Eq. (1) does not hold, we can no longer speak of the physical quantity A having a particular value. This is the case, for example, with the coordinate of the particle. The operator corresponding to it, say q, is the operator of multiplication by the independent variable. Thus,

$$q\psi = x\psi \neq a\psi. \tag{5}$$

In accordance with quantum mechanics we can only say that the relative probability that a measurement of the coordinate will give a result lying between a and b is

$$P(a, b) = \int_a^b \bar{\psi}\psi\, dx = \int_a^b dx = b - a. \tag{6}$$

Since this probability is independent of a, but depends only upon the difference $b - a$, we see that all values of the coordinate are equally probable.

A definite value of the coordinate, for a particle in the state given by Eq. (2), is thus not predictable, but may be obtained only by a direct measurement. Such a measurement however disturbs the particle and thus alters its state. After the coordinate is determined, the particle will no longer be in the state given by Eq. (2). The usual conclusion from this in quantum mechanics is that *when the momentum of a particle is known, its coordinate has no physical reality.*

More generally, it is shown in quantum mechanics that, if the operators corresponding to two physical quantities, say A and B, do not commute, that is, if $AB \neq BA$, then the precise knowledge of one of them precludes such a knowledge of the other. Furthermore, any attempt to determine the latter experimentally will alter the state of the system in such a way as to destroy the knowledge of the first.

From this follows that either (1) *the quantum-mechanical description of reality given by the wave function is not complete or* (2) *when the operators corresponding to two physical quantities do not commute the two quantities cannot have simultaneous reality*. For if both of them had simultaneous reality – and thus definite values – these values would enter into the complete description, according to the condition of completeness. If then the wave function provided such a complete description of reality, it would contain these values; these would then be predictable. This not being the case, we are left with the alternatives stated.

In quantum mechanics it is usually assumed that the wave function *does* contain a complete description of the physical reality of the system in the state to which it corresponds. At first sight this assumption is entirely reasonable, for the information obtainable from a wave function seems to correspond exactly to what can be measured without altering the state of the system. We shall show, however, that this assumption, together with the criterion of reality given above, leads to a contradiction.

2.

For this purpose let us suppose that we have two systems, I and II, which we permit to interact from the time $t = 0$ to $t = T$, after which time we suppose that there is no longer any interaction between the two parts. We suppose further that the states of the two systems before $t = 0$ were known. We can then calculate with the help of Schrödinger's equation the state of the combined system I + II at any subsequent time; in particular, for any $t > T$. Let us designate the corresponding wave function by Ψ. We cannot, however, calculate the state in which either one of the two systems is left after the interaction. This, according to quantum mechanics, can be done only with the help of further measurements, by a process known as the *reduction of the wave packet*. Let us consider the essentials of this process.

Let $a_1, a_2, a_3, \ldots$ be the eigenvalues of some physical quantity A pertaining to system I and $u_1(x_1), u_2(x_2), u_3(x_3), \ldots$ the corresponding eigenfunctions, where x_1 stands for the variables used to describe the first system. Then Ψ, considered as a function of x_1, can be expressed as

$$\Psi(x_1, x_2) = \sum_{n=1}^{\infty} \psi_n(x_2) u_n(x_1), \tag{7}$$

where x_2 stands for the variables used to describe the second system. Here $\psi_n(x_2)$ are to be regarded merely as the coefficients of the expansion of Ψ into a series of orthogonal functions $u_n(x_1)$. Suppose now that the quantity A is measured and it is found that it has the value a_k. It is then concluded that after the measurement the first system is left in the state given by the wave function $u_k(x_1)$, and that the second system is left in the state given by the wave function $\psi_k(x_2)$. This is the process of reduction of the wave packet; the wave packet given by the infinite series (7) is reduced to a single term $\psi_k(x_2)u_k(x_1)$.

The set of functions $u_n(x_1)$ is determined by the choice of the physical quantity A. If, instead of this, we had chosen another quantity, say B, having the eigenvalues $b_1, b_2, b_3, \ldots$ and eigenfunctions $v_1(x_1), v_2(x_1), v_3(x_1), \ldots$ we should have obtained, instead of Eq. (7), the expansion

$$\Psi(x_1, x_2) = \sum_{s=1}^{\infty} \varphi_s(x_2) v_s(x_1), \tag{8}$$

where φ_s's are the new coefficients. If now the quantity B is measured and is found to have the value b_r, we conclude that after the measurement the first system is left in the state given by $v_r(x_1)$ and the second system is left in the state given by $\varphi_r(x_2)$.

We see therefore that, as a consequence of two different measurements performed upon the first system, the second system may be left in states with two different wave functions. On the other hand, since at the time of measurement the two systems no longer interact, no real change can take place in the second system in consequence of anything that may be done to the first system. This is, of course, merely a statement of what is meant by the absence of an interaction between the two systems. Thus, *it is possible to assign two different wave functions* (in our example ψ_k and φ_r) *to the same reality* (the second system after the interaction with the first).

Now, it may happen that the two wave functions, ψ_k and φ_r, are eigenfunctions of two non-commuting operators corresponding to some physical quantities P and Q, respectively. That this may actually be the case can best be shown by an example. Let us suppose that the two systems are two particles, and that

$$\Psi(x_1, x_2) = \int_{-\infty}^{\infty} e^{(2\pi i/h)(x_1 - x_2 + x_0)p} dp, \tag{9}$$

where x_0 is some constant. Let A be the momentum of the first particle; then, as we have seen in Eq. (4), its eigenfunctions will be

$$u_p(x_1) = e^{(2\pi i/h)px_1} \tag{10}$$

corresponding to the eigenvalue p. Since we have here the case of a continuous spectrum, Eq. (7) will now be written

$$\Psi(x_1, x_2) = \int_{-\infty}^{\infty} \psi_p(x_2) u_p(x_1) dp, \tag{11}$$

where

$$\psi_p(x_2) = e^{-(2\pi i/h)(x_2 - x_0)p}. \tag{12}$$

This ψ_p however is the eigenfunction of the operator

$$P = (h/2\pi i)\partial/\partial x_2, \tag{13}$$

corresponding to the eigenvalue $-p$ of the momentum of the second particle. On the other hand, if B is the coordinate of the first particle, it has for eigenfunctions

$$v_x(x_1) = \delta(x_1 - x), \tag{14}$$

corresponding to the eigenvalue x, where $\delta(x_1 - x)$ is the well-known Dirac delta-function. Eq. (8) in this case becomes

$$\Psi(x_1, x_2) = \int_{-\infty}^{\infty} \varphi_x(x_2) v_x(x_1) dx, \tag{15}$$

where

$$\varphi_x(x_2) = \int_{-\infty}^{\infty} e^{(2\pi i/h)(x - x_2 + x_0)p} dp = h\delta(x - x_2 + x_0). \tag{16}$$

This φ_x, however, is the eigenfunction of the operator

$$Q = x_2 \tag{17}$$

corresponding to the eigenvalue $x + x_0$ of the coordinate of the second particle. Since

$$PQ - QP = h/2\pi i, \tag{18}$$

we have shown that it is in general possible for ψ_k and φ_r to be eigenfunctions of two non[-]commuting operators, corresponding to physical quantities.

Returning now to the general case contemplated in Eqs. (7) and (8), we assume that ψ_k and φ_r are indeed eigenfunctions of some non-commuting operators P and Q, corresponding to the eigenvalues p_k and q_r, respectively. Thus, by measuring either A or B we are in a position to predict with certainty, and without in any way disturbing the second system, either the value of the quantity P (that is p_k) or the value of the quantity Q (that is q_r). In accordance with our criterion of reality, in the first case we must consider the quantity P as being an element of reality, in the second case the quantity Q is an element of reality. But, as we have seen, both wave functions ψ_k and φ_r belong to the same reality.

Previously we proved that either (1) the quantum-mechanical description of reality given by the wave function is not complete or (2) when the operators corresponding to two physical quantities do not commute the two quantities cannot have simultaneous reality. Starting then with the assumption that the wave function does give a complete description of the physical reality, we arrived at the conclusion that two physical quantities, with non[-]commuting operators, can have simultaneous reality. Thus the negation of (1) leads to the negation of the only other alternative (2). We are thus forced to conclude that the quantum-mechanical description of physical reality given by wave functions is not complete.

One could object to this conclusion on the grounds that our criterion of reality is not sufficiently restrictive. Indeed, one would not arrive at our conclusion if one insisted that two or more physical quantities can be regarded as simultaneous elements of reality *only when they can be simultaneously measured or predicted.* On this point of view, since either one or the other, but not both simultaneously, of the quantities P and Q can be predicted, they are not simultaneously real. This makes the reality of P and Q depend upon the process of measurement carried out on the first system, which does not disturb the second system in any way. No reasonable definition of reality could be expected to permit this.

While we have thus shown that the wave function does not provide a complete description of the physical reality, we left open the question of whether or not such a description exists. We believe, however, that such a theory is possible.

10

Discussion of Probability Relations Between Separated Systems*

ERWIN SCHRÖDINGER

1. When two systems, of which we know the states by their respective representatives, enter into temporary physical interaction due to known forces between them, and when after a time of mutual influence the systems separate again, then they can no longer be described in the same way as before, viz. by endowing each of them with a representative of its own. I would not call that *one* but rather *the* characteristic trait of quantum mechanics, the one that enforces its entire departure from classical lines of thought. By the interaction the two representatives (or ψ-functions) have become entangled. To disentangle them we must gather further information by experiment, although we knew as much as anybody could possibly know about all that happened. Of either system, taken separately, all previous knowledge may be entirely lost, leaving us but one privilege: to restrict the experiments to one only of the two systems. After re-establishing one representative by observation, the other one can be inferred simultaneously. In what follows the whole of this procedure will be called *the disentanglement*. Its sinister importance is due to its being involved in every measuring process and therefore forming the basis of the quantum theory of measurement, threatening us thereby with at least a *regressus in infinitum*, since it will be noticed that the procedure itself involves measurement.

Another way of expressing the peculiar situation is: the best possible knowledge of a *whole* does not necessarily include the best possible knowledge of all its *parts*, even though they may be entirely separated and therefore virtually capable of being 'best possibly known', i.e. of possessing, each of them, a representative of its own. The lack of knowledge is by no means due to the interaction being insufficiently known – at least not in the way that it could possibly be known more completely – it is due to the interaction itself.

* [The main text follows the original publication in the *Proceedings of the Cambridge Philosophical Society* (Schrödinger 1935e), as communicated by Max Born, received 14 August 1935 and read 28 October. The Appendix reproduces the final section of the subsequent Schrödinger (1936) and follows the original publication, also in the *Proceedings of the Cambridge Philosophical Society*, as communicated by Paul Dirac, received 21 April 1936 and read 26 October of that year (*eds.*).]

Attention has recently[1] been called to the obvious but very disconcerting fact that even though we restrict the disentangling measurements to *one* system, the representative obtained for the *other* system is by no means independent of the particular choice of observations which we select for that purpose and which by the way are *entirely* arbitrary. It is rather discomforting that the theory should allow a system to be steered or piloted into one or the other type of state at the experimenter's mercy in spite of his having no access to it. This paper does not aim at a solution of the paradox, it rather adds to it, if possible. A hint as regards the presumed obstacle will be found at the end.

2. To begin with I wish to establish a simple theorem, which makes it very obvious that the phenomenon in question is a quite general one; that it is the rule and not the exception. The representative arrived at for *one* system depends on the *programme* of observations to be taken with the other one. It is necessary to envisage the dependence on the *programme*. For since one device only can be carried out in every individual case and since, moreover, we cannot tell the result (because after all we are not actually experimenting, but sitting at our desk), there seems to be a certain liberty for presuming that perhaps, after all, there is always or at least in most cases a result possible, which is also possible when other devices are followed, and that perhaps it is this that actually would turn up.

Let x and y stand for all the coordinates of the first and second systems respectively and $\Psi(x, y)$ for the normalized representative of the state of the composed system, when the two have separated again, after the interaction has taken place. What constitutes the entanglement is that Ψ is not a product of a function of x and a function of y. Now suppose that we perform on the second system certain observations in consequence of which its representative, at the moment in which disentanglement is reached, is sure to turn up as one out of the known complete set of normalized orthogonal functions $f_n(y)$. Then, provided that the variables which we have measured all commute, we have to develop $\Psi(x, y)$ into a series with respect to the f_n,

$$\Psi(x, y) = \sum_n c_n g_n(x) f_n(y), \tag{1}$$

in order to come to know the representative of the other system. When the readings on the y-system point to $f_k(y)$, we have to adopt $g_k(x)$ as the representative of the x-system. The c_k have been introduced in order to assume that the g_k are normalized, i.e. that

[1] A. Einstein, B. Podolsky and N. Rosen, *Phys. Rev.* **47** (1935), 777 [Einstein, Podolsky and Rosen 1935; see Chapter 9].

$$\int g_k^*(x)g_k(x)dx = 1. \tag{2}$$

Of course $|c_k|^2$ is the probability of that particular case occurring. The equations

$$c_k g_k(x) = \int f_k^*(y)\Psi(x, y)dy \tag{3}$$

together with (2) determine the c's and the g's, apart from an irrelevant phase-factor in every g and its reciprocal in the corresponding c *and* apart from the possible indeterminateness of g, should the integral for some values of k vanish identically in x.

There is no reason for the g_k to be orthogonal to each other. Let us ask *when* they are, i.e. *how* must the f_k be chosen for that purpose? The condition evidently is

$$c_k^* c_l \delta_{kl} = \int dx \int dy \int dy' f_k(y')\Psi^*(x, y')f_l^*(y)\Psi(x, y). \tag{4}$$

This amounts to saying that, for every k, the function

$$u_k(y) = \int dx \int dy' f_k(y')\Psi^*(x, y')\Psi(x, y) \tag{5}$$

is to be orthogonal to all the $f_l(y)$, with the possible exception of $f_k(y)$. Hence $u_k(y)$ must be a numerical multiple of $f_k(y)$. From (4), with $l = k$, it is seen that the numerical multiplier is $|c_k|^2$. We have therefore

$$|c_k|^2 f_k(y) = \int dx \int dy' f_k(y')\Psi^*(x, y')\Psi(x, y). \tag{6}$$

Introducing the function

$$K(y, y') = \int dx\, \Psi^*(x, y')\Psi(x, y), \tag{7}$$

which has Hermitian symmetry, we see from (6) that the reciprocals of the $|c_k|^2$ and the functions $f_k(y)$ are required to be the eigenvalues and a system of eigenfunctions respectively of the homogeneous integral equation

$$f(y) = \lambda \int K(y, y')f(y')dy'. \tag{8}$$

Provided that the integral in (7) converges, so that K is defined, a complete solution of (8) exists. (It is convenient for our purposes, in order to be concerned with *complete* sets only, to include the functions, orthogonal to K, as eigenfunctions belonging to $\lambda = \infty$, at variance with the custom of mathematicians.) By using this set for the development (1) one easily satisfies oneself that all requirements are fulfilled, in particular that the λ_k^{-1} *are* all non-negative and that their sum *is* unity.

The general case is evidently that all the λ_k^{-1} are different from one another, except maybe for an arbitrary set of them vanishing. Then the *relevant* $f_k(y)$ are uniquely determined and so are the $g_k(x)$. Hence there is always *one* and as a rule only one development of $\Psi(x, y)$ of the type which might suitably be called 'biorthogonal'[.][2]

Whenever (and of course only when) the eigenfunctions of a programme to be carried out on the y-system include the relevant functions $f_k(y)$, or the eigenfunctions properly speaking of (8), the programme will lead to the biorthogonal development and imply the relevant $g_k(x)$ as the other set. Now if for an arbitrarily fixed programme of measurements on the y-system the representative arrived at for the x-system was the same *in all individual cases*, the same $g_k(x)$ would have to turn up (and even with the same probabilities) as in the biorthogonal development; for in two infinite series of repetitions *ab ovo* of one and of the other programme respectively every possible result occurs according to its due probability. Hence the relevant functions $g_k(x)$ would have to be implied whatever programme is carried out. But since, of course, they also determine the biorthogonal development uniquely and thereby require the relevant $f_k(y)$ as the other set, these would have to be included in the eigenfunctions of every programme which cannot be, since the latter are, by principle, an entirely arbitrary complete orthogonal set. Hence the non-invariance is proved.[3]

There must, of course, be cases in which the biorthogonal development refers to a continuous variable (or set of commuting variables), an integral replacing the series (1); and also mixed cases. In our present treatment they would be indicated by the integral (7) diverging and would therefore require a separate treatment, on which I shall not enter here.

The biorthogonal development is the one to give us true insight into the entanglement. If there are no coincidences among the $|c_k|^2$ (excluding also the case, that more than one of them vanish) the relevant f_k's form a well-determined and complete set and so do the g_k's. Then one can say that the entanglement consists in that one and only one observable (or set of commuting observables) of one system is uniquely determined by a definite observable (or set of commuting observables) of the other system. This is the general case. We shall now turn to the opposite extreme, which is the Einstein–Podolsky–Rosen case. It could be characterized by *all* the $|c_k|^2$ being equal and *all* possible developments being biorthogonal. *Every* observable (or set, etc.) of one system is determined by an observable (or set, etc.)

[2] The whole mathematical treatment is familiar to mathematicians in dealing with an 'unsymmetrical kernel' $\Psi(x, y)$. See Courant–Hilbert, *Methoden der mathematischen Physik*, 2nd edition, [vol. 1,] p. 134 [Courant and Hilbert 1930].

[3] In order to adapt this proof to the case when the biorthogonal development is *not* unique, just replace *the* biorthogonal development by a particular one, on which you fix your attention.

of the other one. But the mere fact, that the equality of the $|c_k|^2$ prevents their sum from being normalized to unity, shows us that very improper representatives (in fact much more so than Dirac's δ, δ', δ'', ...) are involved in this case, making it advisable to deal with it on slightly different lines.

3. For simplicity's sake we suppose each of the two systems to have one degree of freedom only. Let the q-numbers x_1, p_1 and x_2, p_2 denote coordinate and momentum of the first and of the second system respectively. The existence of further degrees of freedom would not affect the considerations of this section except for slight alterations in the wording; but for section 4 to hold it would have to be assumed, that within each of the two systems the degree of freedom which we investigate has its Hamiltonian separated from the rest.

The two systems are of course supposed not to interact with each other. The entanglement is to be such that the two *commuting* observables

$$x = x_1 - x_2, \quad p = p_1 + p_2, \tag{9}$$

which we choose to represent the state of the *composed* system, have definite numerical values, say x' and p' respectively, which we suppose to be known. The representative Ψ of the composed system is a function of the eigenvalues of x and p, which involves x' and p' as parameters and vanishes everywhere except in that point where the former are equal to the latter. It is not a δ-function though and can hardly be written explicitly. According to our assumptions Ψ must have the properties

$$x\Psi = x'\Psi \quad \text{and} \quad p\Psi = p'\Psi. \tag{10}$$

We shall use no others.

From (9) the variable x can be observed by observing x_1 and x_2 separately, because the latter commute. The difference of the observed values, x_1' and x_2' say, must be equal to x':

$$x_1' - x_2' = x'. \tag{11}$$

Hence x_1' can be predicted from x_2' and *vice versa.* Similarly

$$p_1' + p_2' = p', \tag{12}$$

so that the result of measuring p_1 serves to predict the result for p_2 and *vice versa.* But of course every *one* of the *four* observations in question, when actually performed, disentangles the systems, furnishing each of them with an independent representative of its own. A *second* observation, whatever it is and on whichever system it is executed, produces no further change in the representative of the *other* system.

Yet since I can predict *either* x_1' or p_1' without interfering with system No. 1 and since system No. 1, like a scholar in examination, cannot possibly know which of the two questions I am going to ask it first: it so seems that our scholar is prepared to give the right answer to the *first* question he is asked, *anyhow*. Therefore he must know both answers; which is an amazing knowledge, quite irrespective of the fact that after having given his first answer our scholar is invariably so disconcerted or tired out, that all the following answers are 'wrong'[.]

Thus far the results of the paper quoted above. Now I wish to point out that system No. 1 (say) has further knowledge. It does not only know these two answers but a vast number of others, and that with no mnemotechnical help whatsoever, at least with none that we know of.

Let us consider an Hermitian operator referring to the first system and given as a 'well-ordered' analytic function of the observables x_1 and p_1:

$$F(x_1, p_1), \tag{13}$$

which we suppose not to contain the $\sqrt{-1}$ *explicitly*. It is an observable of system No. 1. We shall prove that its value is equal to the value of the following observable of system No. 2

$$F(x_2 + x', p' - p_2), \tag{14}$$

so that the result of either observation can be predicted from the other one. That is not trivial, because the equations $x = x'$ and $p = p'$ do not hold, except in the form (10), that is to say they are not identities.

The proof will be produced, if we can show that the difference of the two operators, when applied to Ψ, gives zero:

$$\{F(x_2 + x', p' - p_2) - F(x_1, p_1)\}\Psi = 0. \tag{15}$$

Using (9), we may write this in the form

$$\{F(x_1 + x' - x, p_1 + p' - p) - F(x_1, p_1)\}\Psi = 0. \tag{16}$$

To prove it we observe that any operator which *ends* on its right with a factor $x' - x$ or $p' - p$ reduces Ψ to zero, from (10). Additive terms of this type can therefore be dropped within the curved bracket. Now fix the attention on one of the power products in the minuendus. Its last factor, either $x_1 + x' - x$ or $p_1 + p' - p$, can be replaced by x_1 or p_1, as the case may be, and then this x_1 or p_1 commutes with the rest of the power product and can be removed to its extreme left. The second step consists in applying a similar treatment to the factor ($x_1 + x' - x$ or $p_1 + p' - p$, as the case may be), which has *now* become the last; but this one cannot safely be displaced to the *extreme* left but only to the second place, counting from the left. This procedure is continued until we are left with a power product which differs from the original one in that x_1 and p_1 have replaced $x_1 + x' - x$ and $p_1 + p' - p$ respectively and also that the order of factors has been reversed. But F, owing to its Hermiticity and to the further condition that it should not contain $\sqrt{-1}$ explicitly, must contain the 'reversed' power product too. Hence after applying the same treatment to all of them, we are left with $F(x_1, p_1)$, which cancels with the subtrahendus, and the statement is proved.

If F contains the $\sqrt{-1}$ explicitly, we could replace it by $(x_1p_1 - p_1x_1)/\hbar$. Then the prescription (14) would apply without corollary. It would turn the operator just mentioned into $(p_2x_2 - x_2p_2)/\hbar$ which now can be replaced by $-\sqrt{-1}$. From this follows the corollary to prescription (14), that an explicit $\sqrt{-1}$ has to change sign.

By this theorem all observables are placed on the same footing. Our system, in its virgin state, must know the answers to all of them. One might presume that it avails itself at least of a suggestive mnemotechnical device, viz. that the answer prepared for the variable $F(x_1, p_1)$ is simply $F(x_1', p_1')$, if x_1' and p_1' are those prepared for x_1 and p_1 respectively. But this is not so. For consider, e.g., the *series* of observables

$$F(x_1, p_1, b) = \frac{1}{b}p_1^2 + bx_1^2,$$

where b is to be a positive c-number parameter. With every value for b we are confronted with a new observable, to which an answer must be pending. Moreover the answer must be, irrespective of b, an odd integral multiple of $\hbar$ (though not necessarily independent of b). This shows plainly that all these answers cannot conform to the results which would be obtained by inserting into the expression the same pair of c-numbers, p_1' and x_1'; which, by the way, are simultaneously accessible to experiment in every individual case, one by direct observation, the other one by inference from an observation on the other system.

Our complete lack of insight into the relationship between the different answers in *one* system is all the more bewildering, since we have proved, on the other hand, that the one-to-one correspondence between the answers of the two systems necessarily extends to *all* pairs of observables whenever it holds for two of them.

4. If equations (10) are assumed to hold at time zero, the equations of motion determine what becomes of them as time proceeds. Let the Hamiltonian of the composed system be

$$H = H_1(x_1, p_1) + H_2(x_2, p_2) \tag{17}$$

and let it not contain the time explicitly. We shall use what Dirac calls a Heisenberg representation; then every variable at time t is a function of the variables at time zero, e.g.

$$x_{1t} = e^{itH_1/\hbar}x_1e^{-itH_1/\hbar}. \tag{18}$$

From prescription (14), including the corollary, we can find out what observable of system No. 2 is equivalent to x_{1t}; we call it $[x_{1t}]_2$,

$$[x_{1t}]_2 = e^{-itH_1(x_2+x',p'-p_2)/\hbar}(x_2 + x')e^{itH_1(x_2+x',p'-p_2)/\hbar}. \tag{19}$$

This equation, by its form, indicates the observation on No. 2 *at time zero*, which would serve to predict the coordinate of No. 1 at time t. Solving two equations similar to (18) for x_2 and p_2, we get

$$x_2 = e^{itH_2/\hbar}x_{2t}e^{itH_2/\hbar} \tag{20}$$

and similarly for p_2. Of course H_2 has now to be thought of as written with the arguments x_{2t} and p_{2t}; which does not affect its form, since it is a constant of the motion. With these expressions replacing x_2 and p_2 in (19), the exponentials with H_2 cancel in the interior, leaving just one in front and in the rear. So the final result is

$$[x_{1t}]_2 = e^{-itH_2/\hbar}e^{-itH_1\hbar}(x_{2t} + x')e^{itH_1/\hbar}e^{itH_2\hbar}, \tag{21}$$

where H_1 and H_2 are precisely the functions of equation (17), but written with the arguments

$$x_{2t} + x' \quad \text{and} \quad p' - p_{2t} \quad \text{for } H_1,$$
$$x_{2t} \quad \text{and} \quad p_{2t} \quad \text{for } H_2.$$

This rather complicated function of x_{2t} and p_{2t} is *that* observable of system No. 2 which is equivalent to x_{1t}. Though we have deduced it by means of a Heisenberg representation, the functional connection is of course exactly the same for what Dirac calls the Schrödinger representatives. That is to say, we can take x_{2t} and p_{2t} to have the general meaning of x_2 and p_2 of the preceding section. Regarded as operators they then do not involve the notion of time but work on a Ψ-function, which develops according to the wave equation. This consideration applies to every moment of time. It is therefore correct to say, that the variable which in No. 2 is equivalent to the coordinate in No. 1 undergoes a continuous unitary or contact transformation as time goes on. The transformation is of course the same for every observable, so that we need not write out the formulae for $[p_{1t}]_2$ or for an arbitrary $[F(x_{1t}, p_{1t})]_2$. It is noteworthy that the two exponentials of which the transformation is composed may not be amalgamated, because H_1 and H_2, considering the arguments with which they are written, do not in general commute.

All this is moderately trivial. But it is necessary to consider it lest one should believe that the antinomies could be solved by suggesting or proving that some of the observations must take a certain minimum time. Provided that they *relate* to a definite moment, this will not help us. It cannot be argued that, before the results are reached, the situation to which they refer has passed away. A prediction for time zero does not dissolve into nought as time goes on, but simply transforms into the prediction of another observable. And any desired observable can be predicted for time t by making a *suitable* observation at time zero on the *other* system.

When at time zero a certain observable of system No. 1, say x_1, is inferred from observing x_2, *I am forced* to assign to system No. 1 a representative that makes the observable x_1 precise and tells nothing about its canonically conjugate, *although* I safely infer that system No. 1 *does* know quite a definite (as opposed to a haphazard) answer for the canonical conjugate as well, the only difference being that I know the one while I am ignorant of the other. Now this paradoxical situation is not confined to time zero and could not, therefore, be avoided by my satisfying myself that the result of observing x_2 cannot be known before a certain time has elapsed. From the moment I come to know the result I should be faced with exactly the same situation, only the pair of canonically conjugate observables that is involved changes with time.

The paradox would be shaken, though, if an observation did not *relate* to a definite moment. But this would make the present interpretation of quantum mechanics meaningless, because at present the *objects* of its predictions are considered to be the results of measurements for definite moments of time.

My sincerest gratitude is due to Imperial Chemical Industries, Limited, to whose generosity I owe the leisure for carrying out these studies.

Summary

The probability relations which can occur between two separated physical systems are discussed, on the assumption that their state is known by a representative in common. The *two families* of observables, relating to the first and to the second system respectively, are linked by at least *one match* between two definite numbers, one of either family. The word *match* is short for stating that the *values* of the two observables in question determine each other uniquely and therefore (since the actual labelling is irrelevant) can be taken to be *equal*. In general there is but one match, but there can be more. If, in addition to the first match, there is a second one between canonical conjugates of the first mates, then there are infinitely many matches, every function of the first canonical pair matching with the same function of the second canonical pair.[4] Thus there is a complete one-to-one correspondence between *those* two branches (of the two families of observables) which relate to the two degrees of freedom in question. If there *are* no others,[5] the one-to-one

[4] To make the earlier text conform to the present simplified wording, replace $x_2 + x'$ by P and $p' - p_2$ by X. Then X and P are canonical conjugates. The mating (x with P and p with X) has to be *cross-wise*, though.

[5] In fact it persists anyhow, but as a rule in a very much more complicated form.

correspondence persists as time advances, but the observables of the first system (say) change their mates in the way that the latter, i.e. the observables of the second system, undergo a certain continuous contact-transformation.

Appendix: Section 4 from Schrödinger (1936)**

Indubitably the situation described here is, in present quantum mechanics, a necessary and indispensable feature. The question arises, whether it is so in Nature too. I am not satisfied about there being sufficient experimental evidence for that. Years ago I pointed out[6] that when two systems separate far enough to make it possible to experiment on one of them without interfering with the other, they are bound to pass, during the process of separation, through stages which were beyond the range of quantum mechanics as it stood then. For it seems hard to imagine a complete separation, whilst the systems are still so close to each other, that, from the classical point of view, their interaction could still be described as an unretarded *actio in distans*. And ordinary quantum mechanics, on account of its thoroughly unrelativistic character, really only deals with the *actio in distans* case. The whole system (comprising in our case *both* systems) has to be small enough to be able to neglect the time that light takes to travel across the system, compared with such periods of the system as are essentially involved in the changes that take place.

Though in the mean time some progress seemed to have been made in the way of coping with this condition (quantum electrodynamics), there now appears to be a strong probability (as P. A. M. Dirac[7] has recently pointed out on a special occasion) that this progress is futile.

It seems worth noticing that the paradox could be avoided by a very simple assumption, namely if the situation after separating were described by the [biorthogonal] expansion

$$\left[\Psi(x, y) = \sum_k a_k g_k(x) f_k(y)\right],$$

** [We reproduce here the final section (pp. 451–452) of the paper's companion, which appeared under the similar title 'Probability Relations between Separated Systems' (Schrödinger 1936). The bulk of the paper deals with generalising further the results on steering that Schrödinger has discussed above. Specifically, after reviewing some results on mixed states, Schrödinger shows that for entangled systems, every convex decomposition of one system into states that are linearly independent can be realised by an appropriate measurement on the other. In the generic case, a system can thus be steered into any state with non-zero probability. The final section, however, could follow on quite seamlessly from the conclusion of the first paper: it speculates about the origins of the paradox discussed, about whether it might get resolved in the proper relativistic setting, and on whether there are indeed any EPR correlations in the relativistic regime (*eds.*).]

[6] E. Schrödinger, *Annalen der Physik* (4), 83 (1927), 961 [Schrödinger 1927]. *Collected Papers* (Blackie and Son, 1928), p. 141 [Schrödinger 1928c].

[7] P. A. M. Dirac, *Nature*[,] 137 (1936), 298 [Dirac 1936].

but with the additional statement that the knowledge of the *phase relations* between the complex constants a_k has been entirely lost in consequence of the process of separation. This would mean that not only the parts, but the whole system, would be in the situation of a mixture, not of a pure state. It would not preclude the possibility of determining the state of the first system by *suitable* measurements in the second one or *vice versa*. But it would utterly eliminate the experimenter[']s influence on the state of the system which he does not touch.

This is a very incomplete description and I would not stand for its adequateness. But I would call it a possible one, until I am told, either why it is devoid of meaning or with which experiments it disagrees. My point is, that in a domain which the present theory does not cover, there is room for new assumptions without necessarily contradicting the theory in that region where it is backed by experiment.

11

The Present Situation in Quantum Mechanics*

ERWIN SCHRÖDINGER

1. The Physics of Models

In the second half of the previous century there arose, from the great progress in kinetic theory of gases and in the mechanical theory of heat, an ideal of the exact description of nature that stands out as the reward of centuries-long search and the fulfillment of millennia-long hope, and that is called classical. These are its features.

Of natural objects, whose observed behavior one might treat, one sets up a representation [Vorstellung] – based on the experimental data in one's possession but without handcuffing the intuitive imagination – that is worked out in all details exactly, *much* more exactly than any experience, considering its limited extent, can ever authenticate. The representation in its absolute determinacy resembles a mathematical object[i] or a geometric figure which can be completely calculated from a number of *determining data* [Bestimmungsstücke]; as, e.g., a triangle's one side and two adjoining angles, as determining data, also determine the third angle, the other two sides, the three altitudes, the radius of the inscribed circle, etc. Yet the representation differs intrinsically from a geometric figure in this important respect, that also in *time* as fourth dimension it is just as sharply determined as the figure is in the three space dimensions. Thus it is a question (as is self-evident) always of an object[ii] that changes with time, that can assume different *states*; and if a state

* [Originally published as 'Die gegenwärtige Situation in der Quantenmechanik', *Naturwissenschaften* **23**(48–50) (1935), 807–812, 823–828, 844–849, issued 29 November, 6 December and 13 December 1935, respectively. The translation here reprinted was made by John D. Trimmer and published in *Proceedings of the American Philosophical Society* **124**(5) (1980), 323–338. Occasional inaccuracies in the translation as originally published have been corrected and indicated, mostly with endnotes (of the form 'Trimmer translation'/'German text'). The only amendment we have made systematically throughout the text is to translate 'Bestimmungsstücke' not as 'determining parts' but as 'determining data'. Obvious typos and slight (typographical) inconsistencies have been amended tacitly, the format of the original German for in-text or displayed equations and other such details has been restored, and (as in other reprinted texts) typographical conventions have been generally adapted to fit the rest of the volume (*eds.*). By kind permission of the American Philosophical Society and Springer Nature Customer Service Centre GmbH: Springer Nature, *Naturwissenschaften*, Die gegenwärtige Situation in der Quantenmechanik, Erwin Schrödinger, 1935.]

becomes known in the necessary number of determining data, then not only are all other data also given for this moment (as illustrated for the triangle above), but likewise all data, the complete state, for any given later time; just as the character of a triangle on its base determines its character at the apex. It is part of the inner law of the object[iii] that it should change in a given manner, that is, if left to itself in a given initial state, that it should continuously run through a given sequence of states, each one of which it reaches at a fully determined time. That is its nature, that is the hypothesis, which, as I said above, one postulates on the basis[iv] of intuitive imagination.

Of course one must not think so naively,[v] that in this way one learns how things go in the real world. To show that one does not think this, one calls the precise thinking aid that one has created, an *image* or a *model*. With its uncritical[vi] clarity, which cannot be attained without arbitrariness, one has merely insured that a fully determined [ganz bestimmte] hypothesis can be tested for its consequences, without admitting further arbitrariness during the tedious calculations required for deriving these consequences. Here one has explicit marching orders and actually works out only what a clever fellow could have told directly from the data! At least one then knows where the arbitrariness lies and where improvement must be made in case of disagreement with experience: in the initial hypothesis or model. For this one must always be prepared. If in many various experiments the natural object [really] behaves like the model, one is happy and thinks that the image fits the reality in essential features. If it fails to agree, under novel experiments or with refined measuring techniques, it is not said that one should *not* be happy. For basically this is the means of gradually bringing our [image], i.e., our thinking, closer to the realities.

The classical method of the precise model has as principal goal keeping the unavoidable arbitrariness neatly isolated in the assumptions, more or less as the body isolates the germ plasm,[vii,1] for the historical process of adaptation to continuing experience. Perhaps the method is based on the belief that *somehow* the initial state *really* determines uniquely the subsequent events, or that a *complete* model, agreeing with reality in *complete exactness* would permit predictive calculation of outcomes of all experiments with complete exactness. Perhaps on the other hand this belief is based on the method. But it is quite probable that the adaptation of thought to experience is an infinite process and that 'complete model' is a contradiction in terms, somewhat like 'largest integer'.

A clear idea[viii] of what is meant by classical *model*, its *determining data*, its *state*, is the foundation for all that follows. Above all, a *determinate model* and a

[1] [This is presumably a reference to Weismann's (1892) theory of inheritance, which distinguishes strictly between germ cells and somatic cells, in particular ruling out that acquired somatic characteristics may be inherited (*eds.*).]

determinate state of the same must not be confused. Best consider an example. The Rutherford model of the hydrogen atom consists of two point masses. As determining data one could for example use the two times three rectangular coordinates of the two points and the two times three components of their velocities along the coordinate axes – thus twelve in all. Instead of these one could also choose: the coordinates and velocity components of the *center of mass*, plus the *separation* of the two points, *two angles* that establish the direction in space of the line joining them, and the speeds (= time derivatives) with which the separation and the two angles are changing at the particular moment; this again adds up of course to twelve. It is *not* part of the concept 'R[utherford] model of the H-atom' that the determining data should have particular numerical values. Such being assigned to them, one arrives at a *determinate state* of the model. The clear view over the totality of possible states – yet without relationship among them – constitutes 'the model' or 'the model in *any* state *whatsoever*'. But the concept of the model then amounts to more than merely: the two points in certain positions, endowed with certain velocities. It embodies also knowledge for *every* state how it will change with time in absence of outside interference. (Information on how one half of the determining data will change with time is indeed given by the other half, but how this other half will change must be independently determined.) *This* knowledge is implicit in the assumptions: the points have the masses m, [resp.] M and the charges $-e$, [resp.] $+e$ and therefore attract each other with force e^2/r^2, if their separation is r.

These pieces of information,[ix] with definite numerical values for m, M, and e (but of course *not* for r), belong to the description *of the model* (not first and only to that of a definite state). m, M, and e are *not* determining data. By contrast, separation r is one. It appears as the seventh in the second 'set' of the example introduced above. And if one uses the first, r is not an independent thirteenth but can be calculated from the 6 rectangular coordinates:

$$r = \sqrt{(x_1 - x_2)^2 + (y_1 - y_2)^2 + (z_1 - z_2)^2}.$$

The number of determining data (which are often called *variables* in contrast to *constants of the model* such as m, M, e) is unlimited. Twelve conveniently chosen ones determine all others, or the *state*. No twelve have the privilege of being *the* determining data. Examples of other especially important determining data are: the energy, the three components of angular momentum relative to center of mass, the kinetic energy of center of mass motion. These just named have, however, a special character. They are indeed *variable*[*s*], i.e., they have different values in different states. But in every *sequence* of states, that is actually passed through in the course of time, they retain the same value. So they are also called *constants of the motion* – differing from constants of the model.

2. Statistics of Model Variables in Quantum Mechanics

At the pivot point of contemporary quantum mechanics (Q.M.) stands a doctrine, that perhaps may yet undergo many shifts of meaning but that will not, I am convinced, cease to be the pivot point. It is this, that models with determining data that uniquely determine each other, as do the classical ones, cannot do justice to nature.

One might think that for anyone believing this, the classical models have played out their roles. But this is not the case. Rather one uses precisely *them*, not only to express the negative of the new doctrine, but also to describe the diminished mutual determination[x] remaining afterwards as though obtaining among the same variables of the same models as were used earlier, as follows:

A. The classical concept of *state* becomes lost, in that at most a well-chosen *half* of a complete set of variables can be assigned definite numerical values; in the Rutherford model[xi] for instance the six rectangular coordinates *or* the velocity components (still other groupings are possible). The other half then remains completely indeterminate, while supernumerary data can show highly varying degrees of indeterminacy. In general, of a complete set (for the R[utherford] model twelve [pieces of] data) *all* will be known only uncertainly. One can best keep track of the degree of uncertainty by following classical mechanics and choosing variables arranged *in pairs* of so-called canonically-conjugate ones. The simplest example is a space coordinate x of a point mass and the component p_x along the same direction, its linear momentum (i.e., mass times velocity). Two such constrain each other in the precision with which they may be simultaneously known, in that the product of their tolerance- or variation-widths (customarily designated by putting a Δ ahead of the quantity) cannot fall *below* the magnitude of a certain universal constant,[2] thus

$$\Delta x \cdot \Delta p_x \geqq \hbar.$$

(Heisenberg uncertainty relation.)

B. If even at any given moment not all variables are determined by some of them, then of course neither are they all determined for a later moment by data obtainable earlier. This may be called a break with causality, but in view of A it is nothing essentially new. If a classical state does not exist at any moment, it can hardly change causally [zwangsläufig]. What do change are the *statistics* or *probabilities, these* moreover causally [zwangsläufig]. Individual variables meanwhile may become more, or less, uncertain. Overall it may be said that the total precision of the description does not change with time, because the principle of limitations described under A remains the same at every moment. –

[2] $h = 1.041 \cdot 10^{-27}$ erg sec. Usually in the literature the 2π-fold of this ($6.542 \cdot 10^{-27}$ erg sec) is designated as h and for *our* h an h with a cross-bar is written. [Transl. Note (J. Trimmer): In conformity with the now universal usage, $\hbar$ is used in the translation in place of h.]

Now what is the meaning of the terms 'uncertain', 'statistics', 'probability'? Here Q.M. gives the following account. It takes over unquestioningly from the classical model the entire infinite roll call of imaginable variables or determining data and proclaims each datum to be *directly measurable*, indeed measurable to arbitrary precision, so far as it alone is concerned. If through a well-chosen, constrained set of measurements one has gained that maximal knowledge of an object which is just possible according to A, then the mathematical apparatus of the new theory provides means of assigning, for the same or for any later instant of time, a fully determined *statistical distribution* to *every* variable, that is, an indication of the fraction of cases it will be found at this or that value, or within this or that small interval (which is also called probability). The idea[xii] is that this is in fact the probability of encountering the relevant variable, if one measures it at the relevant time, at this or that value. By a single trial the correctness of this *probability prediction* can be given at most an approximate test, namely in the case that it is comparatively sharp, i.e., declares possible only a small range of values. To test it thoroughly one must repeat the trial *ab ovo* (i.e., including the orientational or preparatory measurements) *very* often and may use only those cases in which the *preparatory* measurements gave exactly the same results. For these cases, then, the statistics of a particular variable reckoned forward from the preparatory measurements, is to be confirmed by measurement – this is the idea.[xiii]

One must guard against criticizing this idea[xiv] because it is so difficult to express; this is a matter of [our] language. But a different criticism surfaces. Scarcely a single physicist of the classical era would have dared to believe, in thinking about a model, that its determining data are measurable on the natural object. Only much remoter consequences of the picture were actually open to experimental test. And all experience pointed toward one conclusion: long before the advancing experimental arts had bridged the broad chasm, the model would have substantially changed through gradual adaption to new facts. – Now while the new theory calls the classical model incapable of specifying all the details of the *mutual interrelationship of the determining data* (for which its creators intended it), it nevertheless considers the model suitable for guiding us as to just which *measurements*[xv] can in principle be made on the relevant natural object. This would have seemed to those who thought up the picture a scandalous extension of their thought-aid[xvi] and an unscrupulous anticipation of future development.[xvii] Would it not be pre-established harmony of a peculiar sort if the classical-epoch researchers, those who, as we hear today, had no idea of what *measuring* truly is, had unwittingly gone on to give us as legacy a guidance scheme revealing just what is fundamentally measurable for instance about a hydrogen atom!?

I hope later to make clear that the reigning doctrine is born of distress. Meanwhile I continue to expound it.

3. Examples of Probability Predictions

Thus all predictions still pertain[xviii] to determining data of a classical model, to positions and velocities of point masses, to energies, angular momenta, etc. The only unclassical feature is that only probabilities are predicted. Let us have a closer look. The orthodox treatment is always that, by way of certain measurements performed *now* and their results, one gains the best possible probability estimates permitted by nature for the results to be expected of other measurements following thereafter either immediately or at some given time.[xix] Now how does the matter really stand? In important and typical cases as follows.

If one measures the energy of a Planck oscillator, the probability of finding for it a value between E and E' cannot possibly be other than zero unless between E and E' there lies at least one value from the series

$$3\pi\hbar\nu,\ 5\pi\hbar\nu,\ 7\pi\hbar\nu,\ 9\pi\hbar\nu,\ \ldots$$

For any interval containing none of these values the probability is zero. In plain English: other measurement results are excluded. The values are odd multiples of the *constant of the model* $\pi\hbar\nu$ ($[\hbar] =$ (Planck constant)$/2\pi$, $\nu =$ frequency of the oscillator). Two points stand out. First, no account is taken of preceding measurements – these are quite unnecessary. Second, the statement certainly doesn't suffer an excessive lack of precision – quite to the contrary it is sharper than any actual measurement could ever be.

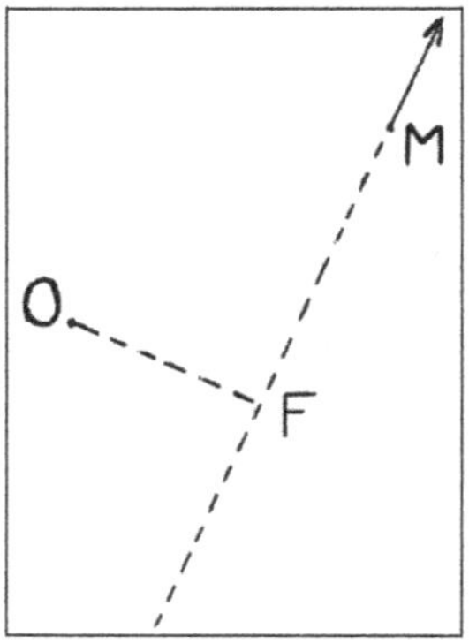

Fig. 11.1 *[Schrödinger's Fig. 1 (eds.).] Angular momentum*: M is a material point, O a geometric reference point. The vector arrow represents the momentum (= mass times velocity) of M. Then the *angular momentum* is the product of the length of the arrow by the length OF.

Another typical example is magnitude of angular momentum. In Fig. 1 let M be a moving point mass, with the vector representing, in magnitude and direction, its momentum (mass times velocity). O is any arbitrary fixed point in space, say the origin of coordinates; thus not a physically significant point, but rather a geometric reference point. As magnitude of the angular momentum of M about O classical mechanics designates the product of the length of the momentum vector by the length of the *normal OF*. In Q.M. the magnitude of angular momentum is governed much as the energy of the oscillator. Again the probability is zero for any interval not containing some value(s) from the following series[xx]

$$0,\ \hbar\sqrt{2},\ \sqrt{2\times 3},\ \hbar\sqrt{3\times 4},\ \hbar\sqrt{4\times 5},\ \ldots$$

that is, only one of these values is allowed. Again, this is true without reference to preceding measurements. And one readily conceives how important is this precise statement, *much* more important than knowing which of these values, or what probability for each of them, would actually pertain to a given case. Moreover it is also noteworthy here that there is no mention of the reference point: however it is chosen one will get a value from the series. This assertion is nonsensical for the model,[xxi] because the normal OF changes *continuously* as the point O is displaced, [and] the momentum vector remains unchanged. In this example we see how Q.M. does indeed use the model to read off those quantities which one can measure and for which it makes sense to predict results, but finds the classical model inadequate for explicating relationships among these quantities.

Now in both examples does one not get the feeling that the essential content of what is being said can only with some difficulty be forced into the Spanish boot of a prediction of probability of finding this or that measurement result for a variable of the classical model? Does one not get the impression that here one deals with fundamental properties of *new* classes of characteristics, that keep only the name in common with classical ones? And by no means do we speak here of exceptional cases, rather it is precisely the truly valuable statements of the new theory that have this character. There are indeed problems more nearly of the type for which the mode of expression is suitable. But they are by no means equally important. Moreover of no importance whatever are those that are naively set up as class exercises. 'Given the position of the electron in the hydrogen atom at time $t=0$, find the statistics of its position at a later time'. No one cares about that.

Taken literally[xxii] all statements pertain to the intuitive model. But the useful statements are scarcely intuitive to it, and its intuitive aspects are of little worth.

4. Can One Base the Theory on Ideal Ensembles?

The classical model plays a Protean role in Q.M. Each of its determining data can under certain circumstances become an object of interest and achieve a certain reality. But never all of them together – now it is these, now those, and indeed always at most *half* of [a] complete set of variables that would allow a full picture of the momentary state.[xxiii] Meantime, how about the others? *Have* they then no reality, perhaps (pardon the expression) a blurred reality; or are all of them always real and is it merely, according to Theorem A of Sect. 2, that simultaneous *knowledge* of them is ruled out?

The second conception[xxiv] is especially appealing to those acquainted with the *statistical viewpoint* that came up in the second half of the preceding century; the more so, considering that on the eve of the new century quantum theory was born *from it*, from a central problem in the statistical theory of heat (Max Planck's theory of heat radiation, December, 1899).[3] The essence of this line of thought is precisely this, that one practically never knows all the determining data of the system, but rather *much* fewer. To describe an actual body at a given moment one relies therefore not on *one* state of the model but on a so-called *Gibbs ensemble*. By this is meant an ideal, that is, merely imagined ensemble of states, that accurately reflects our limited knowledge of the actual body. The body is then considered to behave as though in a single state *arbitrarily chosen from this ensemble*. This conception[xxv] had the most extensive results. Its highest triumphs were in those cases for which *not* all states appearing in the ensemble led to *the same* observable behavior. Thus the body's conduct is now this way, now that, just as foreseen (thermodynamic fluctuations). At first thought one might well attempt likewise to refer back the always uncertain statements of Q.M. to an ideal ensemble of states, of which a quite specific one applies in any concrete instance – but one does not know which one.

That this won't work is shown by the *one* example of angular momentum, as one of many. Imagine in Fig. 1 the point M to be situated at various positions relative to O and fitted with various momentum vectors, and all these possibilities to be combined into an ideal ensemble. Then one can indeed so choose these positions and vectors that in every case the product of vector length by length of normal OF yields one or the other of the acceptable values – relative to the particular point O. But for an arbitrary different point O', of course, unacceptable values occur. Thus appeal to the ensemble is no help at all.

Another example is the oscillator energy. Take the case that it has a sharply determined value, e.g., the lowest, $3\pi\hbar\nu$. The separation of the two point masses (that

[3] [In the translation this is italicised as *Theory of Heat Radiation*, but Planck's book on the subject is from 1906 (Planck 1906). Schrödinger might be referring to Planck's paper 'Über irreversible Strahlungsvorgänge' ['On Irreversible Radiation Processes'] (Planck 1900a), summarising his research of the preceding few years, received by *Annalen der Physik* on 7 November 1899 and published 16 January 1900, or he may be misremembering the date of 14 December 1900 when Planck presented his derivation of the blackbody radiation formula (Planck 1900b) (*eds.*).]

constitute the oscillator) then appears very *unsharp*. To be able to refer this statement to a statistical collective of states would require the distribution of separations to be sharply limited, at least toward large values, by that separation for which the *potential energy* alone would equal or exceed the value $3\pi\hbar\nu$. But that's not the way it is – arbitrarily large separations occur, even though with markedly reduced probability. And this is no mere secondary calculation result, that might in some fashion be circumvented, without striking at the heart of the theory: along with many others, the quantum mechanical treatment of radioactivity (Gamow[4]) rests on this state of affairs. – One could go on indefinitely with more examples. One should note that there was no question of any time-dependent changes. It would be of no help to permit the model to vary quite 'unclassically', perhaps to 'jump'. Already for the single instant things go wrong. At no moment does there exist an ensemble of classical states of the model that squares with the totality of quantum mechanical statements of this moment. The same can also be said as follows: if I wish to ascribe to the model at each moment a definite (merely not exactly known to me) state, or (which is the same) to *all* determining data definite (merely not exactly known to me) numerical values, then there is no supposition as to these numerical values *to be imagined* that would not conflict with some portion of quantum theoretical assertions.

That is not quite what one expects, on hearing that the pronouncements of the new theory are always uncertain compared to the classical ones.

5. Are the Variables Really Blurred?

The other alternative consisted of granting reality only to the momentarily sharp determining data – or in more general terms to each variable a sort of realization just corresponding to the quantum mechanical statistics of this variable at the relevant moment.

That it is in fact not impossible to express the degree and kind of blurring of *all* variables in *one* perfectly *clear* concept follows at once from the fact that Q.M. as a matter of fact has and uses such an instrument, the so-called wave function or ψ-function, also called system vector. Much more is to be said about it further on. That it is an abstract, unintuitive mathematical construct is a scruple that almost always surfaces against new aids to thought and that carries no great message. At all events it is an imagined entity that images the blurring of all variables at every moment just as clearly and faithfully as the classical model does its sharp numerical values. Its equation of motion too, the law of its time variation, so long as the system is left undisturbed, lags not one iota, in clarity and determinacy, behind the

[4] [E.g. Gamow (1928).]

equations of motion of the classical model. So the latter could be straight-forwardly replaced by the ψ-function, so long as the blurring is confined to atomic scale, not open to direct control. In fact the function has provided quite intuitive and convenient ideas, for instance the 'cloud of negative electricity' around the nucleus, etc. But serious misgivings arise if one notices that the uncertainty affects macroscopically tangible and visible things, for which the term 'blurring' becomes[xxvi] simply wrong. The state of a radioactive nucleus is presumably blurred in such degree and fashion that neither the instant of decay nor the direction, in which the emitted α-particle leaves the nucleus, is well-established. Inside the nucleus, blurring doesn't bother us. The emerging particle is described, if one wants to explain intuitively, as a spherical wave that continuously emanates in all directions from the nucleus and that impinges continuously on a surrounding luminescent screen over its full expanse. The screen however does not show a more or less constant dull[xxvii] surface glow, but rather flashes[xxviii] up at *one* instant at *one* spot – or, to honor the truth, it flashes[xxix] up now here, now there, for it is impossible to do the experiment with only a single radioactive atom. If in place of the luminescent screen one uses a spatially extended detector, perhaps a gas that is ionised by the α-particles, one finds the ion pairs arranged along rectilinear columns,[5] that project backwards on to the bit of radioactive matter from which the α-radiation comes (C. T. R. Wilson's cloud chamber tracks, made visible by drops of moisture condensed on the ions).

One can even set up quite ridiculous cases. A cat is penned up in a steel chamber, along with the following diabolical device (which must be secured against direct interference by the cat): in a Geiger counter there is a tiny bit of radioactive substance, *so* small, that *perhaps* in the course of one hour one of the atoms decays, but also, with equal probability, perhaps none; if it happens, the counter tube discharges and through a relay releases a hammer which shatters a small flask of hydrocyanic acid. If one has left this entire system to itself for an hour, one would say that the cat still lives *if* meanwhile no atom has decayed. The first atomic decay would have poisoned it. The ψ-function of the entire system would express this by having in it the living and the dead cat (pardon the expression) mixed or smeared out in equal parts.

It is typical of these cases that an indeterminacy originally restricted to the atomic domain becomes transformed into macroscopic indeterminacy, which can then be *resolved* by direct observation. That prevents us from so naively accepting as valid a 'blurred model' for representing reality. In itself it would not embody anything

[5] For illustration see Fig. 5 or 6 on p. 375 of the 1927 volume of this journal; or Fig. 1, p. 734 of [last] year's volume (1934), though these are proton tracks. [The references are to Meitner (1927) and Meitner (1934), respectively. It is perhaps noteworthy that against the background of Berliner's dismissal from *Naturwissenschaften* Schrödinger is here citing papers by a Jewish physicist (*eds.*).]

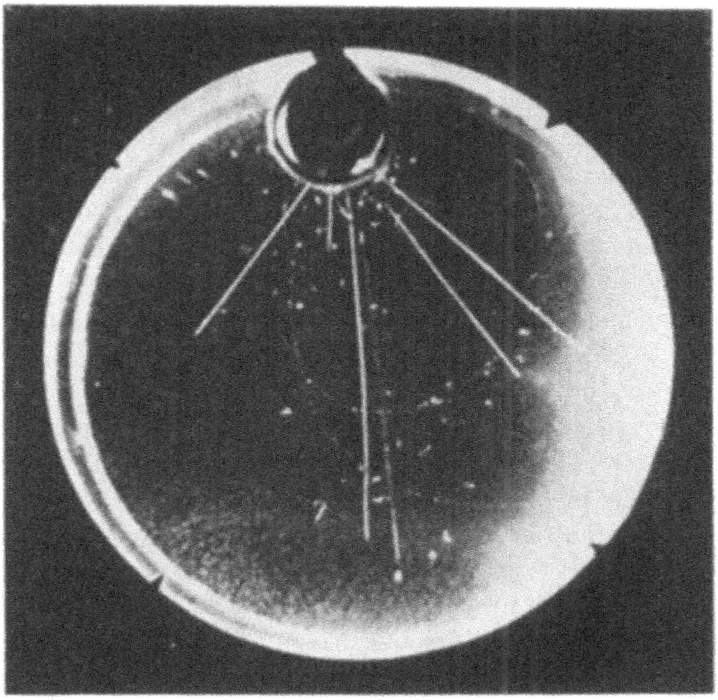

Fig. 1. Aluminiumzertrümmerung durch α-Strahlen.

Fig. 11.2 [Photograph from Meitner's 1934 paper (one of Schrödinger's suggestions). The caption reads: 'Disintegration of Aluminium with α-rays'. Reprinted by permission from Springer Nature Customer Service Centre GmbH: Springer Nature, *Naturwissenschaften*, Atomkern und periodisches System der Elemente, Lise Meitner, 1934 (*eds.*).]

unclear or contradictory. There is a difference between a shaky or out-of-focus photograph and a snapshot of clouds and fog banks.

6. The Deliberate Change in Epistemological Viewpoint[xxx]

In the fourth section we saw that it is not possible smoothly to take over models and to ascribe, to the momentarily unknown or not exactly known variables, nonetheless determinate values, that we simply don't know. In Sect. 5 we saw that the indeterminacy is not either[xxxi] an actual blurring, for there are in fact[xxxii] cases where an easily executed observation provides the missing knowledge. So what is left?

From this very hard dilemma the reigning doctrine rescues itself or us by having recourse to epistemology. We are told that no distinction is to be made between the state of a natural object and what I know about it, or perhaps better, what I can know about it if I go to some trouble. What is *real* – so they say – are actually[xxxiii] only awareness, observation, measurement. If through them I have procured at a given moment the best knowledge of the state of the physical object that is possibly attainable in accord with natural laws, then I can turn aside as *meaningless* any further questioning about the 'real state',[xxxiv] inasmuch as I am convinced that no further observation can extend my knowledge of it – at least, not without an equivalent diminution in some other respect (namely by changing the state, see below).

Now this sheds some light on the origin of the claim[xxxv] that I mentioned at the end of Sect. 2 as something very far-reaching: that all model quantities are measurable in principle. One can hardly get along without this article of belief if one sees himself constrained, in the interests of physical methodology, to call in as dictatorial help the above-mentioned philosophical principle, which no knowledgeable person[xxxvi] can fail to esteem as the supreme protector of all empiricism.

Reality resists imitation [in thought] through a model. So one lets go of naive realism and leans directly on the indubitable proposition that what is *real* (for the physicist) after all is said and done is only observation, measurement.[xxxvii] Then all our physical thinking thenceforth has as sole basis and as sole object the results of measurements which can in principle be carried out, for we must now explicitly *not* relate our thinking any longer to any other kind of reality or to a model. All numbers arising in our physical calculations must be interpreted as measurement results [Maßzahlen]. But since we didn't just now come into the world and start to build up our science from scratch, but rather have in use a quite definite scheme of calculation, from which in view of the great success of[xxxviii] Q.M. we would less than ever want to be parted, we see ourselves forced to dictate from the writing-table which measurements are in principle possible, that is, must be possible in order to support adequately our reckoning system. This allows a sharp value for each single variable of the model (indeed for a whole 'half set') and so each single variable must be measurable to arbitrary exactness. We cannot be satisfied with less, for we have lost our naively realistic innocence. We have nothing but our reckoning scheme to specify where Nature draws the ignorabimus-line, i.e., what is a *best possible* knowledge of the object. And if we couldn't do that, then indeed would our measurement reality become highly dependent on the diligence or laziness of the experimenter, how much trouble he takes to inform himself. We must go on to tell him how far he could go if only he were clever enough. Otherwise it would be seriously feared that just there, where we forbid further questions, there might well still be something worth knowing that we might ask about.

7. The ψ-function as Expectation-catalog

Continuing to expound the official teaching, let us turn to the already (Sect. 5) mentioned ψ-function. It is now the means for predicting [the] probability of measurement results. In it is embodied the momentarily-attained sum of theoretically based future expectation, somewhat as laid down in a *catalog*. It is the relation- and conditioning-bridge[xxxix] between measurements and measurements, as in the classical theory the model and its [momentary] state were. With this latter the ψ-function moreover has much in common. It is, in principle, determined by a finite number of suitably chosen measurements on the object, half as many as were required in the classical theory. Thus the catalog of expectations is initially compiled. From then on it changes with time, just as the state of the model of classical theory, in constrained and unique fashion [zwangsläufig und eindeutig] ('causally') – the evolution of the ψ-function is governed by a partial differential equation (of first order in time and solved for $\partial\psi/\partial t$). This corresponds to the undisturbed motion of the model in classical theory. But this goes on only until one again carries out [some] measurement. For each measurement one is required to ascribe to the ψ-function (= the prediction-catalog) a peculiar,[xl] quite sudden change, which depends *on the measurement result obtained*, and so *cannot be foreseen*; from which alone it is already quite clear that this second kind of change of the ψ-function has nothing whatever in common with its orderly development *between* two measurements. The abrupt change by measurement ties in closely with matters discussed in Sect. 5 and will occupy us further at some length; it is the most interesting point of the entire theory. It is precisely *the* point that demands the break with naive realism. For *this* reason one can *not* put the ψ-function directly in place of the model or of the real thing.[xli] And indeed not because one might never dare impute abrupt unforeseen changes to a real thing[xlii] or to a model, but because in the realism point of view observation is a natural process like any other and cannot *per se* bring about an interruption of the orderly flow of natural events.

8. Theory of Measurement, Part One

The rejection of realism has logical consequences. In general, a variable *has* no definite value before I measure it; then measuring it does *not* mean ascertaining the value that it *has*. But then what does it mean? There must still be some criterion as to whether a measurement is correct or wrong,[xliii] a method is good or bad, accurate, or inaccurate – whether it deserves the name of measurement process at all. Any old playing around with an indicating instrument in the vicinity of another body, whereby at any old time one then takes a reading, can hardly be called a measurement on this body. Now it is fairly clear; if reality does not determine the measured value, then at least the measured value must determine reality – it must actually be present *after* the measurement in *that* sense which alone is still recognized.[xliv] That is, the desired criterion can be merely this: repetition of the measurement must give the same result. By many repetitions I can prove the accuracy of the procedure and show that I am not just playing. It is agreeable that this prescription[xlv] matches exactly the method of the experimenter, to whom likewise the 'true value' is not known beforehand. We formulate the essential point as follows:

The systematically arranged interaction of two systems (measured object and measuring instrument) is called a measurement on the first system, if a directly-sensible variable feature of the second (pointer position) is always reproduced within certain error limits when the process is immediately repeated (on the same object, which in the meantime must not be exposed to any additional influences).

This statement will require considerable added comment: it is by no means a faultless definition. Empirics is more complicated than mathematics and is not so easily captured in polished sentences.

Before the first measurement there might have been an arbitrary quantum-theory prediction *for* it. *After* it the prediction *always* runs: within error limits again the same result. The expectation-catalog ($=\psi$-function) is therefore changed by the measurement in respect to the variable being measured. If the measurement procedure is known from beforehand to be *reliable*, then the first measurement at once reduces the theoretical expectation within error limits on to the value found, regardless of whatever the prior expectation may have been. This is the typical abrupt change of the ψ-function discussed above. But the expectation-catalog changes in unforeseen manner not only for the measured variable itself, but also for others, in particular for its 'canonical conjugate'. If for instance one has a rather sharp prediction for the *momentum* of a particle and proceeds to measure its *position* more exactly than is compatible with Theorem A of Sect. 2, then the *momentum* prediction must change. The quantum mechanical reckoning scheme moreover takes care of this automatically; there is no ψ-function whatsoever that would contradict Theorem A when one reads off it the expectations in the agreed way.[xlvi]

Since the expectation-catalog changes radically during measurement, the object is then no longer suited for testing, in their full extent, the statistical predictions made earlier; [in] the very least for the measured variable itself, since for it now the (nearly) same value would occur over and over again. *That* is the reason for the prescription already given in Sect. 2: one can indeed test the probability predictions completely, but for this one must repeat the entire experiment *ab ovo*. One's prior treatment of the measured object (or one identical to it) must be exactly the same as that given the first time, in order that the same expectation-catalog ($=\psi$-function) should be valid as before the first measurement. Then one 'repeats' it. (This repeating now means of course something quite other than earlier!) All this one must do not twice but very often. Then the predicted statistics will result – that is the idea.[xlvii]

One should note the difference between the error limits and the error distribution of the *measurement*, on the one hand, and [of] the theoretically predicted statistics, on the other hand. They have nothing to do with each other. They [result from] the two quite different types of *repetition* just discussed.

Here there is opportunity to deepen somewhat the above-attempted delimitation of *measuring*. There are measuring instruments that remain fixed on the reading given by the measurement just made. Or the pointer could remain stuck because of a defect. One would then repeatedly make exactly the same reading, and according to our instruction that would be a spectacularly accurate measurement. Indeed it is, just not on the object, but on the instrument itself![xlviii] As a matter of fact there is still missing from our exposition an important point, but one which could not readily be stated earlier, namely what it is that truly makes the difference between *object* and *instrument* (that it is the latter on which the reading is made, is more or less superficial). We have just seen that the instrument under certain circumstances, as required, must be set back to its neutral initial condition before any control measurement is made. This is well known to the experimentalist. Theoretically the matter may best be expressed by prescribing that on principle the instrument should be subjected to the identical prior treatment before each measurement, so that *for it* each time the same expectation-catalog (= ψ-function) applies, as it is brought up to the object. For the object it is just the other way around, any interference being forbidden when a control measurement is to be made, a 'repetition of the first kind' (that leads to *error* statistics). That is the characteristic difference between object and instrument. It disappears for a 'repetition of the second kind' (that serves for checking the quantum predictions). Here the difference between the two is actually rather insignificant.

From this we gather further that for a second measurement one may [also] use another similarly built and similarly prepared instrument – it need not necessarily be *the same one*; this is in fact sometimes done, as a check on the first one. It may indeed happen that two quite differently built instruments are so related to each other that if one measures with them one after the other (repetition of the first kind!) their two indications are in one-to-one correlation with each other. They then measure on the object essentially the same variable – i.e., the same for suitable calibration of the scales.

9. The ψ-function as Description of State

The rejection of realism also imposes obligations. From the standpoint of the classical model the momentary statement content of the ψ-function is far from complete;

it comprises only about 50 per cent of a complete description. From the new standpoint it must be complete for reasons already touched upon at the end of Sect. 6. It must be impossible to add on to it additional correct statements, without otherwise changing it; else one would not have the right to call meaningless all questions extending beyond it.

Thence it follows that two different catalogs, that apply to the same system under different circumstances or at different times, may well partially overlap, but never so that the one is entirely contained within the other. For otherwise it would be susceptible to completion through additional correct statements, namely through those by which the other one exceeds it. – The mathematical structure of the theory automatically satisfies this condition. There is no ψ-function that furnishes exactly the same statements as another and in addition several more.

Therefore if a system changes, whether by itself or because of measurements, there must always be statements missing from the new function that were contained in the earlier one. In the catalog not just new entries, but also deletions, must be made. Now knowledge can well be *gained*, but not *lost*. So the deletions mean that the previously correct statements have now become incorrect. A correct statement can become incorrect only if the *object* to which it applies changes. I consider it unimpeachable to express this conclusion as follows:[xlix]

Theorem 1: *If different ψ-functions obtain, the system is in different states.*[l]

If one speaks only of systems for which a ψ-function is in general available, then the inverse of this theorem runs:

Theorem 2: *For the same ψ-function the system is in the same state.*

This inverse does not follow from Theorem 1 but independently of it, directly from *completeness* or *maximality*. Whoever for the same expectation-catalog would yet claim a difference is possible, would be admitting that it (the catalog) does not give information on all justifiable questions. – The language usage of almost all authors implies validity of the above two theorems. Of course, they set up a kind of new reality – in entirely legitimate fashion, I believe. Moreover they are not trivially tautological, not mere verbal [explications] of 'state'. Without presupposed maximality of the expectation-catalog, change of the ψ-function could be brought about by mere collecting of new information.

We must even face up to an objection[li] to the derivation of Theorem 1. One can argue that each individual statement or item of knowledge, under examination there, is after all a probability statement, to which the category of *correct*, or *incorrect* does not apply in any relation to an individual case, but rather in relation to a collective that comes into being from one's preparing the system a thousand times in identical fashion (in order then to allow the same measurement to follow; cf. Sect. 8). That makes sense, but we must declare[lii] all members of this collective to be identically prepared, since to each the same ψ-function, the same statement-catalog applies and we must not admit[liii] differences that are not expressed in the catalog (cf. the justification[liv] of Theorem 2). Thus the collective is made up of identical individual cases. If a statement is wrong for *it*, then [also] the individual case must have changed, or else the collective too would again be the same.

10. Theory of Measurement, Part Two

Now it was previously stated (Sect. 7) and explained (Sect. 8) that any *measurement* suspends the law that otherwise governs continuous time-dependence of the ψ-function and brings about in it a quite different change, not governed by any law but rather dictated by the result of the measurement. But laws of nature differing from the usual ones cannot apply during a measurement, for objectively viewed it is a natural process like any other, and it cannot interrupt the orderly course of natural events. Since it does interrupt that of the ψ-function, the latter – as we said in Sect. 7 – can *not* serve, like the classical model, as a tentative[lv] representation of an objective reality. And yet in the last Section something like that has taken shape.

So, using catchwords for emphasis, I try again to contrast: 1) The discontinuity of the expectation-catalog due to measurement is *unavoidable*, for if measurement is to retain any meaning at all then the *measured value*, from a good measurement, *must* obtain. 2) The discontinuous change is certainly *not* governed by the otherwise valid causal law, since it depends on the measured value, which is not predetermined. 3) The change also definitely includes (because of 'maximality') some *loss* of knowledge, but knowledge cannot be lost, and so the *object must* change – *also* in the case of[lvi] the discontinuous changes and *also*, during these changes, in an unforeseen, *different* way.

How does this add up? Things are not at all simple. It is the most difficult and most interesting point of the theory. Obviously we must try to comprehend objectively the interaction between measured object and measuring instrument. To that end we must lay out a few very abstract considerations.

This is the point. Whenever one has a complete expectation-catalog – a maximum total knowledge – a ψ-function – for two completely separated bodies, or, in better terms, for each of them singly, then one obviously has it also for the two bodies together, i.e., if one imagines that neither of them singly but rather the two of them together make up the object of interest, of our questions about the future.[6]

But the converse is not true. *Maximal knowledge of a total system does not necessarily include total knowledge of all its parts, not even when these are fully*

[6] Obviously. We cannot fail to have, for instance, statements on the relation of the two to each other. For that would be, at least for one of the two, something in addition to its ψ-function. And such there cannot be.

separated from each other and at the moment are not influencing each other at all. Thus it may be that some part of what one knows may pertain to relations or conditionings[lvii] between the two subsystems (we shall limit ourselves to two), as follows: if a particular measurement on the first system yields *this* result, then for a particular measurement on the second the valid expectation statistics are such and such; but if the measurement in question on the first system should have *that* result, then some other expectation holds for that on the second; should a *third* result occur for the first, then still another expectation applies to the second; and so on, in the manner of a complete disjunction of all possible measurement results which the measurement one happens to contemplate[lviii] on the first system can yield. In this way, any measurement process at all or, what amounts to the same, any variable at all of the second system can be tied to the not-yet-known value of any variable at all of the first, and of course *vice versa* also. If that is the case, if such conditional statements occur in the combined catalog, *then it can not possibly be maximal in regard to the individual systems.* For the content of two maximal individual catalogs would by itself suffice for a maximal combined catalog; the conditional statements could not be added on.

These condition[al] predictions, moreover, are not something that has suddenly fallen in here from the blue. They are in every expectation-catalog. If one knows the ψ-function and makes a particular measurement and this has a particular result, then one again knows the ψ-function, *voilà tout.* It's just that for the case under discussion, because the combined system is supposed to consist of two fully separated parts, the matter stands out as a bit strange. For thus it becomes meaningful to distinguish between measurements on the one and measurements on the other subsystem. This provides to each full title to a private maximal catalog; on the other hand it remains possible that a portion of the attainable combined knowledge is, so to say, squandered on conditional statements, that operate between the subsystems, so that the private expectancies are left unfulfilled – even though the combined catalog is maximal, that is even though the ψ-function of the combined system is known.

Let us pause for a moment. This result in its abstractness actually says it all: Best possible knowledge of a whole does not necessarily include the same for its parts. Let us translate this into terms of Sect. 9: The whole is in a definite state, the parts taken individually are not.

– 'How so? Surely a system must be in some sort of state.'

= 'No. State is ψ-function, is maximal sum of knowledge. I didn't necessarily provide myself with this, I may have been lazy. Then the system is in no state.'

– 'Fine, but then too the agnostic prohibition of questions is not yet in force and in our case I can tell myself: the subsystem is already in some state, I just don't know which.'

= 'Wait. Unfortunately no. There is no "I just don't know". For as to the total system, maximal knowledge is at hand.' – *The insufficiency of the ψ-function as model replacement rests solely on the fact that one doesn't always have it.* If one does have it, then by all means let it serve as description of the state. But sometimes one does not have it, in cases where one might reasonably expect to. And in that case, one [must] not postulate that it 'is actually a particular one, one just doesn't know it'; the above-chosen standpoint forbids this. 'It' is namely a sum of knowledge; and knowledge, that no one knows, is none. –

We continue. That a portion of the knowledge should float in the form of disjunctive conditional statements *between* the two systems can certainly not happen if we bring up the two from opposite ends of the world and juxtapose them without interaction. For then indeed the two 'know' nothing about each other. A measurement on one cannot possibly furnish any grasp of what is to be expected of the other. Any 'entanglement of predictions' that takes place can obviously only go back to the fact that the two bodies at some earlier time formed in a true sense *one* system, that is were interacting, and have left behind *traces* on each other. If two separated bodies, each by itself known maximally, enter a situation in which they influence each other, and separate again, then there occurs regularly that which I have just called *entanglement* of our knowledge of the two bodies. The combined expectation-catalog consists initially of a logical sum of the individual catalogs; during the process it develops causally in accord with known law (there is no question whatever of measurement here). The knowledge remains maximal, but at its end, [when] the two bodies have again separated, it is not again split into a logical sum of knowledges about the individual bodies. What still remains *of that* may have become less than maximal, even very strongly so. – [Note] the great difference [compared to] the classical model theory, where of course from known initial states and with known interaction the individual end states would be exactly known.

The measuring process described in Sect. 8 now fits neatly into this general scheme, if we apply it to the combined system, measured object + measuring instrument. As we thus construct an objective picture of this process, like that of any other, we dare hope to clear up, if not altogether to avoid, the singular jump of the ψ-function. So now the one body is the measured object, the other the instrument. To suppress any interference from outside we arrange for the instrument by means of built-in clockwork to creep up automatically to the object and in like manner creep away again. The reading itself we postpone, as our immediate purpose is to investigate whatever may be happening 'objectively'; but for later use we let the result be recorded automatically in the instrument, as indeed is often done these days.

Now how do things stand, after automatically completed measurement? We possess, afterwards same as before, a maximal expectation-catalog for the total system.

The recorded measurement result is of course not included therein. As to the instrument the catalog is [thus] far from complete, telling us nothing at all about where the recording pen left its trace. (Remember that poisoned cat!) What this amounts to is that our knowledge has evaporated into conditional statements: *if* the mark is at line 1, *then* things are thus and so for the measured object, *if* it is at line 2, then such and such, if at 3, then a third, etc. Now has the ψ-function of the measured *object* made a leap? Has it developed further in accord with natural law [dem zwangläufigen Gesetz] (in accord with the partial differential equation)? No to both questions. It is no more. It has become snarled up [verheddert], in accord with the causal law [dem zwangläufigen Gesetz] for the *combined* ψ-function, with that of the measuring instrument. *The expectation-catalog of the object has split into a conditional disjunction of expectation-catalogs* – like a Baedeker that one has taken apart in the proper manner. Along with each section there is given also the probability that it proves correct – transcribed from the original expectation-catalog of the object. But *which one* proves right – which section of the Baedeker should guide the ongoing journey – that can be determined only by actual inspection of the record.

And what if we *don't* look? Let's say it was photographically recorded and by bad luck light reaches the film before it is developed. Or we inadvertently put in black paper instead of film. Then indeed have we not only not learned anything new from the miscarried measurement, but we have suffered loss of knowledge. This is not surprising. It is only natural that outside interference will in the first place[lix] always spoil the knowledge that one has of a system. The interference, if it is to allow the knowledge to be gained back afterwards, must be carefully planned[lx] indeed.

What have we won by this analysis? *First*, the insight into the disjunctive splitting of the expectation-catalog, which still takes place quite continuously and is brought about through embed[ding] in a combined catalog for instrument and object. From this amalgamation the object can again be separated out only by the living subject actually taking cognizance of the result of the measurement. Some time or other this must happen if that which has gone on is actually to be called a measurement – however dear to our hearts it was to distil[lxi] the process throughout as objectively as possible. And that is the *second* insight we have won: *not until this inspection*, which resolves[lxii] the disjunction, does anything discontinuous, or leaping, take place. One is inclined to call this a *mental* action, for the object is already out of touch [abgeschaltet], is no longer physically affected; what befalls it is already past. But it would not be quite right to say that the ψ-function of the object which changes *otherwise* according to a partial differential equation, independent of the observer, should *now* change leap-fashion because of a mental act. For it had disappeared, it was no more. Whatever is not, no more can it change. It is born anew, is reconstituted, is separated out from the entangled knowledge that one has, through an act of perception, which as a matter of fact is [certainly] not a physical effect on the measured object. From the form in which the ψ-function was last known, to the new in which it reappears, runs no continuous road – it ran indeed through annihilation. Contrasting the two forms, the thing looks like a leap. In truth something of importance happens in between, namely the influence of the two bodies on each other, during which the object possessed no private expectation-catalog nor had any claim thereto, because it was not independent.

11. Resolution of the 'Entanglement'. Result Dependent on the Experimenter's Intention

We return to the general case of 'entanglement', without having specifically in view the special case, just considered, of a measurement process. Suppose the expectation-catalogs of two bodies A and B have become entangled through transient interaction. Now let the bodies be again separated. Then I can take one of them, say B, and by successive measurements bring my knowledge of it, which had become less than maximal, back up to maximal. I maintain: just as soon as I succeed in this, and not before, then first, the entanglement is immediately resolved and, second, I will also have acquired maximal knowledge of A through the measurements on B, making use of the conditional relations that were in effect.

For in the first place the knowledge of the total system remains always maximal, being in no way damaged by good and exact measurements. In the second place: conditional statements of the form 'if for A . . . , then for B . . . ' can no longer exist, as soon as we have reached a maximal catalog on B. For it is *not* conditional and to it nothing at all can be added on relevant to B. Thirdly: conditional statements in the inverse sense ('if for B . . . , then for A . . . ') can be transformed into statements about A alone, because all probabilities for B are already known unconditionally. The entanglement is thus completely put aside, and since the knowledge of the total system has remained maximal, it can only mean that along with the maximal catalog for B came the same thing for A.

And it cannot happen moreover,[lxiii] that A becomes maximally known indirectly, through measurements on B, before B is. For then all conclusions work in the reversed direction – that is, B is too. The systems become simultaneously maximally known, as asserted. Incidentally, this would also be true if one did not limit the measurement to just one of the two systems. But the interesting point is precisely this, that one *can* limit it to one of the two; that thereby one reaches his goal.

Which measurements on B and in what sequence they are undertaken, is left entirely to the arbitrary choice of the experimenter. He need not pick out specific variables, in order to be able to use the conditional statements. He is free to formulate a plan that would lead him to maximal knowledge of B, even if he should know nothing at all about B. And it can do no harm if he carries through this plan to the end. If he asks himself after each measurement whether he has perhaps already reached his goal, he does so only to spare himself from further, superfluous labor.

What sort of A-catalog comes forth in this indirect way depends obviously on the measured values that are found in B (before the entanglement is entirely resolved; no more on any later ones, in case the measuring is continued superfluously[lxiv]). Suppose now that in this way I derived an A-catalog in a particular case. Then I can reflect[lxv] and consider whether I might perhaps have found a *different* one if I had put into action a *different* measuring plan for B. But since after all I neither have actually touched the system A, nor in the imagined other case would have touched it, the statements of the other catalog, whatever it might be, must all *also* be correct. They must therefore be entirely contained within the first, since the first is maximal. But so is the second. So it must be identical with the first.

Strangely enough, the mathematical structure of the theory by no means satisfies this requirement automatically. Even worse, examples can be set up where the requirement is necessarily violated. It is true that in any experiment one can actually carry out only *one* group of measurements (always on B!), for once that has happened the entanglement is resolved and one learns nothing more about A from further measurements on B. But there are cases of entanglement in which *two definite programs* are specifiable, of which each 1) must lead to resolution of the entanglement, and 2) must lead to an A-catalog to which the *other can* not possibly lead – whatsoever measured values may turn up in one case or the other. It is simply like this, that the *two series* of A-catalogs, that can possibly arise from the one or the other of the programs, are sharply separated and have in common not a single term.

These are especially pointed cases, in which the conclusion lies so clearly exposed. In general one must reflect more carefully. If two programs of measurement on B are proposed, along with the two series of A-catalogs to which they can lead, then it is by no means sufficient that the two series have one or more terms in common in order for one to be able to say: well now, surely one of these will always turn up – and so to class the requirement as[lxvi] 'presumably fulfilled'. That's not enough. For *indeed one knows* the probability of every measurement on B, considered as measurement on the total system, and under many *ab-ovo*-repetitions each one must occur with the frequency assigned to it. Therefore the two series of A-catalogs would have to agree, member by member, and furthermore the probabilities in each series would have to be the same. And that not merely for these two programs but also for each of the infinitely many that one might think up. But this is utterly out of the question. The requirement that the A-catalog that one gets should

always be the same, regardless of what measurements on B bring it into being, this requirement is plainly and simply never fulfilled.

Now we wish to discuss a simple 'pointed' example.

12. An Example[7]

For simplicity, we consider two systems with just *one* degree of freedom. That is, each of them shall be specified through a *single* coordinate q and its canonically conjugate momentum p. The classical picture would be a point mass that could move only along a straight line, like the spheres of those playthings on which small children learn to calculate. p is the product of mass by velocity. For the second system we denote the two determining data by capital Q and P. As to whether the two are 'threaded on the same wire' we shall not be at all concerned, in our abstract consideration. But even if they are, it may in that case be convenient not to reckon q and Q from the same reference point. The equation $q = Q$ thus does not necessarily mean coincidence. The two systems may in spite of this be fully separated.

In the cited paper it is shown that between these two systems an entanglement can [obtain], which *at a particular moment, to which everything following is referred*, can be compactly shown in the two equations:

$$q = Q \quad \text{and} \quad p = -P.$$

That means: *I know, if* a measurement of q on the [first] system yields a certain value, that a Q-measurement performed immediately thereafter on the second will give the *same* value, and vice versa; *and I know*, if a p-measurement on the first system yields a certain value, that a P-measurement performed immediately thereafter will give the opposite value, and vice versa.

A single measurement of q *or* p *or* Q *or* P resolves the entanglement and makes *both* systems maximally known. A second measurement on the same system modifies only the statements about *it*, but teaches nothing more about the other. So one cannot check both equations in a single experiment. But one can repeat the experiment *ab ovo* a thousand times; each time set up the same entanglement; according to whim check one or the other of the equations; and find confirmed that one which one is momentarily pleased to check. We assume that all this has been done.

If for the thousand-and-first experiment one is then seized by the desire to give up further checking and instead measure q on the first system and P on the second, and one obtains

$$q = 4; \qquad P = 7;$$

[7] A. Einstein, B. Podolsky, and N. Rosen, *Phys. Rev.* **47**: 777 (1935) [Einstein, Podolsky and Rosen 1935; see Chapter 9]. The appearance of this work motivated the present – shall I say [report] or general confession?

can one then doubt that

$$q=4; \qquad p=-7;$$

would have been a correct prediction for the first system, or

$$Q=4; \qquad P=7$$

a correct prediction for the second? Not verifiable in full in a single experiment – quantum predictions are never such – but correct, because whoever made them[lxvii] suffered no disillusion, whichever half he decided to check.

There's no doubt about it. Every measurement is for its system the first. Measurements on separated systems cannot directly influence each other – that would be magic. Neither can these be randomly generated numbers,[lxviii] if from a thousand experiments it is established that virginal measurements agree.

The prediction-catalog $q=4$, $p=-7$ would of course be hypermaximal.

13. Continuation of the Example: All Possible Measurements are Entangled Unequivocally

Now a *prediction* of this extent is [also] utterly impossible according to the teaching of Q.M., which we here follow out to its last consequences. Many of my friends feel soothed by this[lxix] and declare: what answer a system *would have given* to the experimenter *if*..., – has nothing to do with an actual measurement and so, from our epistemological standpoint, does not concern us.

But let us once more make the matter very clear. Let us focus attention on the system labeled with small letters p, q and call it for brevity the 'small' one. Then things stand as follows. I can direct *one* of two questions to the small system, either that about q or that about p. Before doing so I can, if I choose, procure the answer to *one* of these questions by a measurement on the fully separated other system (which we shall regard as auxiliary apparatus), or I may intend to [procure] this afterwards. My small system, like a schoolboy under examination, *cannot possibly know* whether I have done this or for which questions, or whether and for which I intend to do it later. From arbitrarily many pretrials I know that the pupil will correctly answer the first question that I put to him. From that it follows that in every case he *knows* the answer to *both* questions. That the answering of the first question, that it pleases me to put to him, so tires or confuses the pupil that his further answers are worthless, changes nothing at all of this conclusion. No school principal [Gymnasialdirektor] would judge otherwise, if this situation repeated itself with thousands of pupils of similar provenance, however much he might wonder *what* makes all the scholars so dim-witted or obstinate after the answering of the first question. He would not come to think that his, the teacher's, consulting a workbook[lxx] [Hilfsbuch] suggests to the pupil the correct answer [in the first place], or even, in the cases when the teacher chooses to consult it only after ensuing answers by the pupil, that the pupil's answer has changed the text of the notebook [Notizbuch] in the pupil's favor.

Thus my small system holds a quite definite answer to the q-question and to the p-question in readiness for the case that one or the other is the first to be put directly to it. Of this preparedness not an iota can be changed if I should perhaps measure the Q on the auxiliary system (in the analogy: if the teacher looks up one of the questions in his notebook and thereby indeed ruins with an inkblot *the* page where the other answer stands). The quantum mechanician maintains that after a Q-measurement on the auxiliary system my small system has a ψ-function in which 'q is fully sharp, but p fully indeterminate'. And yet, as already mentioned, not an iota is changed of the fact that my small system also has ready an answer to the p-question, and indeed the same one as before.

But the situation is even worse yet. Not only to the q-question and to the p-question does my clever pupil have a definite answer ready, but rather also to a thousand others, and indeed without my having the least insight into the memory technique by which he is able to do it. p and q are not the only variables that I can measure. Any combination of them whatsoever, for example

$$p^2 + q^2$$

also corresponds to a fully definite measurement according to the viewpoint[lxxi] of Q.M. Now it may be shown[8] that also for this the answer can be obtained by a measurement on the auxiliary system, namely by measurement of $P^2 + Q^2$, and indeed the answers are just the same. By general rules of Q.M. this sum of squares can only take on a value from the series

$$\hbar,\ 3\hbar,\ 5\hbar,\ 7\hbar,\ \ldots .$$

The answer that my small system has ready for the $(p^2 + q^2)$-question (in case this should be the first it must face) must be a number from this series. – It is very much the same with measurement of

$$p^2 + a^2 q^2$$

where a is an arbitrary positive constant. In this case the answer must be, according to Q.M., a number from the following series

$$a\hbar,\ 3a\hbar,\ 5a\hbar,\ 7a\hbar,\ \ldots .$$

For each numerical value of a one gets a different question, and to each my small system holds ready an answer from the series (formed with the a-value in question).

[8] E. Schrödinger, *Proc. Cambridge Phil. Soc.* (in press) [Schrödinger 1935e].

Most astonishing is this: these answers cannot possibly be related to each other in the way given by the formulas! For let q' be the answer held ready for the q-question, and p' that for the p-question, then the relation

$$\frac{p'^2 + a^2 q'^2}{a\hbar} = \text{an odd integer}$$

cannot possibly hold for given numerical values q' and p' and for *every*[lxxii] *positive number a*. This is by no means an operation with imagined numbers, that one can not really ascertain. One can in fact get two of the numbers, e.g., q' and p', the one by direct, the other by indirect measurement. And then one can (pardon the expression) convince himself that the above expression, formed with the numbers q' and p' and an arbitrary a, is not an odd integer.

The lack of insight into the relationships among the various answers held in readiness (into the 'memory technique' of the pupil) is a total one, a gap not to be filled by some new kind of algebra[lxxiii] of Q.M. The lack is all the stranger, since on the other hand one can show: the entanglement is already uniquely determined by the requirements $q = Q$ and $p = -P$. If we know that the coordinates are equal and the momenta equal but opposite, then there follows by quantum mechanics a *fully determined* one-to-one pairing[lxxiv] of *all possible* measurements on both systems. For *every* measurement on the 'small' one the numerical result can be procured by a suitably arranged measurement on the 'large' one, and each measurement on the large [one] indicates[lxxv] the result that a particular measurement on the small [one] would give or has given. (Of course in the same sense as always heretofore: only the virgin measurement on each system counts.) As soon as we have brought the two systems into the situation where they (briefly put) coincide in coordinate and momentum, then they (briefly put) coincide also in regard to all other variables.

But as to how the numerical values of all these variables of *one* system relate to each other we know nothing at all, even though for each the system must have a quite specific one in readiness, for if we wish we can learn it from the auxiliary system and then find it always confirmed by direct measurement.

Should one now think that because we are so ignorant about the relations among the variable-values held ready in *one* system, that none exists, that far-ranging arbitrary combination can occur? That would mean that such a system of '*one* degree of freedom' would need not merely *two* numbers for adequately describing it, as in classical mechanics, but rather many more, perhaps infinitely many. It is then nevertheless strange that *two* systems always agree in *all* variables if they agree in two. Therefore one would have to make the second assumption, that this is due to our awkwardness; would have to think that as a practical matter we are not competent to bring two systems into a situation such that they coincide in reference to two variables, without *nolens volens* bringing about coincidence also for all other variables, even though that would not in itself be necessary. One would have to make these *two* assumptions in order not to perceive as a great embarrassment[lxxvi] the complete lack of insight into the interrelationship of variable-values within one system.

14. Time-dependence of the Entanglement. Worries about the Special Role of Time[lxxvii]

It is perhaps not superfluous to recall that everything said in Sections 12 and 13 pertains to a single instant of time. The entanglement is not constant in time. It does continue to be a one-to-one entanglement of *all* variables, but the pairing[lxxviii] changes. That means the following. At a later time t one can very well again learn the values of q or of p that *then* obtain, by a measurement on the auxiliary system, but the measurements, that one must undertake thereto on the auxiliary system, are *different*. Which ones they should be, one can easily see in simple cases. It now of course becomes a question of the forces at work within each of the two systems. Let us assume that no forces are working. For simplicity we will set the mass of each to be the same and call it m. Then in the classical model the momenta p and P would remain constant, since they are just[lxxix] the masses multiplied by the velocities; and the coordinates at time t, which we shall distinguish by giving them subscripts t, (q_t, Q_t), would be calculated from the initial ones, which we still[lxxx] designate q, Q, thus:

$$q_t = q + \frac{p}{m}t$$

$$Q_t = Q + \frac{P}{m}t.$$

Let us first talk about the small system. The most natural way of describing it classically at time t is in terms of coordinate and momentum *at this time*, i.e., in terms of q_t and p. But one may do it differently. In place of q_t one could specify q. It too is a 'determining datum at time t', and indeed at every time t, and in fact one that does not change with time. This is similar to the way in which I can specify a certain determining datum of my own person, namely my *age*, either through the number 48, which changes with time and in the system corresponds to specifying q_t, or through the number 1887, which is usual in documents and corresponds to specifying q. Now according to the foregoing:

$$q = q_t - \frac{p}{m}t.$$

Similarly for the second system. So we take as determining data

$$\text{for the first system } q_t - \frac{p}{m}t \text{ and } p$$

$$\text{for the second system } Q_t - \frac{P}{m}t \text{ and } P.$$

The advantage is that *among these the same entanglement goes on indefinitely*:

$$q_t - \frac{p}{m}t = Q_t - \frac{P}{m}t$$
$$p = -P$$

or solved:

$$q_t = Q_t - \frac{2t}{m}P; \qquad p = -P.$$

So that what changes with time is just this: the coordinate of the 'small' system is not ascertained simply by a coordinate measurement on the auxiliary system, but rather by a measurement of the aggregate

$$Q_t - \frac{2t}{m}P.$$

Here however, one must not get the idea that maybe he measures Q_t *and* P, because that just [isn't possible]. Rather one must suppose, as one always must suppose in Q.M., that there is a direct measurement procedure for this aggregate. Except for this change, *everything* that was said in Sections 12 and 13 applies at any point of time; in particular there exists at all times the one-to-one entanglement of *all* variables together with its evil consequences.

It is just this way too, if within each system a force works, except that then q_t and p are entangled with variables that are more complicated combinations of Q_t and P.

I have briefly explained this in order that we may consider the following. That the entanglement should change with time makes us after all a bit thoughtful. Must perhaps all measurements, that were under discussion, be completed in very short time, actually *instantaneously*, in zero time, in order that the inexorable[lxxxi] consequences be vindicated? Can the ghost be banished by reference to the fact that measurements take time? No. For each single experiment one needs just *one* measurement on each system; only the virginal one matters, further ones apart from this would be without effect. How long the measurement lasts need not therefore concern us, since we have no second one following on. One must merely be able to so arrange the two virgin measurements that they yield variable-values for the same definite *point* of time, known to us in advance – known in advance, because [in fact] we must direct the measurements at a pair of variables that are entangled at just this point of time.

– 'Perhaps it is not possible so to [set up] the measurements?'

= 'Perhaps. I even presume so. Merely: *today's* Q.M. must require this. For it is now so set up that its predictions are always made for a *point* of time. Since they are supposed to relate to measurement results, they would be entirely without content if the relevant variables were not measurable *for* a definite point of time, whether the measurement itself lasts a long or a short while.'

When we *learn* the result is of course quite immaterial. Theoretically that has as little weight as for instance the fact that one needs several months to integrate the differential equations of the weather for the next three days. – The drastic analogy with the pupil examination misses the mark in a few points of the law's letter, but it fits the spirit of the law. The expression 'the system knows' will perhaps no longer carry the meaning that the answer comes forth from an instantaneous situation; it may perhaps derive from a succession of situations, that occupies a finite length of time. But even if it be so, it need not concern us so long as the system somehow brings forth the answer from within itself, with no other help than that we tell it (through the experimental arrangement) *which* question we would like to have answered; and so long as the answer itself is uniquely tied to a *moment* of time; which for better or for worse must be presumed for every measurement [of] which contemporary Q.M. speaks, for otherwise the quantum mechanical predictions would have no content.

In our discussions, however, we have stumbled across a possibility. If the idea could be worked out[lxxxii] that the quantum mechanical predictions did not or did not always pertain to a quite sharply defined point of time, then one would also be freed from requiring this of the measurement results. Thereby, since the entangled variables change with time, setting up the antinomical assertions would become much more difficult.

That prediction for sharply-defined time is a blunder, is probable also on other grounds. The numerical value of time is like any other the result of observation. Can one make exception [precisely] for measurement with a clock? Must it not like any other pertain to a variable that in general has no sharp value and in any case cannot have it simultaneously with *every*[lxxxiii] other variable? If one predicts the value of *another* for a particular *point of time*, must one not fear that both can [not] be sharply known together? Within contemporary Q.M. one can hardly deal with this apprehension. For time is always considered a priori as known precisely, although one would have to admit that every look[ing]-at-the-clock disturbs the clock's motion in uncontrollable fashion.

Permit me to repeat that we do not possess a Q.M. whose statements should *not* be valid for sharply fixed points of time. It seems to me that this lack manifests itself directly in the former antinomies. Which is not to say that it is the only lack which manifests itself in them.

15. Principle of Nature[lxxxiv] or Calculating Device?

That 'sharp time' is an anomaly in Q.M. and that besides, more or less independent of that, the special treatment of time forms a serious hindrance to adapting Q.M. to the *relativity principle*, is something that in recent years I have brought up again and again, unfortunately without being able to make the shadow of a useful counterproposal.[9] In [overviewing] the entire contemporary situation, [which] I have tried to sketch here, there comes up, in addition, a quite different kind of remark in relation to the so ardently sought, but not yet actually attained, 'relativisation' of Q.M.

The [peculiar] theory of measurement, the apparent jumping around of the ψ-function, and finally the 'antinomies of entanglement', all derive from the simple manner in which the calculation methods of quantum mechanics allow two separated systems conceptually to be combined together into a single one; for which the methods seem plainly predestined. When two systems interact, their ψ-functions, as we have seen, do not come into interaction but rather they immediately cease to exist and a single one, for the combined system, takes their place. It consists, to mention this briefly, at first simply of the *product* of the two individual functions; which, since the one function depends on quite different variables from the other, is a function of all these variables, or 'acts in a space of much higher dimension number' than the individual functions. As soon as the systems begin to influence each other, the combined function ceases to be a product and moreover does not again divide up, after they have again become separated, into factors that can be assigned individually to the systems. Thus one disposes provisionally (until the entanglement is resolved by an actual observation) of only a *common* description of the two in that space of higher dimension. This is the reason that knowledge of the individual systems can decline to the scantiest, even to zero, while that of the combined system remains continually maximal. Best possible knowledge of a whole does *not* include best possible knowledge of its parts – and that is what keeps coming back to haunt us.

Whoever reflects on this must [indeed] be left fairly thoughtful by the following fact. The conceptual joining of two or more systems into *one* encounters great difficulty as soon as one attempts to introduce the principle of special relativity into Q.M. Already seven years ago P. A. M. Dirac found a startlingly simple and elegant relativistic solution to the problem of a single electron.[10] A series of experimental confirmations, marked by the key terms electron spin, positive electron, and pair creation, can leave no doubt as to the basic correctness of the solution. But in the first place it does nevertheless very strongly transcend the conceptual plan of Q.M.

[9] *Berl. Ber.* 16 Apr. 1931; *Annales de l'Institut Henri Poincaré*, p. 269 (Paris, 1931); *Cursos de la universidad internacional de verano en Santander*, **1**: p. 60 (Madrid, Signo, 1935) [Schrödinger 1931b,1932a,1935h].
[10] *Proc. Roy. Soc. Lond.* **A117**: p. 610 (1928) [Dirac 1928].

(that which I have attempted to picture *here*),[11] and in the second place one runs into stubborn resistance as soon as one seeks to go forward, according to the prototype of non-relativistic theory, from the Dirac solution to the problem of several electrons. (This shows at once that the solution lies outside the general plan, in which, as mentioned, the combining together of sub-systems is extremely simple.) I do not presume to pass judgment on the attempts which have been made in this direction.[12] That they have reached their goal, I must doubt [already] because the authors make no such claim.

Matters stand much the same with another system, the electromagnetic field. Its laws are 'relativity theory personified', a *non*-relativistic treatment being in general impossible. Yet it was this field, which in terms of the classical model of heat radiation provided the first impetus[lxxxv] for quantum theory, that was the first system to be 'quantized'. That this could be successfully done with simple means comes about because here one has things a bit easier, in that the photons, the 'atoms of light', do not in general interact directly with each other,[13] but only via the charged particles. Today we do not as yet have a truly unexceptionable quantum theory of the electromagnetic field.[14] One can go a long way with *building up out of subsystems* according to the pattern of the non-relativistic theory (Dirac's theory of light[15]), yet without quite reaching the goal.

The simple procedure provided for this by the non-relativistic theory is perhaps after all only a convenient calculational trick, but one that today, as we have seen, has attained influence of unprecedented scope over our basic attitude toward nature.

My warmest thanks to Imperial Chemical Industries, London, for the leisure to write this [report].

[11] P. A. M. Dirac, *The Principles of Quantum Mechanics*, 1st ed. p. 239; 2nd ed. p. 252. Oxford: Clarendon Press, 1930 [resp.] 1935 [Dirac 1930, 1935a].

[12] Herewith a few of the more important references: G. Breit, *Phys. Rev.* **34**: p. 553 (1929) and **39**: p. 616 (1932) [Breit 1929, 1932]; C. Møller, *Z. Physik* **70**: p. 786 (1931) [Møller 1931]; P. A. M. Dirac, *Proc. Roy. Soc. Lond.* **A136**: p. 453 (1932) and *Proc. Cambridge Phil. Soc.* **30**: p. 150 (1934) [Dirac 1932, 1934]; R. Peierls, *Proc. Roy. Soc. Lond.* **A146**: p. 420 (1934) [Peierls 1934]; W. Heisenberg, *Z. Physik.* **90** p. 209 (1934) [Heisenberg 1934].

[13] But this holds, probably, only approximately. See M. Born and L. Infeld, *Proc. Roy. Soc. Lond.* **A144**: p. 425 and **A147**: p. 522 (1934); **A150**: p. 141 (1935) [Born and Infeld 1934a, b, 1935]. This is the most recent attempt at a quantum electrodynamics.

[14] Here again the most important works, partially assignable, [due] to their contents, also to the [previous-but-one reference]: P. Jordan and W. Pauli, *Z. Physik* **47**: p. 151 (1928) [Jordan and Pauli 1928]; W. Heisenberg and W. Pauli, *Z. Physik* **56**: p. 1 (1929); **59**: p. 168 (1930) [Heisenberg and Pauli 1929, 1930]; P. A. M. Dirac, V. A. Fock, and B. Podolsky, *Physik. Z. Sowjetunion* **6**: p. 468 (1932) [Dirac, Fock and Podolsky, 1932]; N. Bohr and L. Rosenfeld, *Danske Videnskabernes Selskab* (math.-phys. Comm.) **12**: p. 8 (1933) [Bohr and Rosenfeld 1933].

[15] An excellent [report]: E. Fermi, *Rev. Mod. Phys.* **4**: p. 87 (1932) [Fermi 1932].

Notes on the Translation

i concept/Gebilde
ii concept/Gebilde
iii concept/Gebilde
iv builds on a foundation/setzt auf Grund
v literally/einfältig
vi hindsight-free/nachsichtslos
vii body cells isolate the nucleoplasm/der Körper das Keimplasma isoliert hält
viii presentation/Vorstellung
ix results/Angaben
x determinacy/Determinierung
xi example/Modell
xii doctrine/Meinung
xiii doctrine/Meinung
xiv doctrine/Meinung
xv measurements/*Messungen*
xvi thought-pattern/Denkbehelf
xvii proscription against future development/Vorwegnahme künftiger Entwicklung
xviii All of the foregoing pertains/Also alle Voraussagen beziehen sich nach wie vor
xix by way of certain measurements performed *now* and by way of their resulting prediction of results to be expected of other measurements following thereafter either immediately or at some given time, one gains the best possible probability estimates permitted by nature/vermittels einiger *jetzt* angestellter Messungen und ihrer Resultate über die zu erwartenden Resultate anderer Messungen, die entweder augenblicklich oder zu bestimmter Zeit darauf folgen sollen, die bestmöglichen Wahrschenlichkeitsangaben gewonnen werden, welche die Natur zuläßt
xx The translation omits the value 0.
xxi This assertion seems unreasonable for the model/Am Modell ist diese Behauptung unsinnig
xxii The big idea seems to be that/Dem Wortlaut nach
xxiii allowed by a full picture of the momentary state/der ein klares Bild von dem augenblicklichen Zustand erlauben würde
xxiv interpretation/Auffassung
xxv interpretation/Auffassung
xxvi seems/wird
xxvii uniform/matt
xxviii lights/blitzt
xxix lights/blitzt
xxx The Deliberate About-face of the Epistemological Viewpoint/Der bewußte Wechsel des erkenntnistheoretischen Standpunkts
xxxi not even/auch nicht
xxxii always/jedenfalls
xxxiii Actually – so they say – there is intrinsically/*Wirklich* – so sagt man – sind ja eigentlich
xxxiv about the 'actual state'/nach dem 'wirklichen Zustand'
xxxv proposition/Behauptung
xxxvi sensible person/Verständiger
xxxvii that *actually* (for the physicist) after all is said and done there is only observation, measurement/daß *wirklich* (für den Physiker) letzten Endes nur die Beobachtung, die Messung ist
xxxviii progress in/Erfolgen der
xxxix relation- and-determinacy-bridge/Beziehungs- und Bedingtheitsbrücke
xl characteristic/eigenartig
xli physical thing/Realding
xlii physical thing/Realding
xliii true or false/richtig oder falsch

xliv will be recognized again/noch anerkannt wird
xlv program/Anweisung
xlvi when one deduces from it the combined expectations/wenn man vereinbarungsgemäß die Erwartungen an ihr abliest
xlvii Then the predicted statistics are established – that is the doctrine./Dann wird sich die vorausgesagte Statistik einstellen – das ist die Meinung.
xlviii Moreover that would be true not merely for the object but also for the instrument itself!/Das ist es auch, bloß nicht am Objekt, sondern am Instrument selbst!
xlix I consider it acceptable to express this reasoning sequence as follows/Ich halte es für einwandfrei, diese Schlußfolgerung so auszudrücken
l If different ψ-functions are under discussion the system is in different states./*Wenn verschiedene ψ-Funktionen vorliegen, befindet sich das System in verschiedenen Zuständen.*
li We must face up to yet another objection/Wir müssen sogar noch einem Einwand begegnen
lii specify/erklären
liii dare not specify/nicht zugeben dürfen
liv foundation/Begründung
lv experimentally verifiable/versuchsweise
lvi *both* along with/*auch* bei
lvii stipulations/Bedingtheiten
lviii which one specifically contemplated measurement/die eine gerade ins Auge gefaßte Messung
lix almost/zunächst
lx circumspect/behutsam organisiert
lxi prepare/herauspräparieren
lxii determines/entscheidet
lxiii And it cannot happen the other way around/Es kann aber auch nicht etwa vorkommen
lxiv not on more, on any later ones, in case the measuring goes on superfluously/von den späteren, falls überflüssigerweise weitergemessen wird, nicht mehr
lxv look back/nachdenken
lxvi set forth the requirements as/die Forderung hinstellen als
lxvii Quantum predictions are indeed not subject to test as to their full content, ever, in a single experiment; yet they are correct, in that whoever possessed them/Nicht vollinhaltlich im Einzelversuch prüfbar, das sind Quantenvoraussagen ja nie, aber richtig, weil, wer sie besessen hätte
lxviii Neither can it be by chance/Zufallszahlen können es auch nicht sein
lxix remain reassured in this/halten sich dadurch beruhigt
lxx textbook/Hilfsbuch
lxxi formulation/Auffassung
lxxii *any/jede beliebige*
lxxiii perhaps by a new kind of algebra/nicht etwa durch eine neuartige Algebra
lxxiv arrangement/Zuordnung
lxxv stipulates/orientiert über
lxxvi dilemma/Verlegenheit
lxxvii Consideration of the Special Role of Time/Bedenken gegen die Sonderstellung der Zeit
lxxviii arrangement/Zuordnung
lxxix still/doch
lxxx henceforth/auch weiterhin
lxxxi unwelcome/unerbittlich
lxxxii If the formulation could be so carried out/Wenn sich die Auffassung durchfüren ließe
lxxxiii *any/jeder*
lxxxiv Natural Law/Naturprinzip
lxxxv hurdle/Anstoß

12

Note on the Quantum-Mechanical Theory of Measurement*

WENDELL H. FURRY

In recent notes by Einstein, Podolsky and Rosen and by Bohr, attention has been called to the fact that certain results of quantum mechanics are not to be reconciled with the assumption that a system has independently real properties as soon as it is free from mechanical interference. We here investigate in general, and in abstract terms, the extent of this disagreement. When suitably formulated, such an assumption gives to certain types of questions the same answers as does quantum mechanics; this is true of the formulas usually given in discussions of the theory of measurement. There exists, however, a general class of cases in which contradictions occur. That such contradictions are not restricted to the abstract mathematical theory, but can be realized in the commonest physical terms, is shown by the working out of an example.

1. Introduction

Some time ago there appeared a paper by Einstein, Podolsky and Rosen[1] entitled 'Can Quantum-Mechanical Description of Physical Reality be Considered Complete?' The writers concluded that the answer must be negative, on the ground that quantum mechanics forbids simultaneous measurement of two noncommuting variables even when both variables simultaneously possess 'physical reality', in the sense that either might be measured 'without in any way disturbing the system'. Recently Bohr[2] has upheld the view that quantum-mechanical description of nature can be considered complete, by demonstrating how the restrictions on simultaneous measurement which it imposes are inherent in the character and use of the measuring instruments. These measuring instruments must always be included as part of the physical situation from which our experience is obtained, and by

* [The text follows the original publication in *Physical Review* **49** (Furry 1936a), which was received 12 November 1935 and published 1 March 1936. The author's affiliation is given as *Department of Physics, Harvard University*. Reprinted with permission from: W. H. Furry, *Physical Review* **49**, 393–399. Copyright 1936 by the American Physical Society (*eds.*).]

[1] A. Einstein, B. Podolsky and N. Rosen, *Phys. Rev.* **47**, 777 (1935) [Einstein, Podolsky and Rosen 1935; see Chapter 9]. Referred to as EPR.

[2] N. Bohr, *Phys. Rev.* **48**, 696 (1935) [Bohr 1935b; see Chapter 16].

doing this one sees that quantum mechanics provides a complete and peculiarly apt interpretation of experience.

Bohr has again clearly called attention to this circumstance, and has remarked that one must be careful not to suppose that a system is an independent seat of 'real' attributes simply because it has ceased to interact dynamically with other systems. The paper of Einstein, Podolsky and Rosen has shown the sort of situations in which this characteristic of quantum mechanics may become especially prominent. This indicates an extension of the usual discussions of the theory of measurement.[3] In the present note a discussion more comprehensive in this respect will be summarized, and some further consideration will be given to the possibility of illustrating the point in question in concrete physical terms.

We shall have to make use of the concepts and results presented in von Neumann's rigorous and detailed discussion[4] of the theory of measurement by means of an instrument. Since the mathematical language of von Neumann's work is not that most current among physicists, it is desirable first to explain the meaning of these concepts in more usual terms. A similar rephrasing of the proofs of important results is omitted in the interests of brevity, and the results are simply stated.

2. Possible Types of Statistical Information about a System

Our statistical information about a system may always be expressed by giving the expectation values of all observables.[5] Now the expectation value of an arbitrary observable F, for a state whose wave function is φ, is

$$\bar{F} = (\varphi, F\varphi). \tag{1}$$

If we do not know the state of the system, but know that w_i (with $\sum_i w_i = 1$) are the respective probabilities of its being in states whose wave functions are φ_i, then we must assign as the expectation value of F the weighted average of its expectation values for the states[6] φ_i. Thus

$$\bar{F} = \sum_i w_i(\varphi_i, F\varphi_i). \tag{2}$$

[3] Cf. W. Heisenberg, *The Physical Principles of the Quantum Theory*, particularly pp. 55ff. [Heisenberg 1930b]; J. von Neumann, *Mathematische Grundlagen der Quantenmechanik*, Chapter VI [von Neumann 1932, 1955]; W. Pauli, *Handbuch der Physik*, vol. 24, no. 1 [Pauli 1933], pp. 143ff., p. 165 [Pauli 1990, pp. 88ff., 110–111].

[4] [See note 3 (*eds.*).]

[5] V Neumann, reference 3, p[p. 162–]163 [von Neumann 1955, pp. 306–307]. We confine ourselves throughout to the case of discrete spectra. (φ, ψ) means the Hermitian inner product, $\int \varphi^* \psi d\tau$. The same letters, $A, B \ldots$, are used to denote observables and the corresponding operators.

[6] For convenience we shall often refer to 'the state ψ' instead of 'the state whose wave function is ψ'. All functions are normalized unless otherwise stated.

This formula for $\bar{F}$ is the appropriate one when our system is one of an ensemble[7] of systems of which numbers proportional to w_i are in the states φ_i. It must not be confused with any such formula as

$$\bar{F} = \Big(\sum_i (w_i)^{\frac{1}{2}}\varphi_i,\, F\sum_i (w_i)^{\frac{1}{2}}\varphi_i\Big),$$

which corresponds to the system's having a wave function which is a linear combination of the φ_i. This last formula is of the type of (1), while (2) is an altogether different type.

An alternative way of expressing our statistical information is to give the probability that measurement of an arbitrary observable F will give as result an arbitrary one of its eigenvalues, say δ. When the system is in the state φ, this probability is

$$|(\varphi, \chi_\delta)|^2\,, \tag{1'}$$

where χ_δ is the eigenfunction of F corresponding to the eigenvalue δ. When we know only that w_i are the probabilities of the system's being in the states φ_i, the probability in question is

$$\sum_i w_i|(\varphi_i, \chi_\delta)|^2. \tag{2'}$$

Formula (2′) is not the same as any special case of (1′) such as

$$\Big|\Big(\sum_i (w_i)^{\frac{1}{2}}\varphi_i,\, \chi_\delta\Big)\Big|^2.$$

It differs generically from (1′) as (2) does from (1).

When such equations as (1), (1′) hold, we say that the system is in the 'pure state' whose wave function is φ. The situation represented by Eqs. (2), (2′) is called a 'mixture' of the states φ_i with the weights w_i. It can be shown[8] that the most general type of statistical information about a system is represented by a mixture. A pure state is a special case, with only one nonvanishing w_i. The term 'mixture' is usually reserved for cases in which there is more than one nonvanishing w_i. It must again be emphasized that a mixture in this sense is essentially different from any pure state whatever.

[7] The usefulness of this concept has recently been remarked upon by Kemble (*Phys. Rev.* **47**, 973 (1935)) [Kemble 1935; see Chapter 17].

[8] V Neumann, reference 3, pp. 167–168 [von Neumann 1955, pp. 313–319].

3. Reduction of Wave Packets

Let $\Psi(x_1, x_2)$ be the wave function for two systems I and II which have at some previous time interacted and have now ceased to interact. One can show[9] that there always exists an expansion, which is in general unique, in the form

$$\Psi(x_1, x_2) = \sum_k (w_k)^{\frac{1}{2}} \varphi_{\lambda_k}(x_1) \xi_{\rho_k}(x_2), \tag{3}$$

where the φ_{λ_k} are eigenfunctions of an observable L corresponding to eigenvalues λ_k, and the ξ_{ρ_k} are eigenfunctions of an observable R corresponding to eigenvalues ρ_k. The λ_k are all distinct, and so are the ρ_k. It can be shown[10] that, so far as system II alone is concerned, the statistical information available when (3) is the wave function of the combined systems is represented by a mixture of the states ξ_{ρ_k} with the weights w_k. A similar result holds, of course, for system I. A measurement of L and R on the total system can never give any other value than ρ_i for R to correspond to the value λ_i of L. Thus a measurement of L suffices to predict the value of R and the state of system II, which, *after*[11] such a measurement, is always in one of the pure states ξ_{ρ_k}. Eq. (3) shows that the coupling between the systems has been such as to make system I a suitable instrument for measuring the observable R on system II, the quantity L serving as a 'pointer reading'.

Now the conclusions to be drawn on the basis of these developments are just those we should expect if we ascribed 'real' characteristics to system II as soon as it ceased to interact with system I; this will be shown explicitly later. The contradictions we wish to investigate can be brought out only by going beyond the considerations given in connection with Eq. (3). We may either look for particular cases in which the expansion (3) is not unique – e.g., the example given by EPR and the one we shall give in Section 5 – or develop a way of interpreting expansions of a less special type.

The second alternative brings us directly to the general method of 'reducing the wave packet'. This procedure is commonly known and accepted among physicists, and is applied by EPR; but the writer has been unable to find in the literature an explicit description of its application to the present case. Such a description is briefly as follows: If M is any observable of system I, and ψ_μ its eigenfunction

[9] V Neumann, reference 3, pp. 225ff [von Neumann 1955, pp. 422ff.]. In the following arguments we use the word 'observable' to mean 'complete set of commuting observables', in the sense of Dirac, *Principles of Quantum Mechanics*, §17 (1st edition) [Dirac 1930]. In like fashion a set of eigenvalues of such a set of observables is referred to as an 'eigenvalue' of the 'observable'.

[10] [See note 9 (*eds.*).]

[11] I.e., *immediately* after. The wave functions used are in general not stationary solutions of the wave equation, but in our discussion we can abstract from the time, because our statistical information about a system at one time can be calculated from that at another time according to a definite differential equation (v. Neumann, reference 3, p. 186) [von Neumann 1955, pp. 350–351].

corresponding to the eigenvalue μ, then we can express $\Psi(x_1, x_2)$ as a series in the orthogonal functions $\psi_\mu(x_1)$, with coefficients which are functions[12] of x_2:

$$\Psi(x_1, x_2) = \sum_\mu \psi_\mu(x_1)\zeta_\mu(x_2), \tag{4}$$

where
$$\zeta_{\mu'}(x_2) = \int \psi^*_{\mu'}(x_1)\Psi(x_1, x_2)dx_1. \tag{5}$$

The statistical information which one will have about system II after[13] a measurement of M on system I has given the value μ' may be obtained by the following process: We suppose a large number of measurements made on combined systems I + II prepared so that their wave functions are given by (4), each of these measurements consisting in a determination of the values of M for system I and some observable F for system II. We then obtain the relative numbers of times the different values δ are found for F, counting only those measurements in which the value μ' is found for M. These relative numbers are by definition proportional to the quantities $|(\Psi(x_1, x_2), \psi_{\mu'}(x_1)\chi_\delta(x_2))|^2$ (cf. Eq. (1′)), and these quantities are, by (5), just equal to $|(\zeta_{\mu'}, \chi_\delta)|^2$. Since this is true for all observables F, we see that after a measurement on system I has given the value μ' for M, system II is in the pure state with wave function given – apart from normalization – by (5).

4. Probability Calculations and Their Results

We are now ready to discuss in detail the degrees of agreement and disagreement between the results of quantum-mechanical calculations and those to be expected on the assumption that a system once freed from dynamical interference can be regarded as possessing independently real properties. For we can give a definite form to this assumption, and base on it a method for answering all questions which can be asked about the probabilities of finding different results by measurements on system II. This we call Method A:

Assumption and method A. We assume that during the interaction of the two systems each system made a transition to a definite state, in which it now is, system I being in one of the states φ_{λ_k} and system II in one of the states ξ_{ρ_k}. These transitions are not causally determined, and there is no way of finding out *which* transitions occurred, except by making a suitable measurement. In the absence of measurements we know only that the probabilities of the different transitions are respectively w_k, and that if system I is in the state φ_{λ_i} system II is in the state ξ_{ρ_i}.

[12] Unnormalized, and in general not orthogonal to each other.
[13] [See note 11 (*eds.*).]

This provides a sufficient basis for making all needed calculations of probabilities, the methods being those of ordinary probability theory.

Method B. We shall compare with the results of Method A those obtained by quantum-mechanical calculations, using the facts explained in connection with Eqs. (3) and (4), (5).

There are four types of questions for which answers may be required. The notation used in discussing them is the same as that previously described; in particular, the reader is reminded that the observables L and R have a special significance through their connection with the expansion (3), while M and S are arbitrary observables. The questions and their answers are as follows:

(a) If S, having eigenvalues σ and eigenfunctions η_σ is measured on system II without any measurements having been made on system I, what is the probability of obtaining the result σ'? Both methods give the same result

$$\sum_k w_k |(\xi_{\rho_k}, \eta_{\sigma'})|^2, \tag{6}$$

as is at once evident. (Cf. remarks following Eq. (3).)

(b) If L has been measured on system I and the value λ_i obtained, what is the probability of finding the value σ' for S in II? Both methods at once give the answer

$$|(\xi_{\rho_i}, \eta_{\sigma'})|^2. \tag{7}$$

When $S = R$ and $\sigma' = \rho'$, we get in particular the value

$$|(\xi_{\rho_i}, \xi_{\rho'})|^2 = \delta_{\rho_i \rho'}, \tag{7'}$$

so that a definite result is predicted. The possibility of such definite predictions was taken by EPR as a 'criterion of the physical reality' of the observable R; it is, *par excellence*, the bit of evidence which might incline one to believe Assumption A to be true.

(c) If M has been measured on system I and the value μ' obtained, what is the probability of finding the value ρ_i for R in II? Some calculations are required for this case, but are omitted, being analogous to those which will be given in detail for (d). Both methods give the result

$$w_i |(\varphi_{\lambda_i}, \psi_{\mu'})|^2 \Big/ \Big[\sum_k w_k |(\varphi_{\lambda_k}, \psi_{\mu'})|^2\Big]. \tag{8}$$

(d) If M has been measured on system I and the value μ' obtained, what is the probability of finding the value σ' for S in II?

Method A: If the measurements are carried out on a large number of similarly prepared pairs of systems, the fraction giving the value μ' for M is $\sum_k w_k |(\varphi_{\lambda_k}, \psi_{\mu'})|^2$. The fraction giving this value and having system II in state

ξ_{ρ_k} is $w_k|(\varphi_{\lambda_k}, \psi_{\mu'})|^2$. Then the fraction giving the values μ' for M and σ' for S is $\sum_k w_k|(\varphi_{\lambda_k}, \psi_{\mu'})|^2 \times |(\xi_{\rho_k}, \eta_{\sigma'})|^2$. Dividing this by the fraction giving the value μ' for M, we find as the required *a posteriori* probability,

$$\Big[\sum_k w_k|(\varphi_{\lambda_k}, \psi_{\mu'})|^2|(\xi_{\rho_k}, \eta_{\sigma'})|^2\Big]\Big/\Big[\sum_k w_k|(\varphi_{\lambda_k}, \psi_{\mu'})|^2\Big]. \tag{9A}$$

Method B: The wave function from which we must calculate this probability is, by (5), (3):

$$\int \psi^*_{\mu'}(x_1)\Psi(x_1, x_2)dx_1 = \sum_k (w_k)^{\frac{1}{2}}(\psi_{\mu'}, \varphi_{\lambda_k})\xi_{\rho_k}(x_2).$$

On normalizing this function and taking the square of its inner product with $\eta_{\sigma'}$ one gets for the required probability

$$\Big[\sum_k (w_k)^{\frac{1}{2}}(\varphi_{\lambda_k}, \psi_{\mu'})(\xi_{\rho_k}, \eta_{\sigma'})|^2\Big]\Big/\Big[\sum_k w_k|(\varphi_{\lambda_k}, \psi_{\mu'})|^2\Big], \tag{9B}$$

where the denominator comes from normalization.

The difference between (9A) and (9B) comes from the well-known phenomenon of ‘interference’ between probability amplitudes. The absence of such an effect in case (a) is usually stressed in discussions of the theory, since it shows plainly the effect which the mere attaching of an instrument must in general have on the behavior of a system. Since case (d) is not mentioned, it is possible for a reader to form the impression that the theory is consistent with Assumption A.[14]

The formal discrepancy between (9A) and (9B) is a consequence of the fact that, according to the remarks following (4) and (5), after a measurement of M on system I has been made system II is in a pure state, which is in general not one of the ξ_{ρ_k}. Now no possible manipulation of the w_k will produce from the statistics of the mixture those of any pure state other than one of the ξ_{ρ_k}. Thus not only is Method A inconsistent with Method B, but also there is no conceivable modification of Method A which could produce consistency between Assumption A and Method B.

The contradiction here, like that between quantum mechanics and the classical doctrine of causality, indicates a radical change in concept rather than a mere change in the details of a mechanism. The idea which is found to be untenable may, roughly, be said to be that of the independent existence of two entities, the state of system II and one’s knowledge of its state, only the latter being affected by

[14] Cf. Pauli, reference 3, p. 89 [Pauli 1990, pp. 31–32]. The remarks there given are entirely correct, but liable to be misleading to an unwary reader with a predilection for Assumption A. The same is to some extent true of the remarks in v. Neumann, reference 3, p. 232 [von Neumann 1955, pp. 435–436], and Heisenberg, reference 3, pp. 59–62 [Heisenberg 1930b].

measurements made on system I. Quantum theory shows that this is not an adequate concept of the relation between subject and object.

5. A Physical Example

The inconsistency of quantum mechanics with the point of view which finds its definite formulation in Assumption A has been demonstrated mathematically. One may still wish to inquire whether it can be realized in a concrete physical example; for we may certainly suppose that not all mathematical operators, even though subject to the proper formal requirements,[15] correspond to experimentally measurable quantities.

An example has been outlined in mathematical form by EPR. As Bohr has remarked, and as is evident from the mathematics, the physical realization of this example involves certain difficulties, in particular the necessity of abstracting altogether from the time in circumstances in which this is not really permissible. We shall now outline an example in physical terms, in which full account is taken of the time-dependence of the quantities involved.

In order to give a physical example of the point in question, it would suffice to describe any case in which one system (I) is used as a measuring instrument in observing another system (II), and in which, after interaction has ceased, some other observable besides the one suited to serve as a 'pointer reading' can be measured on system I. The resulting inferences about system II would in the main fall under case (d), in which the contradiction between quantum mechanics and Assumption A is evident. A variety of such examples could doubtless be given.[16] There is, however, a particularly neat and striking special type of example, to which EPR have directed attention; and the one we shall describe is of this type. Before describing it in detail, we shall indicate the nature of this sort of example and its connection with the argument of the preceding section.

The characteristic feature of such an example is obtained by choosing a case in which the expansion (3) is not unique.[17] An assumption in the form of Assumption

[15] V Neumann, reference 3, pp. 75ff. [von Neumann 1955, pp. 145ff.].

[16] E.g., the example we shall discuss could be stated in this way. In the construction and discussion of such examples a difficulty arises owing to the fact that in experimental practice the observables used are almost always incomplete (cf. note [9]), and that the theory of measurements of incomplete observables is not altogether free from ambiguity (cf. v. Neumann, reference 3, pp. 184–185 [von Neumann 1955, pp. 347–349]). For this reason it is much easier and more satisfactory to treat our example – which itself involves the use of incomplete observables – in the way we have here used.

[17] The condition for this is that the w_k be not all distinct (cf. v. Neumann, reference 3, p. 232 and p[p]. 175[–176] [von Neumann 1955, p. 436 and p. 332]). In the example of EPR, the w_k are all equal and *all* states of system II are included. This means that there are an infinite number of different expansions of the form (3), and that any measurement made is as likely to give any one result as any other. Measurement in quantum mechanics has in general a twofold aspect: it gives information (of a statistical nature) about the properties of the state of the system *before* the measurement, and it enables us to predict the state of the system *after* the measurement.

A can be stated corresponding to each of two expansions of type (3). Each of these assumptions is consistent with a number of quantum-mechanical results of the forms (6), (7), (8), and particularly of the form (7′); thus both of them are to be accepted as true, according to the general attitude underlying Assumption A. Now each assumption asserts that system II has made a transition to a definite (though unknown) state, and that it is a state for which a certain observable has a well-defined value. In the case chosen for the example the two requirements thus imposed on the state of system II cannot be fulfilled simultaneously by any quantum-mechanical state, because of the limitations imposed by the uncertainty principle. The point of view expressed in Assumption A is accordingly found to conflict with quantum mechanics.

With this outline of the argument in mind, we proceed to the detailed discussion of the example. In this we shall use the language of experimental physics rather than that of the mathematical formalism.

We consider the determination of the position of a heavy particle of mass M, say a proton, by the use of a 'microscope'. This microscope is designed for use, not with γ-rays, but with light charged particles of mass m, say electrons. If we shine soft light in from the side of the 'barrel' and let it be scattered from an electron which is on its way to the 'lens', a transverse component[18] of the electron's momentum can be calculated from the Doppler effect on the scattered light. Under suitable circumstances this will enable us to infer the momentum[19] of the proton quite accurately. On the other hand, we have only to let the electron travel undisturbed to the photographic plate in order to be able to infer with considerable accuracy the corresponding coordinate[20] of the proton. The 'Assumption A' which corresponds to the first experiment is: During the collision between the electron and the proton, the proton made a transition into some state with well-defined momentum; *which* state this is can be determined by a measurement made on the electron. Corresponding to the second experiment one has an 'Assumption A' which reads exactly the same except that the word 'momentum' is replaced by the word 'coordinate'. Since one is still free to choose which experiment is to be performed *after* the electron has been scattered into the microscope[21] and ceased to interact with the proton, these

In order that it may serve the first purpose, care must be used in choosing a suitable coupling of object and instrument. In the example of EPR, the coupling has been so violent that all trace of the original state of system II is lost, so that the word 'prediction' is the only correct one to apply to one's conclusions about system II.

18 I.e., transverse to the axis of the microscope. This component of the momentum will be called simply the 'momentum', and the corresponding coordinate will be called the 'coordinate' x.

19 [See note 18. (*eds.*)]

20 [See note 18. (*eds.*)]

21 Of course the electron need not have been scattered into the microscope; in any given case we can only go on with the experiment in good faith, hoping that it will have been so scattered. The experiment can fail; and from the fact that it has succeeded when one choice was made, quantum mechanics of course offers no such inference as that it 'would have' succeeded if the other choice had been made. But under the proper

assumptions must both be true at once, if one accepts the point of view on which they are based. But we shall see that their simultaneous truth can be in conflict with the uncertainty principle.

Before deriving the actual expressions for the uncertainties in the two alternative predictions, let us consider briefly their physical origin. In principle the Doppler effect experiment can be made with arbitrary precision (cf. latter part of note [21]). The uncertainty Δp may accordingly be regarded as fixed by the original uncertainties in the momenta of the particles; to make it small, we have only to prepare them properly beforehand. Now the use of particles so prepared will in some degree limit the accuracy of our prediction of x. This comes about through the fact that the predictions we want must refer to a definite time, and that the prediction of x read directly from the photographic plate refers to the moment of the collision: this is not precisely known, because our long wave trains take a finite time to pass over each other. In making a prediction for a definite moment, which we choose to be that at which the wave trains cease to overlap, we must allow for the distance the proton may have moved since the collision. This leads to the existence of a lower limit for the product $\Delta p \Delta x$; but since the electron, with its small mass, moves rapidly across the region where the proton may be found, whereas the proton's motion after the collision is comparatively sluggish, this limit will be found to be not h, but about $(m/M)h$.

To show this, we shall discuss the possible sources of uncertainty in the prediction of x. These are: (1) The finite resolving power of the microscope; (2) inaccuracy of focus; (3) allowance for the proton's motion between the moment of the collision and the moment to which the predictions refer.

(1) *Finite resolving power.* This gives an uncertainty

$$\Delta_1 x \sim (\lambda/\varepsilon) \sim (h/p\varepsilon), \tag{10}$$

where ε is the numerical aperture and λ, p are wave-length and momentum of the electron.

(2) *Inaccuracy of focus.* Let the electron[22] be sent in through a slit of width s. At the far side of the field of view the half-width of the beam will have become,

circumstances, such an inference does follow from the point of view of Assumption A. The circumstances required are, that in the Doppler effect experiment enough quanta must be sent in so that 'if the electron is in the barrel', several are sure to be scattered back into a suitably small solid angle. The quanta must be extremely soft, so that scattering a great many of them will not change the electron's momentum too much; and their energy must be measured very accurately. This means that the electron is in the barrel a long time; and it turns out that, to secure the accuracy represented by Eq. (14), the length of the microscope must be much greater than $(M/m)^{11/3}\Delta x$, while the lateral distance to the source and analyzer of the light must be much greater than $(M/m)^{13/3}\Delta x$. These requirements make in principle no difficulty.

22 Only one beam need be narrow. Under (3) we shall see that it is expedient to make the wave-lengths of electron and proton originally equal, so that $\Delta_2 x$ will be the same whichever beam is limited. By admitting the proton through a slit much broader than s, we can assure that the electron escapes unless it is scattered through a fairly large angle, thus avoiding the prevalence of spurious effects.

through diffraction, about $(s/2) + (L\lambda/2s)$, where L is the breadth of the field of view. By proper choice of s, this expression takes on its minimum value, $(\lambda L)^{\frac{1}{2}}$. Now L must be equal to $(h/\Delta p)$, the whole length of the wave train; otherwise the restriction of the field of view would cause enough diffraction of the scattered waves to spoil the significance of the contemplated Doppler effect experiment. Thus we get for the uncertainty in x from this source

$$\Delta_2 x \sim \varepsilon(\lambda h/\Delta p)^{\frac{1}{2}} = \varepsilon h/(p\Delta p)^{\frac{1}{2}}. \tag{11}$$

(3) *Allowance for proton's motion.* The time available for this motion is of the order of magnitude

$$\Delta t \sim (h/\Delta p)/(p/m),$$

which is the time required for the electron to travel the length of such a wave train as must be used. To make the proton's velocity after the collision as small as possible, we can send the two particles into the field of view with equal and opposite momenta. Then if the electron were scattered exactly at a right angle, the resulting x component of the proton's velocity would be zero. But on account of the finite aperture we must take it to be roughly

$$v'_x \sim \varepsilon(p/M).$$

We then get

$$\Delta_3 x \sim v'_x \Delta t \sim \varepsilon(m/M)(h/\Delta p). \tag{12}$$

Thus we have finally

$$\Delta p\Delta x \sim \Delta p(\Delta_1 x + \Delta_2 x + \Delta_3 x) \sim h\{(\Delta p/\varepsilon p) + \varepsilon(\Delta p/p)^{\frac{1}{2}} + \varepsilon(m/M)\}. \tag{13}$$

By making Δp extremely small compared to p, we can make

$$\Delta p\Delta x \sim [\varepsilon](m/M)h. \tag{14}$$

By taking ε also to be small, we can in principle make $\Delta p\Delta x$ *arbitrarily* small; also we could in principle dispense with the advantage we obtained by using the disparity in mass of two known particles.

As explained in the preliminary discussion of the example, the comparison of a result such as (14) with the uncertainty principle shows that Assumption A is inconsistent with quantum mechanics.

Concluding Remark

Both by mathematical arguments and by discussion of a conceptual experiment we have seen that the assumption that a system when free from mechanical interference necessarily has independently real properties is contradicted by quantum mechanics. This conclusion means that a system and the means used to observe it are to be regarded as related in a more subtle and intimate way than was assumed in classical theory. It does not mean that quantum mechanics is not to be regarded as a satisfactory way of correlating and describing experience; it does illustrate the difficulty, often remarked upon by Bohr, which is inherent in the problem of the distinction between subject and object.

13

Remarks on Measurements in Quantum Theory*

WENDELL H. FURRY

Since the completion of my note[1] on a paper of Einstein, Podolsky and Rosen,[2] there have appeared two articles by Schrödinger[3] on the same subject. Mathematically the content of Schrödinger's papers and mine is essentially the same, although the emphasis is quite different. The interpretations given are exact opposites. I wish here to add a few words to the discussion.

Bohr[4] has emphasized that the essential difficulty with the argument of EPR lies in the assumption that whenever a system is not in mechanical interaction with other systems it may be regarded as having independently real properties. This assumption depends on the ascription of an undue significance to the fact that the systems are not connected in any way which finds a reflection in our description of the situation by means of a model. If it were tenable, it would provide an easy solution, quite acceptable to our habitual ways of thinking, of the problem of the relation of subject and object. In fact, however, the difficulties of this problem as they appear in quantum theory are not to be resolved by any such facile bit of analysis.

To make this assumption a definite basis for arguments, one must add some assertion as to the nature of the 'real' properties of the 'free' system. Being in agreement with Bohr's point of view, and wishing only to make clearer an interesting characteristic of quantum mechanics, I took as my additional assumption one which suggests itself from the content and manner of the usual discussions of the theory of measurement. My 'Assumption A' is that a 'free' system which was formerly coupled to an instrument is 'really' in some one of a certain set of quantum-mechanical

* [The text follows the original publication in the 'Letters to the Editor' section of *Physical Review* (Furry 1936b), which was received 2 March 1936 and published 15 March. Reprinted with permission from: W. H. Furry, *Physical Review* **49**, 476. Copyright 1936 by the American Physical Society (*eds.*).]

1 W. H. Furry, *Phys. Rev.* **49**, 393 (1936) [Furry 1936a; see Chapter 12]. Referred to as F.

2 A. Einstein, B. Podolsky and N. Rosen, *Phys. Rev.* **47**, 777 (1935) [Einstein, Podolsky and Rosen 1935; see Chapter 9]. Referred to as EPR.

3 E. Schrödinger, *Proc. Camb. Phil. Soc.* **31**, 555 (1935) [Schrödinger 1935e; see Chapter 10] (referred to as C); and *Naturwiss.* **23**, 807–812, 823–828, 844–849 (1935) [Schrödinger 1935f; see Chapter 11] (referred to as N).

4 N. Bohr, *Phys. Rev.* **48**, 696 (1935) [Bohr 1935b; see Chapter 16].

states, which are the eigenstates of the observable which the instrument is suited to measure. Assumption A has three important properties:

(1) The corresponding picture of the situation is in full accord with our habitual attitudes, and is the one we use in ordinary practice.

(2) The predictions derived from Assumption A for the sorts of cases which *actually occur* agree exactly with those of quantum mechanics.

(3) In more general cases, *realizable in principle* according to the postulates of the theory, there is flat contradiction between the formulas given by Assumption A and those of quantum mechanics. Therefore Assumption A is actually untrue.

The corresponding postulate chosen by Schrödinger is altogether different, on account of his different point of view. Since he agrees with the underlying assumption of EPR, he is careful to introduce no *a priori* 'doubtful' element into his thesis. Thus he rejects (N, p. 827/193) Assumption A, and ends with taking as his criterion of 'reality' just that of EPR: The 'real' properties of the 'free' system are the values of those observables whose values could be predicted 'without in any way disturbing the system'. On this basis one can give interesting considerations only about certain degenerate situations, such as that chosen as an example by EPR. Schrödinger accordingly devotes most of his mathematical discussion to a complete elaboration of this example. As was to be expected on general principles (F, footnote [14]), the resulting picture of the situation is much more complicated than that originally drawn by EPR, and could scarcely be traced to so simple a source as an 'incompleteness' of the theory; Schrödinger implies that it indicates some serious defect.

We may note two properties of this assumption:

(1′) It is, in the opinion of EPR and of Schrödinger, indisputably true *a priori*.

(2′) None of its assertions – it is innocent of actual predictions – ever comes into direct contradiction with the results of quantum mechanics; hence it cannot be either proved or disproved objectively.

According to the prevailing attitude in theoretical physics, (2′) is enough to consign all such assertions to the limbo of 'meaningless statements'.[5] Against this verdict (1′) is of no avail; for, as asserted already in the second paragraph of this letter, the assumption underlying such an opinion is definitely untenable: it is not correct to assume that the only physically significant relations are those which are directly obvious from a classical model. One striking experimental example is the collision of two similar particles. No matter how far apart the particles are when we try to collect one of them, the relative probabilities of finding it in different places are strongly affected by the 'interference term' in the cross section; it is not really 'free'. (In this case, as in the one under consideration, there is of course nothing

[5] Cf. W. Heisenberg, *The Physical Principles of the Quantum Theory*, p. 15, footnote [Heisenberg 1930b].

'magical' about the affair. The interference effect does not come in unless there has been an actual opportunity for the two particles to get interchanged,[6] just as in the case in hand there is never any relation between the systems without the existence of an actual dynamical interaction to start it.)

Thus there can be no doubt that quantum mechanics requires us to regard the realistic attitude as in principle inadequate. But it need not disturb at all our habit of taking such an attitude as a matter of practical convenience in the interpretation of experiments. In any measuring process actually used the 'biorthogonal' expansion (F, Eq. (3)) is unique, and the experimenter concerns himself only with the observable and 'pointer reading' which belong to this expansion. Thus Assumption A is a perfectly safe working hypothesis.

Schrödinger's misgivings (C, p. 555/164) about the fact that measurement involves actually a chain of object-instrument relations accordingly seem groundless. This question was discussed in full detail some years ago by Neumann,[7] and in my opinion his treatment is altogether adequate.

W. H. FURRY

Department of Physics,
Harvard University,
March 2, 1936.

[6] W. Pauli, *Handbuch der Physik* **24**/1 [Pauli 1933], pp. 192–193 [Pauli 1990, pp. 138–139].

[7] J. v. Neumann, *Mathematische Grundlagen der Quantenmechanik*, chapter VI [von Neumann 1932].

14

Is a Deterministic Completion of Quantum Mechanics Possible?*

WERNER HEISENBERG

The laws of quantum mechanics contain in many cases statements about the probability for a definite event taking place. Although this statistical character of quantum mechanics is intimately connected to the general features of the description of nature investigated by Bohr[1] that have come to light in atomic physics and are described by the concept of 'complementarity', yet, in discussions on[i] atomic theory the question is raised again and again whether quantum mechanics might not be amended later to a deterministic theory through new physical results.[2] It is natural to consider quantum mechanics at first to be an incomplete description of nature; for it seems to follow from entirely general principles that wherever an exact prediction of what happens in the future has not succeeded, this failure should be seen as signalling the presence of a problem that is yet unsolved and needs[iii] to be

* [A typescript of this essay is to be found in the Albert Einstein Archives (document EA 5-207, 11pp., in German). Formulas are written in by hand, and single letters and words that were not clearly legible have been inked in by hand (presumably by Heisenberg); there are also a few marginal notes in Einstein's handwriting, which we reproduce in the footnotes. The typescript is virtually identical (except for more or less complete referencing and a couple of instances of using the form 'welche' instead of the form 'die' for the feminine relative pronoun) to the original manuscript from Heisenberg's Nachlaß in the Werner Heisenberg Archive: AMPG, III. Abt., Rep. 93, Nr. 668 (formerly in Munich, now in Berlin), which is reproduced in the Archive for the History of Quantum Physics (AHQP 45, section 11-001, 29pp., in German), and on which we have based our translation. A transcription of the manuscript was published in vol. 2 of Pauli's scientific correspondence (Pauli 1985, pp. 409–418), but there are a number of discrepancies between the transcription and the original. We have endnoted these (giving first the reading of the transcription, then that of the typescript and manuscript), except when they constitute merely differences in spelling or punctuation with no implications for the meaning of the sentence (*eds.*).]

[1] N. Bohr, Solvay Conference 1927; *Naturwiss.* **16**, 245 (1928) [Bohr 1928a]; *Atomtheorie und Naturbeschreibung*, Berlin 1931 [Bohr 1931]; Faraday-lecture, *Journ. of the Chem. Soc.* 1932, p. 349 [Bohr 1932]; Lys og liv, *Natur. Verden* **17**, 49 (193[3]) [Bohr 1933a]; *Phys. Rev.* Forthcoming [Bohr 1935b; see Chapter 16]. [The paper by Bohr published in the proceedings of the 1927 Solvay conference is a French translation of Bohr (1928a). Notes and a partial transcription of Bohr's actual remarks at the conference are included and translated in Bacciagaluppi and Valentini (2009) (*eds.*).]

[2] E.g. M. v. Laue, *Naturw.* **20**, 915 (1932) [Laue 1932];[ii] *Naturw.* **22**, 439 (1934) [Laue 1934]; E. Schrödinger, *Naturw.* **22**, 518 (1934) [Schrödinger 1934]; A. Einstein, *Phys. Rev.* **47**, 777 (1935) [Einstein, Podolsky and Rosen 1935; see Chapter 9].

solved in the future.[iv] An argument of this kind, in light of the experimental success of quantum mechanics, assumes in general that quantum mechanics gives a *correct* description of nature. However, it ties this assumption to the hope that later research will bring to light a hitherto hidden network of causal connections behind the statistical connections of quantum mechanics – rather like the way classical mechanics is hidden behind the temperature and entropy concepts of thermodynamics. These causal connections need not refer at all to anschaulich classical variables of physical systems. Rather, it is inferred from the validity of the indeterminacy relations that classical concepts allow no adequate description of atomic phenomena, that therefore one needs to develop new concepts to be perhaps associated with hitherto unknown physical properties of atomic systems. For instance, apart from the physical properties that are exhaustively specified by determining its stationary state, the nucleus of the radium atom could, say, possess still other hitherto unknown physical properties, knowledge of which would make possible an exact prediction of the time of radioactive decay. The following considerations are meant to show under very general assumptions that such a deterministic completion of quantum mechanics is impossible, thus that one might maintain the hope for a description of nature that is deterministic in the traditional sense only if one chose to consider the most important experimental successes of quantum mechanics to be accidental. The essential content of the following trains of thought is already to be found in the earliest discussions of the fundamental interpretation of quantum mechanics;[3] presenting it anew is perhaps nonetheless justified by the criticism of quantum mechanics repeatedly expressed in the most recent literature.

§1

Quantum mechanics represents a physical system by a wavefunction in a 'configuration' space[v] whose number of dimensions is determined by the number of degrees of freedom of the system in question. The square of the absolute value of the wavefunction at a specific point of this space gives the probability that the anschaulich physical quantities denoted by the coordinates of the space take on the specific values corresponding to that point, if the system is observed with respect to these values. The formalism of quantum mechanics is thus based on the assumption that

[3] N. Bohr, *l*[*oc*]. *c*[*it*].; W. Heisenberg, *Z. f. Phys.* **43**, 172 (1927) [Heisenberg 1927a]. Cf. also W. Heisenberg: *Die physik*[*alischen*] *Prinzip*[*ien*] *der Quantenth*[*eorie*] Leipzig 1930 [Heisenberg 1930a]; J. v. Neumann: *Die mathematischen Grundlagen der Q*[*uanten*]*M*[*echanik*] 1932 [von Neumann 1932]; W. Pauli: *Die* [*allgemeinen*] *Prinzip*[*ien*] *d*[*er*] *Wellenmech*[*anik*] Hand[buch der Physik] [Pauli 1933]. [There are slight discrepancies between the three versions of this footnote, which are immaterial with the possible exception of the typescript adding '2nd ed., vol. 24/1, p. 257, 1933'. The page reference, however, seems spurious: on that page Pauli is discussing the paper by Landau and Peierls (1931) on the measurability of fields (Pauli 1990, pp. 210–211) (*eds*.).]

a physical system can be represented by classical-anschaulich variables, and that, as in classical theory, there can be an objective sense independent of the processes of observation in speaking of the actual value of a specific physical quantity, e.g. of the 'position of the electron'.

The quantum mechanical way of describing nature thus begins with a peculiar rift: on the one hand, one proceeds from the assumption that the task of physics is the description and synthesis in terms of laws of anschaulich, objective processes in space and time; on the other hand, one uses for the mathematical description of physical processes these wavefunctions[vi] in multi-dimensional configuration spaces, which in no way can be seen simply as representatives of the objective happening in space and time, as can, say, the coordinates of a point-mass in classical mechanics. This rift manifests itself in an arbitrariness in the application of quantum mechanics: should only the atomic system to be observed be represented by a wavefunction, and the devices used for its[vii] observation be treated according to the laws of classical physics, or should also the devices be represented by wavefunctions according to the laws of quantum mechanics, where in the end only the observation of the measuring device, say, the observation of a line on a photographic plate, is taken to be a classical-anschaulich process? At what place should one draw the cut between the description by wavefunctions and the classical-anschaulich description?

The answer to this question is: *the quantum mechanical predictions about the outcome of an arbitrary experiment are independent of the location of the cut just*[viii] *discussed.* Since this theorem is of crucial importance for the internal coherence [für den inneren Zusammenhang] of quantum mechanics, its[ix] proof shall be given here in detail.

Let an atomic system A be given, from which information reaches the observer by means of the measuring devices B, C, etc. The systems B, C ... shall be capable of being regarded as measuring devices in the sense of classical physics, i.e. from, say, a specific reaction of system C, it shall be possible to infer univocally to a reaction of system B, and from[x] this to a specific behaviour of A.

$$\text{Atomic System } A \rightarrow B \rightarrow C \rightarrow \ldots \rightarrow \text{Observer.}$$

For this to be possible, a certain physical interaction between the systems A, B, C etc. must be present. For example, it could be that the interaction energy between A and B is non-zero only when the coordinates $q_{A'}$ of system A possess the quite specific values q'_A. From the occurrence of a reaction in B one could then infer that the coordinates of A have in fact assumed these particular values q'_A. Let us make this special assumption about the interaction energy between A and B, since the essential content of the proof to be carried out can already be made clear in this special case.

A physical process taking place in system A can, on the one hand, be described by the time-dependent wavefunction $\psi_A(q_A, t)$ of system A. $|\psi_A(q'_A, t')|^2$ gives then the probability that at time $t = t'$, the[xi] coordinates q_A assume the values q'_A. One will thus assume with probability $|\psi_A(q'_A, t')|^2$ (up to a constant factor depending on the sensitivity of the measuring instruments) that the measuring devices react when the interaction between A and B is switched on at time t'. To the measuring instruments B, C etc. one applies in this treatment just the laws of classical physics.

But on the other hand, one can incorporate also system B into the quantum mechanical formalism by starting with a wavefunction[xii] for the total system $A + B$. From this wavefunction, the probability that system B reacts can then be inferred, and it is to be proved that this probability, up to a constant factor, is again given by $|\psi_A(q'_A, t')|^2$.

Before switching on the interaction between A and B, the wavefunction of system $A + B$ is given by the product $\psi_A(q_A, t) \cdot \psi_B(q_B, t)$, where $\psi_B(q_B, t)$ represents the state of system B before the onset of the reaction. We now separate the Hamiltonian function of the total system $A + B$ into the three parts H_A, H_B and H_{AB}, the first two of which refer to the systems A and B, while H_{AB} denotes the interaction between A and B, which differs from zero only at the point $q_A = q'_A$ and becomes effective only from time $t = t'$. The Schrödinger equation of the system $A + B$ then reads[xiii]

$$\left(\frac{\hbar}{i}\frac{\partial}{\partial t} + H_A + H_B\right)\psi(q_A, q_B, t) = -H_{AB}\,\psi(q_A, q_B, t). \tag{1}$$

If one is interested only in the behaviour of the system shortly after time $t = t'$, one can substitute for $\psi(q_A, q_B, t)$ on the right-hand side of this equation the 'undisturbed' value $\psi_A(q_A, t')\psi_B(q_B, t')$. Further, since the interaction energy H_{AB} differs from zero only at the point $q_A = q'_A$, the right-hand side can be replaced by H_{AB} $\psi_A(q'_A, t')\psi_B(q_B, t')$. From this it transpires that the solution of the Schrödinger equation for times shortly after $t = t'$ can be represented in the form

$$\psi(q_A, q_B, t) = \psi_A(q_A, t)\,\psi_B(q_B, t) + \psi_A(q'_A, t')\,\varphi(q_A, q_B, t, t'),$$

where the function $\varphi(q_A, q_B, t, t')$ no longer depends on the behaviour of system A before time t'. The probability that system B has undergone a reaction, that it is therefore no longer in the initial state represented by $\psi_B(q_B, t)$, is given by the deviation of the wavefunction from $\psi_A(q_A, t)\,\psi_B(q_B, t)$, and thus, according to the rules of quantum mechanics, essentially given by

$$\int dq_A\, dq_B |\psi_A(q'_A, t')\,\varphi(q_A, q_B, t, t')|^2 = |\psi_A(q'_A, t')|^2 \int dq_A\, dq_B |\varphi(q_A, q_B, t, t')|^2.$$

Thus this[xiv] probability turns out to be proportional to $|\psi_A(q'_A, t')|^2$ also according to this second method of calculation, where the remaining constant factor depends only on system B and its interaction with A, but not on the prior history of system A. One could, in the same way, incorporate also system C into the quantum mechanical formalism, and one would arrive at the same end result by very similar calculations. For the prediction of a given physical event, it is thus indeed indifferent at which place one draws the cut between the classical and quantum mechanical treatment.

The content of this proof can be presented in a somewhat more general form as follows. System B is meant to serve as the measuring device for the coordinates q_A of A. For this it is necessary that, when the interaction between A and B is switched on, a specific value q'_A corresponds to some specific reaction in B, e.g. the transition of the coordinate q_B from the value q'_B to the value q''_B. In the language of wave mechanics, this means: when the interaction is switched on, the evolution of the specific solution to the Schrödinger equation with the wavefunction initially differing from zero only in the neighbourhood of q'_A and q'_B, is such that[xv] the projection of the wavepacket onto the space of the q_B describes a motion from q'_B to q''_B. This is (in this example) the necessary condition for B being a measuring device. Now, before switching on, thus as long as the systems are still independent, the wavefunction is given by the product of the wavefunctions of the subsystems. One can track the temporal evolution of the total wavefunction[xvi] most easily by constructing it as a superposition of the wavepackets just[xvii] described. The total wavefunction then appears (for a short time after switching on the interaction) as a product of two factors, one of which is given by the wavefunction of the system A to be observed at the time when the interaction is switched on, while the other represents the reaction of the measuring device B. From this follows again the result[xviii] discussed above.

The most essential properties of the quantum mechanical formalism[xix] this proof relies on are: first, the fact that any enlargement of the physical system one is describing is represented by an increase in the number of dimensions of the configuration space; second, the possibility of obtaining new solutions through superposition of different solutions to the Schrödinger equation. In addition, in the quantum mechanical representation of the measuring device we have relied on the known[xx] relation between geometrical optics and wave optics. However, it follows from this that the classical-theoretical causal connections that are employed in the measuring devices can be reproduced in quantum mechanics only to the degree of precision to which the anschaulich classical variables of the measuring devices can be represented in the wave picture. However, in all practical cases the fundamental indeterminacy brought about in this way by the indeterminacy relations in the formulation of those causal connections is much smaller than the practical uncertainty that must be allowed for in every measuring device, even the best. Devices in which the said fundamental uncertainty plays a larger role would not be capable of being regarded as measuring devices in the ordinary sense, and of being treated according to classical theory.[4] The claim made earlier, that it is indifferent at which location the cut between the[xxii] parts of the system to be treated quantum mechanically and the classical measuring devices should be drawn, should thus be made more precise in the sense that this cut may indeed be shifted arbitrarily far in the direction of the observer in the region that is otherwise described according to the laws of classical physics; but that this cut cannot be shifted arbitrarily in the direction of the atomic system. Rather, there are physical systems – and all atomic systems belong among these – that the classical concepts are unsuitable to describe, and whose behaviour can therefore be expressed correctly only in the language of wavefunctions.

Wave mechanics is capable of reproducing classical causal connections in the sense just discussed only because its formalism – exactly like that of classical mechanics – is completely self-consistent [in sich eindeutig zusammenhängt]; in its formalism, the temporal evolution of the wavefunction can be inferred univocally from the initial values of this function with the help of the Schrödinger equation. The statistical element enters quantum mechanical statements only through the 'cut', where the partly uncontrollable[xxiii] disturbance necessarily tied to every observation can be held responsible for the appearance of statistical connections.

§2

After these preliminaries we return to the question posed at the beginning, whether the statistical predictions of quantum mechanics are capable of a deterministic completion. We shall thus investigate the assumption that the physical systems about which quantum mechanics makes statistical statements are bearers of hitherto unknown physical properties that determine univocally the behaviour hitherto known only statistically. For example, let the nucleus of the radium atom possess, apart from the properties that are fixed by the knowledge of its stationary state, still other hitherto unknown properties whose knowledge makes possible an exact prediction of when the atomic nucleus will emit an α-particle. And we assume specifically: this statement should hold independently of the means of observation that are used to detect the α-particle.[xxiv]

[4] [Footnote in the typescript, marked 'Anm.' in the manuscript, but part of the main text in the transcription (*eds.*):] At this point, attention must be drawn also to a third category of measuring devices, which employ univocal causal connections[xxi] that, however, cannot be followed in detail within classical theory. As an example, one can take the devices with which one detects the presence of neutrons in a specific position by generating artificial radioactivity. The physical processes that take place between the capture of the neutron and the creation and emission of the electron cannot be described in the framework of classical theory. Nevertheless, we have here a measuring device that allows for univocal inferences and can for instance play the role of system *B* in the proof we have discussed.

One might at first be tempted to drop this last special assumption, and consider that the radioactive emission may also depend on the properties of the means of observation. But this gain in generality is only apparent. For instance, one could reckon as part of the system to be observed the counter device and the photographic plate that registers its counts.[xxv] Then the question of the blackening of the plate replaces the question of the emission of the α-particle. And here, one is surely forced to assume that whether or not the blackening of the plate occurs is entirely independent of how the observer later takes a look at the plate. For our description of nature is based in the end on the assumption that one can speak of objective events in space and time. It would be altogether impossible to speak of the correctness of the predictions of any theory if one did not assume that the occurrence of a particular event is an objective fact that does not depend on our observation of this event. Classical physics defines the domain[xxvi] within which we can objectify our perceptions unproblematically. – If one considers the blackening of a photographic plate to be an objective fact in this way, then there is actually no further reason not to consider also the emission of an α-particle to be an objective fact; except, that is, if at some place between the macroscopic and the microscopic events one assumes there is a discontinuity for which not the slightest indications are present either in experience or in the quantum mechanical formalism.

We consequently need to discuss the special assumption: a physical system, for which quantum mechanics allows one to predict only[xxvii] the probability for the occurrence of a specific event, possesses still other hitherto unknown properties, knowledge of which would make precise statements possible about the occurrence of this event – *independently* of how the occurrence of the event is observed. It shall now be shown that this assumption runs into contradiction with the statements of quantum mechanics. And in fact not just with the statistical statements of quantum mechanics; this would not be surprising, for the observation of the hitherto[xxviii] unknown properties could make it possible to select specific systems from a quantum mechanical ensemble, and one cannot expect that the same statistics hold for individual systems selected according to specific points of view as for the ensemble. Rather, the said assumption also comes into contradiction with *specific* statements of quantum mechanics.

The reason for this contradiction lies in the already discussed fact that the formalism of quantum mechanics itself establishes a univocal connection between the quantities that it relates to each other, and that the quantum mechanical statements acquire their statistical character only because at the location separating the observer with his devices from the system to be observed, a fundamentally[xxix] uncontrollable disturbance of the system by the means of observation prevents us from following the causal connections.[5] Let us thus place[xxx] the cut in the above discussed scenario between, say, the atomic system A and the device B. Then we would have to ascribe to system A hitherto unknown properties that complement with specific statements the statistical statements that are to be made about A according to quantum mechanics. But now there is no reason within the framework of quantum mechanics for not also placing the cut between systems B and C rather than between A and B. Then the systems A and B are univocally connected in the formalism of quantum mechanics, and every statement about A not already contained in this univocal connection can run into contradiction with what this connection implies.

In detail, the contradiction arises in the following way: on the one hand, it is meant to follow from the knowledge of the hitherto hidden physical properties of system A, that A in its interaction with[xxxi] any physical system, thus in particular with B, reacts as if the coordinate q_A had the value q'_A. But on the other hand B can be so chosen that according to quantum mechanics its reaction to A proceeds differently. For instance (as opposed to the above discussed example) B can measure a property of A complementary to q_A. Then the possibility of a reaction of system B, as determined by quantum mechanics, rests precisely on the freedom that still obtains in the value of q_A; thus B's reaction to A is different to what would follow from the coordinate q_A having the value q'_A. More generally:[xxxii] the quantum mechanical formalism that is applied to the total system $A + B$ contains no freedom whatsoever, and leaves no place for additional assumptions regarding the effect of A on B; just as little as the mathematical framework of classical mechanics would. There would be room for complementing the quantum mechanical statements only at the location of the cut; but this location cannot be determined physically, indeed, it is precisely the arbitrariness in the choice of the location of the cut that is crucial for the application of quantum mechanics. Any hitherto unknown physical properties, which would then be necessarily tied to their specific physical system, are thus fundamentally unsuitable for complementing the quantum mechanical statements.

One can easily illustrate this state of affairs with the above mentioned example of the radioactive atomic nucleus. The α-particles leave the atomic nucleus with an energy that is fixed up to quantities of the order $\hbar/T$, if one takes T to be the average decay time of the atomic nucleus. One can let this α-radiation impinge on a diffraction grating of very high resolution, and according to the laws of quantum mechanics, one can be certain that (if the decay time is not too short) the α-radiation is reflected by the grating[xxxiii] only in one of the extraordinarily sharply determined directions that according to the laws of interference correspond to the de Broglie

[5] [Starting at 'and that the quantum', this passage is emphasised in the margin in the typescript, possibly by Einstein (*eds.*).]

wavelength of the α-rays. Now, if there were hitherto unknown physical properties of the atomic nucleus allowing one to predict in which direction and at what moment the α-particle is emitted, then one could calculate what part of the diffraction grating is hit by the α-particle, and when it is hit. The direction of reflection of the α-particle could then (up to negligibly small corrections) only depend on the properties, at the time of the impact, of that part of the grating *that is hit*,[6] because *whatever* the interaction, the α-particle ought to be found at the predicted[xxxiv] position in question – that was the assumption made above –, thus it cannot interact with the distant parts of the grating. On the other hand, according to quantum mechanics,[7] it is precisely the distant parts of the grating that are crucial in determining the direction of reflection. Thus a contradiction arises. One recognises immediately that this contradiction does not arise only in thought experiments that are hard to verify, if one replaces the radioactive atomic nucleus in the above discussed example with a sodium atom emitting light belonging to the D-line. A definite prediction of the position at which the light quantum hits the diffraction grating would be in contradiction with the occurrence of a sharp diffraction phenomenon.[8]

The strength of this example,[xxxv] however, rests crucially on the assumption that the localisation of an α-particle or light quantum applies to every kind of interaction, that one can thus speak of the location of the α-particle with the same right as of the location of a macroscopic object. If one does not accept this, then one should apply the same argument to a system for which there can be no doubt about the applicability of classical anschaulich concepts, thus one should include a part of the macroscopic measuring devices[xxxvi] in the system.

With this argument one can also counter the objection that inferences to the statistical character of atomic physics are not conclusive, because the concepts there used of corpuscle and wave, light quantum and field are in fact unsuitable for describing the physical conditions, and therefore will have to be replaced later by other conceptions. This objection overlooks that in the end the application of the quantum mechanical formalism does not require at all the use of these concepts, thus also that nothing is gained by eliminating these concepts. For quantum mechanics can always be treated such that part of the measuring devices can be counted as part of the system to be observed, and represented by wavefunctions. Then, the results of the quantum mechanical calculation include only[xxxvii] anschaulich statements about such things as pointer positions, blackening of plates and the like, in

[6] [Starting at 'The direction of reflection', this passage is emphasised in the margin in the typescript and marked 'nein' ['no'] (*eds.*).]

[7] [Starting at 'position in question', this passage is emphasised in the margin in the typescript and marked – clearly in Einstein's hand – 'Sprechweise im Sinne gewöhnlicher Mechanik' ['Way of speaking as in the usual mechanics'] (*eds.*).]

[8] [Starting at 'to the D-line', this passage is emphasised in the margin in the typescript and marked 'dito.' ['ditto'] (*eds.*).]

the case of which one cannot doubt one is already using the correct concepts for describing the physical conditions. It is a crucial feature of quantum mechanics that its formalism allows us to link organically the physical domains that are in principle beyond the reach of our Anschauung with the macroscopic anschaulich domains, in such a way that the statements of the formalism can be unambiguously expressed through anschaulich concepts.

Of course, quantum mechanics expressly assumes, just as in the argument carried out here, that it is possible after all to objectify our perceptions[xxxviii] at some point, i.e. to speak of things and processes. Classical physics proves that this is possible in a wide domain, and the whole of natural science rests upon this possibility.

§3

The question posed at the beginning, whether a deterministic completion of quantum mechanics is possible, is thus to be answered in the negative. If one wants to express the reasons put forward for this in a simple formula, then according to G. Hermann,[9] one can say that a deterministic completion of quantum mechanics is impossible for the reason that quantum mechanics already allows the complete specification of the causes for the occurrence of a definite measurement result. From this formulation the question immediately arises, which feature of nature – obviously overlooked by classical physics – is actually responsible for the fact that the formalism of quantum mechanics, in itself completely self-consistent [in sich eindeutig zusammenhängend], is not sufficient for calculating in advance all measurement results, that therefore statistical connections appear at the location of the cut. One recognises this feature best from a simple example: in the classical theory one could say of a particle that it moves with a definite velocity at a definite position. If one wants to translate this statement (with the inherent unavoidable unsharpness) into quantum theory, then one will say that a wavepacket in configuration space is moving with a definite velocity at a definite position. Thereby, however, the state is not yet uniquely fixed. Indeed, to fix it uniquely one must supply still further data about the size and form of the wavepacket, *for which there is altogether no analogue in classical theory.* Quantum mechanics has thus revealed to us here a new property of nature that was unknown to classical physics.

Since knowledge of the quantum mechanical wavefunction can be gained from a suitable observation of the system and is enough to determine for later times which observations on the system will have results that can be predicted exactly,[xl] one can speak of an '*observational context*' specified[xli] by knowledge of the wavefunction.

[9] G. Hermann: *Die [natur]philosophischen Grundlagen der Quantenmechanik*, Berlin 1935 [Hermann 1935c].[xxxix]

The example just discussed shows that the same anschaulich process can belong to different observational contexts – as opposed to classical physics, in which there is only a single observational context. The experimental results accumulated in quantum mechanics have further shown that the observation of a system in general leads discontinuously from one observational context to another. The causal flow can be followed only within a given observational context; in the discontinuous transition from one observational context to another (in fact a 'complementary' one in the Bohrian sense), only statistical predictions are possible. The possibility of different complementary observational contexts, unknown to the classical theory, is thus responsible for the occurrence of statistical laws.

After one manages to see that a deterministic completion of quantum mechanics would lead to contradictions, one can pose the further question whether a later modification of quantum mechanics might not open up the possibility for such a deterministic completion. In attempting to answer this further question, one can naturally only call upon the experimental results. One can verify that the conclusions of §2 essentially make use only of the results of quantum mechanics that have been repeatedly confirmed experimentally; the most important example is given there by the experimentally established fact that the radiation emitted by an atom can give rise to interference phenomena, even though the total energy of the radiation is eventually absorbed at a definite point (as a 'light quantum'). Furthermore, one can point to the fact that the statistical[xlii] character of quantum mechanics is very tightly bound to the formal circumstance that its mathematical framework of wavefunctions operates in multi-dimensional configuration space, not in ordinary space, and that precisely this feature of quantum mechanics has been exactly confirmed through the correct reproduction of the more complicated atomic spectra. Yet of course it can never be decided when a sum of experimental results licenses conclusive inferences to a law.

It will have to suffice for the moment here to observe that neither the experimental results nor considerations of principle give one grounds[xliii] to believe that the future description of nature will admit being fitted into the narrow classical framework of an anschaulich and causal description of objective processes in space and time.

Notes on the Translation

i um/über
ii The transcription gives the page number as '115'.
iii müßte/muss
iv in *Zukunft zu lösenden*/in Zukunft zu lösenden
v The transcription omits the inverted commas.
vi Wellenfunktion/Wellenfunktionen

vii dieser/seiner
viii oben/eben
ix mein/sein
x von/aus
xi Missing ‘die’ in the transcription.
xii Zahlenfunktion/Wellenfunktion
xiii The transcription has ‘h’ in the equation.
xiv Die/Diese
xv verändert sich [. . .], so daß/verändert sich [. . .] so, dass
xvi Gesamtwellenfunktionen/Gesamtwellenfunktion
xvii etwa/eben
xviii System/Ergebnis
xix der Quantenmechanik den Formalismus/des quantenmechanischen Formalismus
xx der bestimmte/der bekannte
xxi Clause ‘in denen zwar eindeutige kausale Zusammenhänge benützt werden’ missing in the transcription.
xxii dem/den
xxiii teilweise nur kontrollierbare/teilweise unkontrollierbare
xxiv der α-Teilchen/des α-Teilchens
xxv ihre Anschläge/ihre Ausschläge
xxvi Beweis/Bereich
xxvii und/nur
xxviii beiden/bisher
xxix prinzipielle/prinzipiell
xxx Leben wir also/Legen wir also
xxxi Spurious ‘B’ in the transcription.
xxxii ‘Allgemeiner:’ missing in the transcription.
xxxiii von Gittern/vom Gitter
xxxiv vorhergegangenen/vorhergesagten
xxxv dieser Beispiele/dieses Beispiels
xxxvi des makroskopischen Meßapparates/der makroskopischen Messapparate
xxxvii auch/nur
xxxviii Wechselwirkungen/Wahrnehmungen
xxxix Title amended already in the transcription.
xl nicht/exakt
xli charakteristischen/charakterisierten
xlii natürliche/statistische
xliii Anlaß bilden/Anlass bieten

15

The Natural-Philosophical Foundations of Quantum Mechanics (Excerpt)*

GRETE HERMANN

The physical consequence of quantum mechanics, which has given rise to the unsettling of familiar natural-philosophical conceptions, in particular causal concepts, is that a sharp, insurmountable limit is set to the predictability of future natural processes. The idea of Laplace's demon – who completely knows the present state of nature, has complete insight into all natural laws, and because of this knowledge can foresee the future course of events in full detail – loses thereby every application to nature. And yet this idea was only the expression of the conviction that every natural event in all its features has been caused by preceding events, and thus must be predictable from these causes for one knowledgeable of the natural laws. Together with the faith in the unlimited possibility of such predictions, also the conviction in the pervasive causal connection of natural events thus comes to falter.

The experimental catalyst for considerations culminating in the claim that there are insurmountable limits to predictability was given by the so-called dualism experiments. According to them, the classical distinction between radiation processes that consist in the rapid movement of small material particles and those in which a wave propagates no longer applies.

In classical physics the α- and β-rays emitted by radioactive elements were considered matter rays, since they e.g. leave linear traces behind when passing through saturated water vapour, thereby showing the discrete character of the particles in motion. However, these same rays, when they seep through a grating or are reflected by it, lead to definite interference phenomena, and thus force the researcher to assume he is dealing with a wave process. In a similar way, light rays, which since

* [Originally published in *Naturwissenschaften* in the issue of 18 October 1935 (Hermann 1935b). This is an abridged and slightly adapted version of a much longer paper that had appeared in March of that year (Hermann 1935a). Both papers have recently been reprinted in the volume edited by K. Herrmann (2019). A previous translation of the abridged version was published by D. Lumma (1999). This translation is based on our own translation of the longer paper in Crull and Bacciagaluppi (2017a, chapter 15) (*eds*.).]

the discovery of interference phenomena were unequivocally interpreted as a wave motion, have displayed properties that allow one to infer to their corpuscular nature.

Quantum mechanics does justice to these experiments through the assumption that each atomic process must be representable *also* in the wave picture, each wave process *also* as corpuscular. Given that the two pictures are contradictory, it is impossible that one and the same process have all the traits of a propagating wave as well as all the features of the motion of a corpuscle. The compatibility of the two pictures is thus only possible through each one limiting the applicability of the other.

In the oft-cited uncertainty relations, Heisenberg has determined exactly the limitations that wave and particle picture, applied to the same physical process, impose on each other. The best-known of them forbids the simultaneous sharp determination of the position and the momentum of particles in applying the corpuscle picture: if Δq is the precision with which the position, say of an electron, is determined, and Δp is the precision with which its momentum is determined, then the relation $\Delta q \cdot \Delta p \geq h$ holds, where h is Planck's constant.

For classical physics, the measurability of different quantities is independent of one another [*sic*]. The physical state of a system therefore can be expressed through a mere enumeration of the values of all occurring physical quantities. In contrast, the quantum mechanical formalism employs novel symbols in the description of the state that express the mutual dependence of the determination of different quantities.

These symbols, the wavefunctions of physical systems, and the mathematical formalism that prescribes the correct rules for their combination, follow the classical theory closely because of Bohr's *correspondence principle*. The classical description is compatible with the quantum mechanical one insofar as its quantities remain undetermined to such a degree that the uncertainty relations are fulfilled.

On the other hand, this correspondence entails that the quantum mechanical formalism does not allow one to predict the result of a measurement with arbitrary precision. Instead it allows, depending on the wavefunction characterising the physical system before the measurement, only the derivation of more or less far-reaching probabilistic statements.

However, the insurmountability of the indicated limits is not guaranteed by the proof that this formalism itself provides the basis only for limited predictions. Whoever doubts this thereby need not attack the formalism itself. It may be that this formalism will prove itself also in the future as it has so far. What prevents us, however, from assuming that through an extension of physical knowledge it will be augmented with new formulas and rules which, together with the current formal approach, make precise predictions possible again? Everything here depends on answering this question.

It lies at hand to want to read off the impossibility of such an extension already from the uncertainty relations. If both position and momentum of a particle cannot fundamentally be measured with arbitrary exactness, how then shall one obtain definite statements about the future motion, which is determined precisely by the instantaneous position and momentum of the body?

But this argumentation is based on the notion that, irrespective of the uncertainty relations, the electron, as a corpuscle in the classical sense, has at each time an exact position and an exactly determinate momentum determining its future motion – up to external disturbances –, and that *this cause* of the forthcoming physical development is always hidden from observation. Thus the uncertainty relations are only subjectively interpreted, and appear to say nothing about the nature of physical systems.

This subjective interpretation is incompatible with the derivation of these relations from the duality of the wave and particle conception: the subsumption of every atomic process *also* under the features of the wave picture limits the application of the corpuscle picture in a manner that not all features of point masses in motion – in the classical sense – may also be properties of the electron in flight.

But if according to these considerations the electron does not simultaneously have an exact position and an exact momentum, then its exact position and its exact momentum cannot be determining its further motion. Dropping this assumption opens the door to the question of whether or not one might find other features upon which the course of the motion depends and from which it can be calculated. The formalism of quantum mechanics knows no such features. But from this does not follow the right to declare them impossible.

Discussions of numerous other attempts at proving the limits of predictability as in-principle insurmountable shall turn out to lead to similar considerations. Indeed, all these arguments manifest deep difficulties that prevent attempts at overcoming the existing limits of predictability. But they leave open the crucial question: for any quantum mechanically characterised state of a physical system there are in fact measurements whose results cannot be predicted on the basis of knowledge of this state. But in what way is the physically otherwise usual path – to search for new features, to refine through them the determination of the states of physical systems, and to find in them the reason for previously unpredictable measurement results – supposed to be blocked?

Anyone who denies altogether the possibility of such features comes into conflict with the principle of the openness of experience. There is no other criterion for having captured all the essential circumstances in a natural domain than the possibility of understanding the regularities of the processes occurring in this domain. But whether one has identified these natural-law connections comes from being able to derive from them predictions that are confirmed empirically.

Hence there can be only one sufficient reason for abandoning as fundamentally useless the further search for the causes of an observed process: *that one already knows these causes.*

Accordingly, quantum mechanics, with its claim to be forever limited in the predictability of measurement results, stands before the following dilemma: either it names itself the causes that completely determine these measurement results – then how shall it prevent the researcher from finding these causes in the individual case and predicting the measurement result from them? Or it does not name these causes – then how shall it, without arbitrarily anticipating the investigation of unknown areas of nature, exclude the possibility of future discoveries of these causes?

The quantum mechanical formalism contains a way out of this dilemma. The direction in which the solution to the difficulties is to be found is indicated by Bohr's correspondence principle. This principle permits and requires that, within the domain of application of the classical concepts allowed by quantum mechanics, every consequence that arises classically from the characterisation of the present circumstances, should also serve as a basis for the quantum mechanical posits.

Now this consideration in terms of correspondence gives, in certain cases of quantum mechanically unpredictable events, accurate information concerning those physical variables on which these processes depend in all their essential features. These are the cases in which such an event belongs to a measurement process and contributes to bringing the observed object into relation with the measuring instrument.

Let us take the case where the measuring instrument indicates the result of measurement by a pointer position; then the step from the reading of this pointer position to the quantum mechanical posit of the state of the observed physical system presupposes a theory of interaction between system and measuring instrument. This theory is based solely on classical concepts, and with their help shows that and to what degree the deflection of the pointer depends on the state of the measured object, and hence gives clues to its determination. The use of any electrical, any optical instrument, any balance, is thus based on an inference from the measuring instrument to the object of measurement. In this inference the reading of the measuring instrument is explained as the necessary effect that the system to be measured has imposed on the instrument in the process of measurement.

In the case of a measurement whose result was not quantum mechanically predictable, the same evidently holds also for the pointer position of the measuring instrument that registers the result of the measurement. However, for this unpredictable event, the interpretation of the measurement process itself gives the reasons whereby it has come to be. It would thus be pointless to wish to seek the cause of its occurrence in new physical features hitherto overlooked by research. *The theory of measurement already contains a sufficient basis for explanation.*

But it is evidently no different for the state of the measured system. For it is contingent in the course of a natural process whether it [the measured system] is treated as the object of measurement, or whether it interests the researcher only as a means for measuring other processes.

The possibility of finding new features that strictly determine the result of a measurement is therefore, indeed, excluded in quantum mechanics by the only reason that is cogent in the face of the openness of experience: *the features that determine the measurement result are already given in quantum mechanics itself.*

This seems strange. If quantum mechanics knows how to explain the measurement result completely *after* it has occurred, why does it not offer a handle on calculating it *before* the measurement from the subsequently ascertained basis for its explanation?

The solution to these difficulties lies again in the correspondence principle. The predictions that one obtains from the quantum mechanical characterisation of a system can never exceed those that are derivable from the classical conceptions that have only limited applicability. Even so, if *after* the reading of the measuring instrument one explains the unpredictable position of its needle through a theory of the measurement process, then this [explanation] traces the process of measurement back to states that were not, and could not, have been included in the preceding description of object and measurement instrument. The description of these systems is thus not univocal in quantum mechanics, but (so to speak) reveals only one aspect of the physical system, which the researcher can capture on the basis of the preceding observation. Relative to this observation the system has no sharp values with respect to certain physical quantities, and accordingly, also no features from which the result of a sharp measurement of these quantities can be read off. However, if one makes such a measurement – which disturbs the system and brings it into a different state – then one obtains for this new state precise quantum mechanical statements for the measured quantity, and moreover, causes for precisely this unpredictable value having had to result. Nevertheless, those causes were not utilisable for a prediction of this result, since they determine the system only relative to the observation that was made in the first place with the measurement itself. They could therefore not have been available beforehand to the physicist for purposes of prediction.

This relative character of the quantum mechanical mode of description shines out clearly in an instructive thought experiment.[1]

[1] V[on] Weizsäcker, Ortsbestimmung eines Elektrons durch ein Mikroskop. *Z. Physik* **70**, Nrs. 1 and 2 [Weizsäcker 1931].

Let the position of an electron be determined only by a plane; where it is within the plane is unknown. Then, according to the uncertainty relations, only the component of the electron's momentum in the plane can be given; in the direction orthogonal to it, the momentum remains indeterminate.

A position measurement on this electron is to be undertaken by illumination of the electron. The scattered light shall go through a microscope and then be captured on a photographic plate. In order to have simple conditions, we imagine the intensity of the light employed here to be so reduced, that the whole process involves only a single light quantum. According to the duality of the wave and particle picture, this light quantum is on the one hand to be considered as a corpuscle that collides with the electron according to the classical laws of elastic collision, on the other hand as a wave that, deflected by the electron, propagates in the microscope according to the classical laws of optics.

The conservation of momentum is valid for the collision between the light quantum and the electron: both are deflected in the collision; the changes in their momenta are opposite and equal.

In order to obtain a sharp image of the electron, we place the plate in the image plane of the microscope corresponding to the plane of the object, in which, according to the classical theory, all wave trains proceeding from a point in the object plane are combined again in a single point. We utilise thus the classical conception that from the place of the collision, there spreads in all directions a spherical wave that, insofar as it strikes the aperture of the microscope, penetrates into its lenses. Therefore, in this process the total aperture angle of the microscope is involved, and therefore – now again in the corpuscle picture – it is meaningless to pick out a definite direction in which the light quantum has been reflected by the electron, and has penetrated into the microscope. From this it follows that also the change in momentum that the electron has experienced through the collision cannot be determined exactly. One will thus have to characterise the state of the electron immediately after the collision by a wavefunction that determines a sharp position, but a less sharp momentum, as compared to the previous state.

One arrives at an entirely different description of the collision if one fixes the plate not in the image plane, but rather in the focal plane of the microscope. Also in this case the plate will display a sharp image; for the light quantum has only enough energy to excite a single atom of the plate. This point of the focal plane hit by the light quantum is characteristic for a definite direction from which the light has penetrated into the microscope. The conception of the wave picture that interprets the observed event in this case is accordingly that of a bundle of parallel rays, which through their refraction in the lenses are combined into a single point in the focal plane. The direction in which the light quantum has entered the microscope is therefore fixed, but the position in the object plane from which it had started after the collision with the electron remains indeterminate. If the momentum of the light quantum was known before the collision, then with the specification of the direction of the light quantum after the collision, its change in momentum is also fixed, and thus according to the conservation of momentum also that of the electron. So although in this case nothing different has happened with the electron than in the first one, we must now characterise its state after the collision by a wavefunction with unsharp position and relatively sharp momentum.[2]

The coexistence of these different possibilities evidently means that – as we can say – depending on the relevant observational context, one can obtain different wavefunctions for the same system and for the same instant – namely for the electron at the time immediately after the collision with the light quantum. The quantum-mechanical characterisation, unlike the classical one, does not pertain to the physical system still somehow 'in itself', and this means here: independently of which observation one uses to procure one's knowledge of it.

On the basis of these results, what revision is to be made to the causal principle of classical physics?

Two points from the preceding considerations are crucial for this: the limits of predictability of future events have turned out, indeed, to be insurmountable in principle, however there is no phenomenon for which one cannot find causes within the framework of the quantum mechanical formalism.

The two claims apparently contradict one another. While the first asserts that unavoidable limits are set on the application of causal inferences and the mastery of mankind over nature that they confer, the second emphasises the in-principle unlimited applicability of causal concepts to which fundamentally every natural process can be subordinated – indeed with regard to all physical features characterising it.

The resolution of this contradiction can only come about on the basis of a discussion of those concepts that play a crucial role in the quantum mechanical results mentioned: the concept of the predictability of natural events on the one hand and the causal relations on the other.

We have already touched upon the close link between both concepts. The explanatory value of a physical hypothesis can be verified only by the prediction of future natural events. And without the possibility of such a verification, the assertion of causal connections loses the character of knowledge of nature.

[2] [In the full paper (Hermann 1935a), Hermann includes here the following paragraph (*eds.*):] Finally, if one sets up no photographic plate at all, but allows the light quantum to pursue its path without detecting it, then one obtains yet a third – though not in the same way intuitive – description of the state after the collision. In this case, the physical system composed of the light quantum and the electron is assigned a wavefunction that describes a linear combination: each of its terms is the product of one wavefunction describing the electron and one describing the light quantum. Through this linear combination the light quantum and the electron are thus not described each by itself, but only in their relation to each other. Each state of the one is associated with one of the other.

This relation has often misled one to believe that in a strict sense one has identical concepts here, and that only the linguistic designation feigns a distinction. On this construal the contradiction between the two aforementioned quantum mechanical claims is unavoidable. If the relationship of cause and effect consists in nothing other than that the effect can be predicted if the cause is known, then there are no causes for in-principle unpredictable events. The fact that quantum mechanics presupposes and looks for a natural-law explanation even for unpredictable events, therefore shows that equating the two concepts is based on a mistake. The causal connection directly concerns only the necessary sequence of events itself. The possibility of predicting these on the basis of insight into the causal relationships provides the criterion for the correct application of causal concepts. Quantum mechanics necessitates careful distinction of the two concepts.

The law of causation, formulated independently of the criterion for its applicability, states that nothing in nature occurs that is not in all physically determinable features, caused by (and this means: follows necessarily from) previous processes. In this sense, gapless causality is not only compatible with quantum mechanics, but is even demonstrably presupposed by it.

But how do things stand with the criterion of causality? Quantum mechanics too requires such a criterion and takes it, like classical physics, from the possibility of predicting future events. In contrast to classical physics, however, it has broken with the presupposition that each causal claim may be verified directly by the prediction of the effect. Even for unpredictable events, quantum mechanics gives a causal explanation and verifies this through predictions. But this verification comes about indirectly. One infers backwards from the unpredictable events to their cause, and from the assumption that this cause was present, to predictions of forthcoming events whose occurrence can be verified empirically. So in the discussed example, the darkening of the plate is traced back to the collision of the electron and the light quantum, from which one can then infer forward to the state of the electron, which is still accessible to observation.

That this new possibility of only indirect verification of causal claims was not appreciated by classical physics, is grounded in the fact that the relative trait of the quantum mechanical description of nature is foreign to classical physics. For it [classical physics], the characterisation of any system is unique and independent of the manner in which the observer procures knowledge of it. And hence it [classical physics] necessarily arrives at the assumption that with sufficiently sharp observation and sufficient knowledge of relations of natural law, the investigation of physical systems allows one to determine the causes of their further evolution with arbitrary distinctness and thereby to predict this further evolution.

Seen in this light, the difficulties into which the advocate of the law of causation is plunged by the discoveries of quantum mechanics do not arise from the causal principle itself. Rather they stem from the tacitly added assumption that physical knowledge accounts for natural events adequately and independently of observational context. This assumption finds its expression in the presupposition that any causal link between processes gives rise to a prediction of the effect from the cause – indeed, that the causal link is in fact identical to the possibility of this prediction.

Quantum mechanics requires us to resolve this merging of different natural-philosophical principles – to drop the assumption of the absolute character of the knowledge of nature and to handle the principle of causality independently of it. Consequently, it has in no way refuted the law of causality, but has clarified it and freed it from other principles that are not necessarily connected with it.

16

Can Quantum-Mechanical Description of Physical Reality be Considered Complete?*

NIELS BOHR

It is shown that a certain 'criterion of physical reality' formulated in a recent article with the above title by A. Einstein, B. Podolsky and N. Rosen contains an essential ambiguity when it is applied to quantum phenomena. In this connection a viewpoint termed 'complementarity' is explained from which quantum-mechanical description of physical phenomena would seem to fulfill, within its scope, all rational demands of completeness.

In a recent article[1] under the above title A. Einstein, B. Podolsky and N. Rosen have presented arguments which lead them to answer the question at issue in the negative. The trend of their argumentation, however, does not seem to me adequately to meet the actual situation with which we are faced in atomic physics. I shall therefore be glad to use this opportunity to explain in somewhat greater detail a general viewpoint, conveniently termed 'complementarity', which I have indicated on various previous occasions,[2] and from which quantum mechanics within its scope would appear as a completely rational description of physical phenomena, such as we meet in atomic processes.

The extent to which an unambiguous meaning can be attributed to such an expression as 'physical reality' cannot of course be deduced from *a priori* philosophical conceptions, but – as the authors of the article cited themselves emphasize – must be founded on a direct appeal to experiments and measurements. For this purpose they propose a 'criterion of reality' formulated as follows: 'If, without in any way disturbing a system, we can predict with certainty the value of a physical quantity, then there exists an element of physical reality corresponding to this

* [The text follows the original publication in *Physical Review* (Bohr 1935b), which was received 13 July 1935 and published 15 October. The Niels Bohr Archive contains two typescripts of the paper, one in German and one in English (AHQP-BSM 14, section 1). The English version corresponds to the published one, and the German one appears to be a translation of the English. See Chapter 5 for further details. Reprinted with permission from: N. Bohr, *Physical Review* **48**, 696—702. Copyright 1935 by the American Physical Society (*eds.*).]

[1] A. Einstein, B. Podolsky and N. Rosen, *Phys. Rev.* **47**, 777 (1935) [Einstein, Podolsky and Rosen 1935; see Chapter 9].

[2] Cf. N. Bohr, *Atomic Theory and* [*the*] *Description of Nature*, I (Cambridge, 1934) [Bohr 1934].

physical quantity'. By means of an interesting example, to which we shall return below, they next proceed to show that in quantum mechanics, just as in classical mechanics, it is possible under suitable conditions to predict the value of any given variable pertaining to the description of a mechanical system from measurements performed entirely on other systems which previously have been in interaction with the system under investigation. According to their criterion the authors therefore want to ascribe an element of reality to each of the quantities represented by such variables. Since, moreover, it is a well-known feature of the present formalism of quantum mechanics that it is never possible, in the description of the state of a mechanical system, to attach definite values to both of two canonically conjugate variables, they consequently deem this formalism to be incomplete, and express the belief that a more satisfactory theory can be developed.

Such an argumentation, however, would hardly seem suited to affect the soundness of quantum-mechanical description, which is based on a coherent mathematical formalism covering automatically any procedure of measurement like that indicated.[3] The apparent contradiction in fact discloses only an essential inadequacy of the customary viewpoint of natural philosophy for a rational account of physical phenomena of the type with which we are concerned in quantum mechanics. Indeed the *finite interaction between object and measuring agencies* conditioned by the very existence of the quantum of action entails – because of the impossibility of controlling the reaction of the object on the measuring instruments

[3] The deductions contained in the article cited may in this respect be considered as an immediate consequence of the transformation theorems of quantum mechanics, which perhaps more than any other feature of the formalism contribute to secure its mathematical completeness and its rational correspondence with classical mechanics. In fact, it is always possible in the description of a mechanical system, consisting of two partial systems (1) and (2), interacting or not, to replace any two pairs of canonically conjugate variables (q_1p_1), (q_2p_2) pertaining to systems (1) and (2), respectively, and satisfying the usual commutation rules

$$[q_1p_1] = [q_2p_2] = ih/2\pi,$$
$$[q_1q_2] = [p_1p_2] = [q_1p_2] = [q_2p_1] = 0,$$

by two pairs of new conjugate variables (Q_1P_1), (Q_2P_2) related to the first variables by a simple orthogonal transformation, corresponding to a rotation of angle θ in the planes (q_1q_2), (p_1p_2)

$$q_1 = Q_1 \cos\theta - Q_2 \sin\theta \qquad p_1 = P_1 \cos\theta - P_2 \sin\theta$$
$$q_2 = Q_1 \sin\theta + Q_2 \cos\theta \qquad p_2 = P_1 \sin\theta + P_2 \cos\theta.$$

Since these variables will satisfy analogous commutation rules, in particular

$$[Q_1P_1] = ih/2\pi, \qquad [Q_1P_2] = 0,$$

it follows that in the description of the state of the combined system definite numerical values may not be assigned to both Q_1 and P_1, but that we may clearly assign such values to both Q_1 and P_2. In that case it further results from the expressions of these variables in terms of (q_1p_1) and (q_2p_2), namely

$$Q_1 = q_1 \cos\theta + q_2 \sin\theta, \qquad P_2 = -p_1 \sin\theta + p_2 \cos\theta,$$

that a subsequent measurement of either q_2 or p_2 will allow us to predict the value of q_1 or p_1 respectively.

if these are to serve their purpose – the necessity of a final renunciation of the classical ideal of causality and a radical revision of our attitude towards the problem of physical reality. In fact, as we shall see, a criterion of reality like that proposed by the named authors contains – however cautious its formulation may appear – an essential ambiguity when it is applied to the actual problems with which we are here concerned. In order to make the argument to this end as clear as possible, I shall first consider in some detail a few simple examples of measuring arrangements.

Let us begin with the simple case of a particle passing through a slit in a diaphragm, which may form part of some more or less complicated experimental arrangement. Even if the momentum of this particle is completely known before it impinges on the diaphragm, the diffraction by the slit of the plane wave giving the symbolic representation of its state will imply an uncertainty in the momentum of the particle, after it has passed the diaphragm, which is the greater the narrower the slit. Now the width of the slit, at any rate if it is still large compared with the wave-length, may be taken as the uncertainty Δq of the position of the particle relative to the diaphragm, in a direction perpendicular to the slit. Moreover, it is simply seen from de Broglie's relation between momentum and wave-length that the uncertainty Δp of the momentum of the particle in this direction is correlated to Δq by means of Heisenberg's general principle

$$\Delta p \Delta q \sim h,$$

which in the quantum-mechanical formalism is a direct consequence of the commutation relation for any pair of conjugate variables. Obviously the uncertainty Δp is inseparably connected with the possibility of an exchange of momentum between the particle and the diaphragm; and the question of principal interest for our discussion is now to what extent the momentum thus exchanged can be taken into account in the description of the phenomenon to be studied by the experimental arrangement concerned, of which the passing of the particle through the slit may be considered as the initial stage.

Let us first assume that, corresponding to usual experiments on the remarkable phenomena of electron diffraction, the diaphragm, like the other parts of the apparatus, – say a second diaphragm with several slits parallel to the first and a photographic plate, – is rigidly fixed to a support which defines the space frame of reference. Then the momentum exchanged between the particle and the diaphragm will, together with the reaction of the particle on the other bodies, pass into this common support, and we have thus voluntarily cut ourselves off from any possibility of taking these reactions separately into account in predictions regarding the final result of the experiment, – say the position of the spot produced by the particle on the photographic plate. The impossibility of a closer analysis of the reactions between the particle and the measuring instrument is indeed no peculiarity of the experimental procedure described, but is rather an essential property of any arrangement suited to the study of the phenomena of the type concerned, where we have to do with a feature of *individuality* completely foreign to classical physics. In fact, any possibility of taking into account the momentum exchanged between the particle and the separate parts of the apparatus would at once permit us to draw conclusions regarding the 'course' of such phenomena, – say through what particular slit of the second diaphragm the particle passes on its way to the photographic plate – which would be quite incompatible with the fact that the probability of the particle reaching a given element of area on this plate is determined not by the presence of any particular slit, but by the positions of all the slits of the second diaphragm within reach of the associated wave diffracted from the slit of the first diaphragm.

By another experimental arrangement, where the first diaphragm is not rigidly connected with the other parts of the apparatus, it would at least in principle[4] be possible to measure its momentum with any desired accuracy before and after the passage of the particle, and thus to predict the momentum of the latter after it has passed through the slit. In fact, such measurements of momentum require only an unambiguous application of the classical law of conservation of momentum, applied for instance to a collision process between the diaphragm and some test body, the momentum of which is suitably controlled before and after the collision. It is true that such a control will essentially depend on an examination of the space-time course of some process to which the ideas of classical mechanics can be applied; if, however, all spatial dimensions and time intervals are taken sufficiently large, this involves clearly no limitation as regards the accurate control of the momentum of the test bodies, but only a renunciation as regards the accuracy of the control of their space-time coordination. This last circumstance is in fact quite analogous to the renunciation of the control of the momentum of the fixed diaphragm in the experimental arrangement discussed above, and depends in the last resort on the claim of a purely classical account of the measuring apparatus, which implies the necessity of allowing a latitude corresponding to the quantum-mechanical uncertainty relations in our description of their behavior.

The principal difference between the two experimental arrangements under consideration is, however, that in the arrangement suited for the control of the momentum of the first diaphragm, this body can no longer be used as a measuring instrument for the same purpose as in the previous case, but must, as regards

[4] The obvious impossibility of actually carrying out, with the experimental technique at our disposal, such measuring procedures as are discussed here and in the following does clearly not affect the theoretical argument, since the procedures in question are essentially equivalent with atomic processes, like the Compton effect, where a corresponding application of the conservation theorem of momentum is well established.

its position relative to the rest of the apparatus, be treated, like the particle traversing the slit, as an object of investigation, in the sense that the quantum-mechanical uncertainty relations regarding its position and momentum must be taken explicitly into account. In fact, even if we knew the position of the diaphragm relative to the space frame before the first measurement of its momentum, and even though its position after the last measurement can be accurately fixed, we lose, on account of the uncontrollable displacement of the diaphragm during each collision process with the test bodies, the knowledge of its position when the particle passed through the slit. The whole arrangement is therefore obviously unsuited to study the same kind of phenomena as in the previous case. In particular it may be shown that, if the momentum of the diaphragm is measured with an accuracy sufficient for allowing definite conclusions regarding the passage of the particle through some selected slit of the second diaphragm, then even the minimum uncertainty of the position of the first diaphragm compatible with such a knowledge will imply the total wiping out of any interference effect – regarding the zones of permitted impact of the particle on the photographic plate – to which the presence of more than one slit in the second diaphragm would give rise in case the positions of all apparatus are fixed relative to each other.

In an arrangement suited for measurements of the momentum of the first diaphragm, it is further clear that even if we have measured this momentum before the passage of the particle through the slit, we are after this passage still left with a *free choice* whether we wish to know the momentum of the particle or its initial position relative to the rest of the apparatus. In the first eventuality we need only to make a second determination of the momentum of the diaphragm, leaving unknown forever its exact position when the particle passed. In the second eventuality we need only to determine its position relative to the space frame with the inevitable loss of the knowledge of the momentum exchanged between the diaphragm and the particle. If the diaphragm is sufficiently massive in comparison with the particle, we may even arrange the procedure of measurements in such a way that the diaphragm after the first determination of its momentum will remain at rest in some unknown position relative to the other parts of the apparatus, and the subsequent fixation of this position may therefore simply consist in establishing a rigid connection between the diaphragm and the common support.

My main purpose in repeating these simple, and in substance well-known considerations, is to emphasize that in the phenomena concerned we are not dealing with an incomplete description characterized by the arbitrary picking out of different elements of physical reality at the cost of sacrifying [*sic*] other such elements, but with a rational discrimination between essentially different experimental arrangements and procedures which are suited either for an unambiguous use of the idea of space location, or for a legitimate application of the conservation theorem of momentum.

Any remaining appearance of arbitrariness concerns merely our freedom of handling the measuring instruments, characteristic of the very idea of experiment. In fact, the renunciation in each experimental arrangement of the one or the other of two aspects of the description of physical phenomena, – the combination of which characterizes the method of classical physics, and which therefore in this sense may be considered as *complementary* to one another, – depends essentially on the impossibility, in the field of quantum theory, of accurately controlling the reaction of the object on the measuring instruments, i.e., the transfer of momentum in case of position measurements, and the displacement in case of momentum measurements. Just in this last respect any comparison between quantum mechanics and ordinary statistical mechanics, – however useful it may be for the formal presentation of the theory, – is essentially irrelevant. Indeed we have in each experimental arrangement suited for the study of proper quantum phenomena not merely to do with an ignorance of the value of certain physical quantities, but with the impossibility of defining these quantities in an unambiguous way.

The last remarks apply equally well to the special problem treated by Einstein, Podolsky and Rosen, which has been referred to above, and which does not actually involve any greater intricacies than the simple examples discussed above. The particular quantum-mechanical state of two free particles, for which they give an explicit mathematical expression, may be reproduced, at least in principle, by a simple experimental arrangement, comprising a rigid diaphragm with two parallel slits, which are very narrow compared with their separation, and through each of which one particle with given initial momentum passes independently of the other. If the momentum of this diaphragm is measured accurately before as well as after the passing of the particles, we shall in fact know the sum of the components perpendicular to the slits of the momenta of the two escaping particles, as well as the difference of their initial positional coordinates in the same direction; while of course the conjugate quantities, i.e., the difference of the components of their momenta, and the sum of their positional coordinates, are entirely unknown.[5] In this arrangement, it is therefore clear that a subsequent single measurement either of the position or of the momentum of one of the particles will automatically determine the position or momentum, respectively, of the other particle with any desired accuracy; at least if the wave-length corresponding to the free motion of each particle is sufficiently short compared with the width of the slits. As pointed out by the named authors, we are therefore faced at this stage with a completely free choice

[5] As will be seen, this description, apart from a trivial normalizing factor, corresponds exactly to the transformation of variables described in the preceding footnote if $(q_1 p_1)$, $(q_2 p_2)$ represent the positional coordinates and components of momenta of the two particles and if $\theta = -\pi/4$. It may also be remarked that the wave function given by formula (9) of the article cited corresponds to the special choice of $P_2 = 0$ and the limiting case of two infinitely narrow slits.

whether we want to determine the one or the other of the latter quantities by a process which does not directly interfere with the particle concerned.

Like the above simple case of the choice between the experimental procedures suited for the prediction of the position or the momentum of a single particle which has passed through a slit in a diaphragm, we are, in the 'freedom of choice' offered by the last arrangement, just concerned with a *discrimination between different experimental procedures which allow of the unambiguous use of complementary classical concepts*. In fact to measure the position of one of the particles can mean nothing else than to establish a correlation between its behavior and some instrument rigidly fixed to the support which defines the space frame of reference. Under the experimental conditions described such a measurement will therefore also provide us with the knowledge of the location, otherwise completely unknown, of the diaphragm with respect to this space frame when the particles passed through the slits. Indeed, only in this way we obtain a basis for conclusions about the initial position of the other particle relative to the rest of the apparatus. By allowing an essentially uncontrollable momentum to pass from the first particle into the mentioned support, however, we have by this procedure cut ourselves off from any future possibility of applying the law of conservation of momentum to the system consisting of the diaphragm and the two particles and therefore have lost our only basis for an unambiguous application of the idea of momentum in predictions regarding the behavior of the second particle. Conversely, if we choose to measure the momentum of one of the particles, we lose through the uncontrollable displacement inevitable in such a measurement any possibility of deducing from the behavior of this particle the position of the diaphragm relative to the rest of the apparatus, and have thus no basis whatever for predictions regarding the location of the other particle.

From our point of view we now see that the wording of the above-mentioned criterion of physical reality proposed by Einstein, Podolsky and Rosen contains an ambiguity as regards the meaning of the expression 'without in any way disturbing a system'. Of course there is in a case like that just considered no question of a mechanical disturbance of the system under investigation during the last critical stage of the measuring procedure. But even at this stage there is essentially the question of *an influence on the very conditions which define the possible types of predictions regarding the future behavior of the system.* Since these conditions constitute an inherent element of the description of any phenomenon to which the term 'physical reality' can be properly attached, we see that the argumentation of the mentioned authors does not justify their conclusion that quantum-mechanical description is essentially incomplete. On the contrary this description, as appears from the preceding discussion, may be characterized as a rational utilization of all possibilities of unambiguous interpretation of measurements, compatible with the finite and uncontrollable interaction between the objects and the measuring instruments in the field of quantum theory. In fact, it is only the mutual exclusion of any two experimental procedures, permitting the unambiguous definition of complementary physical quantities, which provides room for new physical laws, the coexistence of which might at first sight appear irreconcilable with the basic principles of science. It is just this entirely new situation as regards the description of physical phenomena, that the notion of *complementarity* aims at characterizing.

The experimental arrangements hitherto discussed present a special simplicity on account of the secondary role which the idea of time plays in the description of the phenomena in question. It is true that we have freely made use of such words as 'before' and 'after' implying time-relationships; but in each case allowance must be made for a certain inaccuracy, which is of no importance, however, so long as the time intervals concerned are sufficiently large compared with the proper periods entering in the closer analysis of the phenomenon under investigation. As soon as we attempt a more accurate time description of quantum phenomena, we meet with well-known new paradoxes, for the elucidation of which further features of the interaction between the objects and the measuring instruments must be taken into account. In fact, in such phenomena we have no longer to do with experimental arrangements consisting of apparatus essentially at rest relative to one another, but with arrangements containing moving parts, – like shutters before the slits of the diaphragms, – controlled by mechanisms serving as clocks. Besides the transfer of momentum, discussed above, between the object and the bodies defining the space frame, we shall therefore, in such arrangements, have to consider an eventual exchange of energy between the object and these clock-like mechanisms.

The decisive point as regards time measurements in quantum theory is now completely analogous to the argument concerning measurements of positions outlined above. Just as the transfer of momentum to the separate parts of the apparatus, – the knowledge of the relative positions of which is required for the description of the phenomenon, – has been seen to be entirely uncontrollable, so the exchange of energy between the object and the various bodies, whose relative motion must be known for the intended use of the apparatus, will defy any closer analysis. Indeed, it is *excluded in principle to control the energy which goes into the clocks without interfering essentially with their use as time indicators*. This use in fact entirely relies on the assumed possibility of accounting for the functioning of each clock as well as for its eventual comparison with other clocks on the basis of the methods of classical physics. In this account we must therefore obviously allow for a latitude in the energy balance, corresponding to the quantum-mechanical uncertainty relation for the conjugate time and energy variables. Just as in the question discussed above of the mutually exclusive character of any unambiguous use in quantum theory of the concepts of position and momentum, it is in the last resort this circumstance which entails the complementary relationship between any detailed time account of atomic phenomena on the one hand and the unclassical features of intrinsic stability of atoms, disclosed by the study of energy transfers in atomic reactions on the other hand.

This necessity of discriminating in each experimental arrangement between those parts of the physical system considered which are to be treated as measuring instruments and those which constitute the objects under investigation may indeed be said to form a *principal distinction between classical and quantum-mechanical description of physical phenomena*. It is true that the place within each measuring procedure where this discrimination is made is in both cases largely a matter of convenience. While, however, in classical physics the distinction between object and measuring agencies does not entail any difference in the character of the description of the phenomena concerned, its fundamental importance in quantum theory, as we have seen, has its root in the indispensable use of classical concepts in the interpretation of all proper measurements, even though the classical theories do not suffice in accounting for the new types of regularities with which we are concerned in atomic physics. In accordance with this situation there can be no question of any unambiguous interpretation of the symbols of quantum mechanics other than that embodied in the well-known rules which allow to predict the results to be obtained by a given experimental arrangement described in a totally classical way, and which have found their general expression through the transformation theorems, already referred to. By securing its proper correspondence with the classical theory, these theorems exclude in particular any imaginable inconsistency in the quantum-mechanical description, connected with a change of the place where the discrimination is made between object and measuring agencies. In fact it is an obvious consequence of the above argumentation that in each experimental arrangement and measuring procedure we have only a free choice of this place within a region where the quantum-mechanical description of the process concerned is effectively equivalent with the classical description.

Before concluding I should still like to emphasize the bearing of the great lesson derived from general relativity theory upon the question of physical reality in the field of quantum theory. In fact, notwithstanding all characteristic differences, the situations we are concerned with in these generalizations of classical theory present striking analogies which have often been noted. Especially, the singular position of measuring instruments in the account of quantum phenomena, just discussed, appears closely analogous to the well-known necessity in relativity theory of upholding an ordinary description of all measuring processes, including a sharp distinction between space and time coordinates, although the very essence of this theory is the establishment of new physical laws, in the comprehension of which we must renounce the customary separation of space and time ideas.[6] The dependence on the reference system, in relativity theory, of all readings of scales and clocks may even be compared with the essentially uncontrollable exchange of momentum or energy between the objects of measurements and all instruments defining the space-time system of reference, which in quantum theory confronts us with the situation characterized by the notion of complementarity. In fact this new feature of natural philosophy means a radical revision of our attitude as regards physical reality, which may be paralleled with the fundamental modification of all ideas regarding the absolute character of physical phenomena, brought about by the general theory of relativity.

[6] Just this circumstance, together with the relativistic invariance of the uncertainty relations of quantum mechanics, ensures the compatibility between the argumentation outlined in the present article and all exigencies of relativity theory. This question will be treated in greater detail in a paper under preparation, where the writer will in particular discuss a very interesting paradox suggested by Einstein concerning the application of gravitation theory to energy measurements, and the solution of which offers an especially instructive illustration of the generality of the argument of complementarity. On the same occasion a more thorough discussion of space-time measurements in quantum theory will be given with all necessary mathematical developments and diagrams of experimental arrangements, which had to be left out of this article, where the main stress is laid on the dialectic aspect of the question at issue. [No such draft or paper appears to be extant. See also Section 5.2, esp. note 9 (*eds.*).]

Part IV

Other Reactions to EPR

17

The Correlation of Wave Functions with the States of Physical Systems*

EDWIN C. KEMBLE

In their recent article, *Can Quantum-Mechanical Description of Physical Reality be Considered Complete?*[1] Einstein, Podolsky and Rosen arrive at the conclusion that the question put by the title of their paper must be answered in the negative. In the writer's opinion their argument is not sound.

The essential feature of their reasoning is the purported demonstration by means of an example that '*it is possible to assign two different wave functions ... to the same reality*'. If this were true, it would mean that quantum-mechanical description is erroneous as well as incomplete, for each different wave function involves a different prediction regarding the future behavior of the system described and the authors of the above paper clearly intend the phrase 'the same reality' to refer to the same system in the same physical state. Actually, however, the demonstration cited is incorrect.

The special problem discussed by Einstein, Podolsky and Rosen is that of a compound system whose two component systems α and β are assumed to interact during a limited time interval $0 < t < T$. (We may suppose that the interaction is a collision between an atom and a free electron which takes place during this time interval.) After the interaction it is possible to measure any physical property of the system α without disturbing β. According to the usual quantum-mechanical procedure the wave function to be assigned to the system β after the observation of α depends on the type of observation made and on its result. But since β has not been disturbed by the observation of α, Einstein, Podolsky and Rosen argue that it cannot be affected by that observation and must in all cases constitute 'the same physical reality'. In other words, they assume it to be in the same 'state' in all

* [The text follows the original publication in the 'Letters to the Editor' section of *Physical Review* (Kemble 1935), which was received 25 May 1935 and published 15 June. Reprinted with permission from: E. C. Kemble, *Physical Review* **47**, 973–974. Copyright 1935 by the American Physical Society (*eds.*).]

[1] A. Einstein, B. Podolsky and N. Rosen, *Phys. Rev.* **47**, 777 (1935) [Einstein, Podolsky and Rosen 1935; see Chapter 9].

cases. Here lies a fallacy, however, for whenever two systems interact for a short time there is a correlation between the subsequent behavior of one system and that of the other. Thus in the case of a pair of systems with a definite total energy or definite total linear momentum we can infer from the measured value of the energy or momentum of one system after a collision what the value of the corresponding quantity for the other system must be. In the Compton effect each photon scattered at a definite angle from a unidirectional monochromatic primary beam is correlated with a recoil electron driven off at a definite corresponding angle. The process of reducing the wave packet[2] is the orthodox procedure for taking this correlation into account and, *of course*, it yields a different wave function for the recoil electrons correlated with photons scattered at one angle from that for recoil electrons whose photons have been scattered at another angle. If our mathematics did not work out that way we should have to revise it.

It is quite true, as pointed out by E., P. and R., that the act of observing the system α can hardly be supposed to change the state of a system β which may be widely separated from it in space, but it can and does reveal something about the state of β which could not have been deduced from the initial wave function of the combination system without such an observation. Thus each possible wave function for system β gives a correct description of the corresponding state of that system.

These remarks would hardly clear up the whole question if no specific mention were made of the interpretation of quantum mechanics as a statistical mechanics of assemblages of like systems. This point of view most clearly formulated by Slater[3] seems the only way of solving the paradoxes of the average presentation of elementary quantum theory. It asserts that the wave functions of the Schrödinger theory have meaning primarily as descriptions of the behavior of (infinite) assemblages of identical systems similarly prepared. When we say that the 'state' of an electron in motion is described by $\psi(x, t)$ we mean that an assemblage of a very large number of similarly prepared electrons would have statistical properties described by this function, and that we cannot know more about an individual electron than the fact that it belongs to a suitably chosen potential assemblage of this character. Every analysis of a wave function into a linear combination of orthogonal functions can be interpreted as a resolution of the complete assemblage into a collection of sub-assemblages. The expansion set up by E., P. and R. of the complete wave function $\Psi(x_\alpha, x_\beta)$ for the combination system into eigenfunctions of a dynamical variable A of the system α shows that in each sub-assemblage we have a correlation between a particular pure state of the variable A and a corresponding state of the system β.

[2] Reference 1, p. 779 [see p. 160].

[3] J. C. Slater, *J. Frank. Inst.* **207**, 449 (1929) [Slater 1929].

The act of measuring the variable A is the physical counterpart of this mathematical expansion and when successful[4] shows which sub-assemblage a particular pair of systems α, β belongs to. The 'reduction of the wave packet' is the process of substituting the wave function of the sub-assemblage for that of the original complete assemblage which must accompany the transfer of one's attention from an arbitrary member or members of the original assemblage to systems known to belong to said sub-assemblage.

There seems no reason to doubt the completeness of the quantum-mechanical description of atomic systems within the frame of our present experimental knowledge.

EDWIN C. KEMBLE

The Physical Laboratories,
Harvard University,
May 25, 1935.

[4] In most measurements of the properties of atomic systems we do not get a definite result for every system in the group under investigation.

18

States and Reality of Physical Systems*

BORIS PODOLSKY

In a letter to the editor[1] Kemble asserts that the argument of Einstein, Podolsky, and Rosen[2] is in his opinion unsound. Not being in a position to discuss Kemble's objections with Einstein and Rosen, I take the liberty of expressing my own reaction to them.

The closer I examine Kemble's letter the more I am convinced that he does not prove his contention. His argument comes to a climax in the third paragraph, where he says: 'since β has not been disturbed by the observation of α, Einstein, Podolsky, and Rosen argue that it cannot be affected by that observation and must in all cases constitute "the same physical reality". In other words, they assume it to be in the same "state" in all cases. Here lies a fallacy, however, for whenever two systems interact for a short time there is a correlation between the subsequent behavior of one system and that of the other'. Thus, it seems, our 'fallacy' consisted in assuming that if a system is not *disturbed* it is also not *affected*, and thus remains in the same physical state.

In the next paragraph, however, Kemble seems to assert that even this is not a fallacy, for he says: 'It is quite true, as pointed out by E., P. and R., that the act of observing the system α *can hardly be supposed to change the state of a system* β which may be widely separated from it in space, but it can and does reveal something about the state of β which could not have been deduced ... without such an observation' (emphasis mine, B. P.). Kemble, therefore, agrees that, in spite of something being revealed by the observation, we are still dealing with the same *state* of the system, and therefore the same *reality*. I fail to see that after this anything remains of the fallacy asserted earlier.

It is probably because of this anticlimax that Kemble adds a paragraph beginning with 'These remarks would hardly clear up the whole question if no specific

* [Here we follow the text as first published in Jammer (1974, pp. 192–193), with small but obvious typos corrected. The format has also been very slightly uniformised to that of the other Letters to the Editor of *Physical Review* reproduced here (*eds.*).]

[1] *Phys. Rev.* **47**, 973 (1935) [Kemble 1935; see Chapter 17].

[2] *Phys. Rev.* **47**, 777 (1935) [Einstein, Podolsky and Rosen 1935; see Chapter 9].

mention were made of the interpretation of quantum mechanics as a statistical mechanics of assemblages of like systems'. This is an entirely different question. I believe that Einstein and Rosen would agree with me that quantum mechanics is, in view of our present knowledge, a correct and a complete statistical theory of such assemblages. A statistical mechanics, however, may not be a complete description of elementary processes – and it is of these that we speak. If its validity as a complete description is restricted to assemblages, then the difficulty mentioned by Kemble in the second paragraph of his paper does not arise; for in that case we are not dealing with the same reality.

Kemble, however, goes further and says: '... we cannot know more about an individual electron than the fact that it belongs to a suitable potential assemblage of this kind'. It is here that the real difference of opinion is revealed; the difference of opinion which led to the discovery of a fallacy where there was none.

If we cannot know more about a physical system than the fact that it belongs to a certain potential (and therefore imaginary) assemblage, then a mere change in our *knowledge* of the system must result in the change of the assemblage, and thus of the *physical state* of the system. Since, as is the case with the system β, this knowledge can be altered without disturbing the system, the physical system loses the essential property of physical reality – that of existence independent of any mind. This point of view is in direct contradiction with the philosophic point of view explicitly stated in the first paragraph of our article.[3] If independent existence of physical reality is not accepted, there is no common ground for the discussion. What is more, the very question used as the title of our article loses all meaning.

BORIS PODOLSKY

Graduate School,
University of Cincinnati,
July 6, 1935.

[3] *Phys. Rev.* **47**, 777 (1935) [Einstein, Podolsky and Rosen 1935; see Chapter 9].

19

Quantum Mechanics as a Physical Theory*

HENRY T. FLINT

The issue of the *Physical Review* dated May 15 contains a contribution by Prof. A. Einstein, B. Podolsky and N. Rosen under the title: 'Can Quantum-Mechanical Description of Physical Reality be Considered Complete?' The authors answer this question in the negative. They begin by pointing out that it is important to draw a distinction between the physical concepts, which are introduced into a theory and form its subject matter, and the objective reality, which is independent of the theory and which it is the purpose of the theory to describe. That is to say, we have to draw a distinction between the results of experiment and our theoretical description of them.

In order to test a theory the authors ask two questions: (1) Is the theory correct? And (2), Is it complete? The content of the paper is a consideration of the answers obtained when these questions are applied to quantum mechanics.

The first question can be answered in the affirmative if the theoretical conclusions agree with experiment. In physics, we test correctness by means of numerical values, and in this respect the authors have no adverse criticism to make of the quantum theory. The second question is a more difficult one, and some explanation of its meaning is required before an answer can be attempted. A complete theory is understood to be one which provides a counterpart to each element of physical reality. The definition of physical reality lies at the root of all that follows in the paper. The authors say: If, without in any way disturbing a system, we can predict with certainty the value of a physical quantity, then there exists an element of physical reality corresponding to this physical quantity. They accept this as a reasonable criterion. They do not regard it as a necessary but as a sufficient condition of reality.

Accepting this definition, the authors go on to show that the quantum-mechanical description of physical reality is not complete. The first step in the argument is to show that the uncertainty principle places two alternatives before us. Suppose that

* [The text follows the original publication in *Nature* (Flint 1935), which was received before 14 June 1935 (see Chapter 3, note 48) and published 22 June. Reprinted by permission from Springer Nature Customer Service Centre GmbH: Springer Nature, *Nature*, Quantum Mechanics as a Physical Theory, H. T. F., 1935 (*eds.*).]

two physical quantities are represented in the theory of quantum mechanics by the operators A and B. If these operators do not commute, that is, $AB \neq BA$, then the theory maintains that precise knowledge of one quantity is incompatible with precise knowledge of the other. The best known example of this is that in which the physical quantities are the co-ordinate of position and the momentum of an electron. We can only obtain precise knowledge of the second quantity by experiment, and in attempting this we destroy our knowledge of the first. But both operators represent objects of physical reality, so that by the definition it must be concluded either that the theory does not contain concepts for both, that is, it is incomplete, or else the two quantities have not physical reality at the same time. In quantum mechanics it is assumed that the wave function contains a complete description of this reality for the state to which it corresponds. This is reasonable, since the information obtainable from this function corresponds exactly to that which can be predicted with certainty without altering the state of the system.

In the final stage of the argument, an appeal is made to the process known as reduction of the wave packet. We must refer to the original paper for the details, but it is deduced that it is possible to predict with certainty two quantities corresponding to physical realities, and moreover that these quantities are represented by non-commuting operators. Thus we have to admit that if the quantum theory is complete and therefore contains concepts for these quantities then, although non-commuting, they correspond simultaneously to physical realities.

We saw that the uncertainty principle presented us with two alternatives; and now we see that if we deny the first of these alternatives, we have also to deny the second. The conclusion is that the first alternative, that the quantum theory is incomplete, has to be accepted.

It is possible that the criticism does not strike at the root of the quantum theory, since so much depends upon a definition of physical reality, which is not of necessity included in the theory. If we could be satisfied that the definition is necessary or that without it two simultaneous conjugate physical quantities are represented by commuting variables, then the quantum theory fails. An interesting letter on this point from Prof. G. Temple appears in *Nature* of June 8 (p. 957).[1]

The paper by Einstein and his collaborators is rather to be regarded as an appeal for a more direct description of the phenomena of physics. The authors seem to prefer the artists' portrayal of the landscape rather than a conventional representation of its detail by symbols which bear no relation to its form and colour. The

[1] [The short note by Temple (1935) argues that the Jordan product $\frac{1}{2}(AB + BA)$ is the quantum mechanical representation of the product *ab* of two physical quantities. He then demands that the representation of the product *abc* be unambiguous. This in turn requires the Jordan product to be associative, which Temple shows to be the case only if all observables commute. According to Temple, '[t]his destroys the whole structure of the modern form of the quantum theory'. We fail to see the connection with EPR (*eds.*).]

expression of a belief that it is possible to provide a complete description of physical reality indicates the nature and purpose of the criticism. Presumably the appeal is for a replacement of the operator calculus of the quantum theory by a method of description which resembles that of the classical and relativistic theories in that, like them, it introduces concepts more directly corresponding to physical realities.

In criticising quantum mechanics in this way, we must remember that it arose out of the inadequacy of existing models, and it has evolved in its particular way because it has had to do without them. The theory has done what, in the face of difficulty, it set out to do. It does not profess to describe actual phenomena but the possible results of measurements, and it is particularly concerned with the influence of the observer on the quantities observed. Much work is being done at present on the notation of the quantum theory. There are difficulties to be removed and with their removal we may find that we have travelled in the direction desired by the three authors. Physicists can scarcely remain content with a situation in which the electron is an equation, and the remark sometimes heard that the school of quantum mechanics is content to establish itself merely as a bureau for ready reckoning is by no means true.

H. T. F.

20

The Observation of Canonically Conjugates*

ERWIN SCHRÖDINGER

Canonical variables have as eigenvalues all numbers from $-[\infty]$ to $+[\infty]$.[1] From this it follows that a variable possessing a canonically conjugate can *in principle* never be determined by a single observation. For an observation always consists in reading a scale. The limited number of its possible results are always known beforehand. They constitute an essentially *discontinuous* and *finite* point set, like e.g. the following twentytwo: 0, 1, 2, 3, 4, ... 20, and 'beyond the scale' or the following fiftythousandandtwo: 0.000, 0.001[,] . . . . . . 2.545, 2.546, ... 50.000, and 'beyond the scale'. One usually expresses this situation by saying that of course a continuous variable cannot be measured with *absolute* accu[racy,] but thinks no harm in imagining that it can be measured to *any desired* accuracy.[2] But for the present purpose it seems that a finite point set, however large and dense, can never be considered a sufficient approximation to the infinite one. For the trace of a commutator $pq - qp$ in the finite case is known to be zero. Therefore if the diagonal element be at all a constant (i.e. all diagonal elements equal), it has to be zero and the variables commute.[3]

These facts are well known. They might prove to be comforting in view of certain paradoxes in quantum mechanics.[4]

E. SCHRÖDINGER

Magdalen College, Oxford.

* [The text follows the typescript in the 'Einstein Paradox' folder in Schrödinger's Nachlaß, AHQP 92, section 2-117, and Alpbach, 3.1, Manuskripte ES, Berliner (1p., typescript, in English) (*eds.*).]

1 P. A. M. Dirac, *The Principles of Quantum Mechanics*, II. ed., Oxford at the Clarendon Press 1935. See p. 94 [Dirac 1935a].

2 J. von Neumann, *Mathematische Grundlagen der Quantenmechanik*, Berlin, by [*sic*] Springer 1932 [von Neumann 1932]. See chapter III.3, page[s] 110[ff.] [von Neumann 1955, pp. 211 ff.].

3 On p. 214 *l.c.* [von Neumann 1955, p. 402] J. von Neumann expresses the opinion that in macroscopic measuring it is essential for everything that is at all measurable *in this way* also to be measurable simultaneously. The italics are mine.

4 A. Einstein, B. Podolsky and N. Rosen, *Phys. Rev.* **47**, 777, 1935 [Einstein, Podolsky and Rosen 1935; see Chapter 9]. – See also *Nature*, June 22, 1935, p. 1025 [Flint 1935; see Chapter 19].

21

Quantum Mechanics and Physical Reality*

NIELS BOHR

In a recent article by A. Einstein, B. Podolsky and N. Rosen, which appeared in the *Physical Review* of May 15,[1] and was reviewed in *Nature* of June 22,[2] the question of the completeness of quantum mechanical description has been discussed on the basis of a 'criterion of physical reality', which the authors formulate as follows: 'If, without in any way disturbing a system, we can predict with certainty the value of a physical quantity, then there exists an element of physical reality corresponding to this physical quantity'.

Since, as the authors show, it is always possible in quantum theory, just as in classical theory, to predict the value of any variable involved in the description of a mechanical system from measurements performed on other systems, which have only temporarily been in interaction with the system under investigation; and since in contrast to classical mechanics it is never possible in quantum mechanics to assign definite values to both of two conjugate variables, the authors conclude from their criterion that quantum mechanical description of physical reality is incomplete.

I should like to point out, however, that the named criterion contains an essential ambiguity when it is applied to problems of quantum mechanics. It is true that in the measurements under consideration any direct mechanical interaction of the system and the measuring agencies is excluded, but a closer examination reveals that the procedure of measurements has an essential influence on the conditions on which the very definition of the physical quantities in question rests. Since these conditions must be considered as an inherent element of any phenomenon to which the term 'physical reality' can be unambiguously applied, the conclusion of the above-mentioned authors would not appear to be justified. A fuller development

* [The text follows the original publication in the 'Letters to the Editor' section of *Nature* (Bohr 1935a), dated 29 June 1935 and published 13 July. Reprinted by permission from Springer Nature Customer Service Centre GmbH: Springer Nature, *Nature*, Quantum Mechanics and Physical Reality, N. Bohr, 1935 (*eds.*).]

1 [Einstein, Podolsky and Rosen (1935); see Chapter 9.]

2 [Flint (1935); see Chapter 19.]

of this argument will be given in an article to be published shortly in the *Physical Review*.[3]

N. BOHR

Institute of Theoretical Physics,
Copenhagen,
June 29 [1935].

[3] [Bohr (1935b); see Chapter 16.]

22

Is the Quantum-Mechanical Description of Physical Reality Complete?*

ARTHUR E. RUARK

In a recent paper Einstein, Podolsky and Rosen[1] conclude that the description of reality given by wave functions is not complete. This conclusion is directly opposed to the view held by many theoreticians, that a physical property of a given system has reality only when it is actually measured, and that wave mechanics gives a faithful and complete description of all that we can learn from measurements. The purpose of this note is to examine the reason for this divergence of conclusion. The reason is that the authors adopt a criterion of reality different from the one just mentioned; they say, that if, without in any way disturbing a system, we can predict with certainty the value of a physical quantity, there exists a corresponding element of physical reality. They consider two systems, I and II, which interact up to a certain time, after which there is no interaction. They show that by measuring the quantities belonging to certain operators A and B, characteristic of system I, it is in general possible, without disturbing system II, to predict with certainty *either* the value of a quantity P, *or* the value of a quantity Q; where P and Q belong to system II, and do not commute. However, there is no question of actually measuring *both* A and B; indeed, we can show that if P and Q do not commute, A and B in general do not, so that the corresponding physical quantities cannot be measured simultaneously. (Doubtless it is possible to extend the scheme to three systems, so that in the sense discussed by Einstein, Podolsky and Rosen either A or B could be predicted by studying the third system; and so on, until an indefinite number of systems is employed; but this does not help matters, nor is it necessary for the purposes of this discussion.)

* [The text follows the original publication in the 'Letters to the Editor' section of *Physical Review* (Ruark 1935), which was received 2 July 1935 and published 1 September. Reprinted with permission from: A. E. Ruark, *Physical Review* **48**, 466–467. Copyright 1935 by the American Physical Society (*eds.*).]

[1] Einstein, Podolsky and Rosen, *Phys. Rev.* **47**, 777 (1935) [Einstein, Podolsky and Rosen 1935; see Chapter 9].

The conclusion drawn by Einstein, Podolsky and Rosen is, that in accordance with their criterion of reality, the quantities corresponding to both P and Q possess reality; so that the wave description, failing to yield both of them, is incomplete.

This conclusion can be attacked by anyone who prefers to say that P and Q could possess reality only if *both* A and B (not merely one or the other) could be simultaneously measured. Indeed, this is clearly stated in the paper under discussion. Whereas, Einstein, Podolsky and Rosen say it is not reasonable to suppose the reality of P and Q can depend on the process of measurement carried out on system I, an opponent could reply: (1) that it makes no difference whether the measurements are direct or indirect; (2) that system I is nothing more than an instrument, and the measurement of A makes this instrument unfit for the measurement of B. Such an opponent will feel that the ingenious method of measurement discussed by Einstein, Podolsky and Rosen suffers from all the essential difficulties common to measurements which result in disturbing system II.

It seems to the writer that in the present state of our knowledge the question cannot be decided by reasoning based on accepted physical principles. The arguments which can be advanced on either side seem to be far from conclusive, and the issue involved appears to be a matter of personal choice or of definition.

ARTHUR E. RUARK

University of North Carolina,
Chapel Hill, North Carolina,
July 2, 1935.

23

Physical Reality and Quantum Mechanics*

EDWIN C. KEMBLE

In a recent 'Letter to the Editor'[1] the writer criticized the paper by Einstein, Podolsky, and Rosen 'Can Quantum-Mechanical Description of Reality be [C]onsidered Complete?'[2] In that communication I stated that their demonstration of the possibility of assigning two different wave functions to the same physical reality was incorrect, because actually the wave functions apply to different physical realities. Subsequent discussion with Ruark[3] has brought me to the rather tardy conclusion that I misinterpreted their phrase 'the same physical reality'. My own argument is relevant to the question, 'Is it desirable to apply the phrase "the physical reality" under the conditions described by E. P. and R.?' This question has been much more fully and adequately discussed by Bohr.[4] If, however, one interprets the phrase 'the same physical reality' in terms of the – to me unsatisfactory – literally interpreted definition of physical reality which forms the starting point of the paper by E. P. and R., my criticism is beside the point.

EDWIN C. KEMBLE

* [The text follows the typescript that Kemble included with his letter to Margenau of 9 November 1935 (AHQP 54, section 3-017); in the microfilm, however, this typescript is appended to Kemble's subsequent letter to Margenau of 20 November (AHQP 54, section 3-019). The typescript does not include the texts of the footnotes, which, however, are likely to contain only the references we supply here (*eds.*).]

1 [Kemble (1935); see Chapter 17.]

2 [Einstein, Podolsky and Rosen (1935); see Chapter 9.]

3 [Ruark (1935); see Chapter 22.]

4 [Bohr (1935b); see Chapter 16.]

24

Quantum-Mechanical Description*

HENRY MARGENAU

It is pointed out that, by removing a single quantum-mechanical postulate commonly accepted, several conceptual difficulties in quantum-mechanical description can be eliminated.

The discussion of a recent paper by Einstein, Podolski [*sic*] and Rosen[1] has brought to light an interesting divergence of opinions as to the meaning of reality. To the writer it seems, however, that the issue at stake is more than the merit or demerit of any particular conception of reality, and that a very fundamental point of quantum-mechanical axiology is involved. For Einstein, Podolski and Rosen have shown, by employing correctly the usual quantum-mechanical postulates, that the state of system 1 which, by hypothesis, is *isolated* from system 2, depends on the type of measurement performed on system 2. This, if true, is a most awkward *physical* situation, aside from any monstrous philosophical consequences it may have. We wish to show that, by the removal of a single postulate commonly accepted, the real difficulty inherent in Einstein–Podolski–Rosen's conclusion disappears.

I

The postulate in question may be stated thus: When a (single) measurement is performed on a physical system, then immediately after the measurement the state of the system is known with certainty. That is to say, if the system is in a state φ, the measurement causes the state function to transform itself into ψ, where ψ is an eigenstate of the operator belonging to the measured observable. This assumption is clearly involved in Einstein–Podolski–Rosen's interpretation of the reduction of the wave packet, and if it be denied that in general a measurement produces an

* [The text follows the original publication in *Physical Review* (Margenau 1936), which was received 16 November 1935 and published 1 February 1936. The author's affiliation is given as *Sloane Physics Laboratory, Yale University*. Reprinted with permission from: H. Margenau, *Physical Review* **49**, 240–242. Copyright 1935 by the American Physical Society (*eds.*).]

[1] A. Einstein, B. Podolski and N. Rosen, *Phys. Rev.* **47**, 777 (1935) [Einstein, Podolsky and Rosen 1935; see Chapter 9]; E. C. Kemble, *ibid.* **47**, 973 (1935) [Kemble 1935; see Chapter 17]; A. E. Ruark. *ibid.* **48**, 466 (1935) [Ruark 1935; see Chapter 22]; N. Bohr, *ibid.* **48**, 696 (1935) [Bohr 1935b; see Chapter 16].

eigenstate, their conclusion fails, and the dilemma disappears. We shall refer to the postulate stated as I.

II

I is not self-evident, or *a priori* true; for there is hardly more justification for supposing that a single measurement determines completely the state *after*, than to suppose it to determine completely the state of the system *before* the act of measurement. The latter is usually denied; to admit the former supposition would introduce a peculiar asymmetry into quantum-mechanical description. But there are stronger arguments against I.

III

I is definitely contradictory to another, in a sense more fundamental, postulate, namely, Schrödinger's time equation. To supply substance to the following simple but formal consideration the reader may think of the system as a particle, of the state function as referring to a pure case with respect to momentum, of M as a coordinate measurement, and therefore of ψ as a δ-function. I implies that, under the act of measurement, $\varphi \to \psi$, or, to use customary terminology, the operator M converts φ into ψ:

$$M\varphi = \psi. \tag{1}$$

We observe that in this equation M cannot be a *unique* operator such as those with which we are usually confronted in quantum mechanics, for ψ depends on the outcome of the measurement and cannot be predicted. That is to say, while we know, in our example, that ψ is a δ-function, the position of its peak is not fixed. ((1) must not, of course, be mistaken for a characteristic value equation, since φ and ψ are different functions.)

Eq. (1) is sometimes interpreted differently. It is supposed that M selects from an assemblage of systems φ a subassemblage ψ and that it is not the total assemblage φ which undergoes a transition from a state to an eigenstate.[2] But there are difficulties in such a view, of which two may be mentioned.

1. If φ truly referred to a larger collection of physical systems than ψ, the two functions would have to be constructed in different configuration spaces, and it would be difficult to give mathematical meaning to (1).
2. It is certainly proper to make a measurement on a single system. In that case the above interpretation, which involves the splitting off of a subassemblage, is meaningless.

[2] I owe this criticism to Professor Kemble.

We thus conclude that (1), if it has significance at all, must imply the actual transformation of the state function for a single system into an eigenstate function for the same single system.

On the other hand, the measurement is certainly a physical operation, describable in the ordinary manner as an interaction between physical systems. As such it is subject to Schrödinger's equation. Let the 'natural' Hamiltonian operator (the form valid for the isolated state) of the system be H_0, and let H_M represent the interaction with the measuring device (in our example coupling terms with the radiation field), so that $H = H_0 + H_M$. Then

$$H\varphi = (h/2\pi i)(\partial\varphi/\partial t).$$

Thus if the measurement takes place in a small time Δt, we have

$$\Delta\varphi = (2\pi i/h)\Delta t H\varphi.$$

Since $\varphi + \Delta\varphi$ is the function into which the measurement converts φ, and which we previously called ψ,

$$\{1 + (2\pi i/h)\Delta t(H_0 + H_M)\}\varphi = \psi. \tag{2}$$

This relation is of the same form as (1). But it is unlikely to make ψ a δ-function even if the correct form for H_M were known. The essential contradiction, however, arises from the fact that, if we put

$$M = 1 + (2\pi i/h)\Delta t(H_0 + H_M)$$

as (1) and (2) require, the left-hand side of this equation is an unspecifiable, nonunique operator, while the right-hand side is unique.[3] Since both (1) and (2) cannot be correct, we abandon (1).

IV

The assumption that a single measurement fixes a state either before or after the act of observation is untenable from the point of view of any reasonable probability theory. In contrast with classical physics, quantum mechanics defines its states in terms of functions, φ, in configuration space in such a way that there exists a unique (aside from arbitrary phases) correspondence between φ and a sequence of probability amplitudes. More explicitly, φ, and hence the state which it represents, is entirely equivalent to a set of numbers a_i associated with a certain operator P. This

[3] This situation is of course known. Von Neumann (*Mathematische Grundlagen der Quantenmechanik*) [von Neumann 1932] deals with it most thoroughly. We find it difficult to accept his explanation for reasons to be discussed elsewhere.

set of numbers defines a *probability distribution*, whose *elements* are the observations on the observable p which belongs to the operator P, and whose *properties* are the eigenvalues of P.[4] A quantum-mechanical state is therefore synonymous with a probability distribution. A probability distribution, however, cannot be fixed by a single measurement, but requires a very large number of observations. This is true even if the distribution consists of zero's [*sic*] for all properties but one (quantum-mechanical pure state).

To illustrate: In throwing a die, the probability distribution, which corresponds to the quantum-mechanical definition of a state, is the sequence of numbers $\frac{1}{6} \ldots \frac{1}{6}$ for the properties $1 \ldots 6$. If a 5 turns up in a single throw, I can conclude that I have thrown a 5, but cannot infer the distribution $\frac{1}{6} \ldots \frac{1}{6}$. This can be done only after a large number of throws. The fact that I have thrown a 5, while it has converted my ignorance with respect to the outcome of this observation into certain knowledge, has not changed the distribution of the probability aggregate from which I started.

V

The removal of I renders impossible the *preparation* of states in the manner advocated by Dirac and others. But the situation is perhaps not altogether unsatisfactory on that account, for it is still possible to *ascertain* a state, namely, by performing a large number of observations. A typical instance of a *complete* set of simultaneous observations which does define a state is the photographic record of a spectrum, where the intensities represent at once the distribution of probabilities among the various pure energy states. It must be admitted, however, that the relation between abstract states and experience is a source of difficulties which weaken the proposal here made and which must be subjected to closer scrutiny.

VI

While the previous arguments seem to show that I is in several respects undesirable and conflicting with other axioms, one may also convince himself that it is *unnecessary*. For no significant quantum-mechanical calculation requires I. Space does not permit an elaboration of this assertion here.[5]

[4] For terminology see v. Mises, *Wahrscheinlichkeitsrechnung* (Deuticke, 1931) [von Mises 1931]. The present point of view has previously been presented in greater detail by the author in *The Monist* **42**, 161 (1932) [Margenau 1932].

[5] The author hopes to return to the entire problem in a paper to be published in *J. Phil. Sci.* [i.e. in *Philosophy of Science*] [Margenau 1937].

VII

Returning to the argument of Einstein–Podolski–Rosen, the solution of their problem in the light of the present considerations is clear. If a great number of measurements of A had been performed on the state of system (1) after the interaction, *all* the coefficients ψ_n in Eq. (7) of Einstein–Podolski–Rosen's paper would have been determined. Conversely, if many measurements had been made on B, all the φ_s in their Eq. (8) would be known. A difficulty can arise only if the expansion of the right of (7), with the experimentally determined ψ_n's, yielded a different function that [*sic*] does the expansion in (8) with experimentally determined φ_s's. There is no indication that this will be the case.

25

Quantum Mechanics and Physical Reality*

HUGH C. WOLFE

The present writer wishes to add a few words to the discussion of this subject in the current literature.[1]

Most of us have in the back of our minds an idea of 'physical reality' which is satisfactory in classical physics and which we tend to carry along into quantum mechanics. Discussion of a single free particle will be sufficient to bring out the essential point. We believe that the particle is somewhere (it has position) and that it is moving in some definite way (it has momentum) and so on for any other physical properties. In classical physics, these quantities would be, in principle, simultaneously measurable and the state of the particle could be described by giving their values. In spite of the fact that the values of these quantities cannot be determined simultaneously (uncertainty principle), we still are inclined to believe in the state in which they all have definite values as an objective physical reality because we can measure any one of these quantities, chosen arbitrarily, and get a definite answer.

Quantum mechanics does not concern itself with the 'state' of the particle in the above sense but only with our knowledge of that 'state'. The 'state' is unknowable to the extent required by the uncertainty relations and any discussion of it belongs in the realm of speculation rather than physics. If we are to continue to employ this idea of the 'state' of the particle, we must be careful not to base any physical arguments on it and we must not confuse it with our quantum-mechanical description of our knowledge about the particle.

In quantum mechanics, our knowle[d]ge about the particle is represented by a wave function, φ, which is the solution of a wave equation. Our knowledge, based on measurement, determines the choice of the wave equation and the boundary conditions. This knowledge, embodied in φ, may be used to predict the results

* [The text follows the original publication in the 'Letters to the Editor' section of *Physical Review* (Wolfe 1936), which was received 30 December 1935 and published 1 February 1936. Reprinted with permission from: H. C. Wolfe, *Physical Review* **49**, 274. Copyright 1936 by the American Physical Society (*eds.*).]

[1] Einstein, Podolsky and Rosen, *Phys. Rev.* **47**, 777 (1935) [Einstein, Podolsky and Rosen 1935; see Chapter 9]. N. Bohr, *Phys. Rev.* **48**, 696 (1935) [Bohr 1935b; see Chapter 16]. E. Schrödinger, *Proc. Camb. Phil. Soc.* **31**, 555 (1935) [Schrödinger 1935e; see Chapter 10].

of further measurements. Every observable, b, has associated with it a quantum-mechanical operator, B, and the predicted values of b are given by $\int \varphi^* B \varphi d\tau$. Usually this prediction takes the form of stating the probabilities for various possible values to be found by measurement.[2]

Viewed in this light, the case discussed by Einstein, Podolsky and Rosen simply dissolves. Two systems have interacted and then separated. Nothing that we do to the first system after this affects the 'state' of the second. But measurements on the first system affect our knowledge of the second and therefore affect the wave function which describes that knowledge. Different measurements on the first system give us different information about the second and therefore different wave functions and different predictions as to the results of measurement on the second system.

The writer believes the ideas set forth here to be compatible with Bohr's point of view and hopes they will prove helpful in the dissemination of that point of view.

HUGH C. WOLFE

College of the City of New York,
December 30, 1935.

[2] No attempt is made here to give an adequate discussion of this topic. The writer wishes to thank Dr. W. H. Furry for a preview of a forthcoming article which deals instructively with this question. [Presumably Furry (1936a) (*eds.*).]

Part V

Correspondence

26

Correspondence on the 'Einstein Paradox'*

Schrödinger to Einstein, 7 June 1935**

[Oxford,] 7th June 1935.

Dear Einstein,[1]

I was very pleased that in the paper just published in *Phys. Rev.* you have caught the dogmatic quantum mechanics also publicly by the scruff of the neck, about which we had talked so much already in Berlin. May I say a few things about it? It looks at first like objections, but they are just points I would like to have formulated more clearly.

To set up the contradiction, it is not yet enough in my opinion that the following can occur in the case of always identical preparation of the system pair: *one* concrete single measurement on the first system gives for the second a definite value of A, another gives a definite value of B, and the simultaneously reality[2] of A and B is ruled out on general grounds. I think that is not enough yet. For one must allow anyway that the same preparation does not always have the same outcome. If it assigns one time the value A' to the quantity A, another time the value A'' to the quantity A, why should it not a third time confer reality not to the quantity A at all but to the quantity B, specifically the value B'?

* [Discrepancies between transcriptions and originals – which may partly derive from the use of different sources (the originals sent vs the carbon copies kept) – are footnoted (often simply as 'published transcription'/'consulted original'), except for differences in spelling, punctuation or emphasis, or for omissions and emendations in headings (*eds.*).]

** [Translation follows AHQP 92, section 2-105, and Alpbach, 6.5, Korrespondenz, Vorordner (carbon copy, 3pp., in German, cigarette mark on p. 2). Equations and addendum from EA 22-044 and EA 22-045, respectively. Transcription in Meyenn (2011, vol. 2, pp. 527–529) (*eds.*).]

[1] [Schrödinger addresses Einstein by surname but uses the familiar second person 'Du' (*eds.*).]

[2] [The transcription misses out the words 'gibt einen bestimmten Wert von *B* und die gleichzeitige Realität' (*eds.*).]

To set up the contradiction it seems to me one needs the following. There must be a wavefunction of the system pair and there must be two quantities A and B whose simultaneous reality is ruled out on general grounds, and to these the following must apply:

1) There exists *one* measurement method that, given this wavefunction, *always* gives a sharply defined value to the quantity A (if not always the same), so that without actually performing the measurement, I can say: when such a wavefunction obtains, A possesses reality – I don't give a hang which value it has;

2) a *different* measurement method must at least *sometimes* impart a sharp value to quantity B (always given the same wavefunction, of course).

One could perhaps weaken these conditions further. It would at any rate be enough if the percentage of cases in which A is sharp, added with the percentage of cases in which B is sharp, is greater than 100. But then one loses the possibility of presenting contradictory single cases, one can only say there are some among them.

As it happens, now you have given an example in which the above-mentioned sum is actually 200 (to put it briefly). And thus my stricter formulation seems redundant. I do, however, set store by it, and this because your example constitutes a very exceptional case. In general, there is only *one* possible way of expanding a function of two variables (or variable groups) as a bilinear series

$$\Psi(x_1, x_2) = \sum_{n=1}^{\infty} c_n \psi_n(x_2) u_n(x_1)$$

such that both the $u_n(x_1)$ and the $\psi_n(x_2)$ form an orthogonal system. Here (compared to your equation (7)[3]) I have added coefficients c_n. This means that I want to think of the individual functions of both systems as normalised. Then it is easy to specify when and to what extent the above expansion of a specific given $\Psi(x_1, x_2)$ is *not* unique. That is in fact the case if and only if among the c_n there are some of equal modulus. Then one may bring about an arbitrary unitary transformation among the $u_n(x_1)$ that belong to the same c_n, if one simultaneously performs the contragradient one with the corresponding $\psi_n(x_2)$. One can do the same with any collection of terms that happen to have the same c_n-modulus, where of course the unitary transformation is arbitrary every time. More briefly: one is free to perform *one* complete unitary transformation, but which operates a rotation only in the subspaces that happen to have constant c_n-modulus. No more. Your case is so particularly exceptional in that *all* c_n are equal. That is why you can rotate entirely arbitrarily, even from the 'Q-setting' to the 'P-setting'. I think one is entitled to

[3] [Equation (7) in the EPR paper is $\Psi(x_1, x_2) = \sum_{n=1}^{\infty} \psi_n(x_2) u_n(x_1)$ (*eds.*).]

consider the *exact* equality of two c_n as such an exceptional occurrence that one may formulate general theorems without regard to it. With this qualification one can say:

In general, a wavefunction of the system pair determines uniquely the pair of physical variables of the two systems – call them A_1 and A_2 – that satisfy the condition that *each* measurement of A_1 on the first system allows a precise inference to the A_2 of the second system, and vice versa. (N.B. The words 'and vice versa' can also be omitted without making the claim incorrect; the 'and vice versa' is automatically fulfilled.)

One's further position depends essentially on whether one shares my concern: it should not be necessary for the construction of the contradiction to make use of the exceptional occurrence of the equality of coefficients. If one does not share it, then your [own] example[4] is fully sufficient.

Until now I have been taking the contradiction as proven, and as I said already at the beginning, I do not consider what is laid out above to be a real objection either. The rhyme I have been spinning until now on the whole issue is this. We do not have a quantum mechanics that takes relativity, i.e. that takes among other things the finite speed of propagation of all influences into account. In the whole scheme we only have the analogue of the old absolute mechanics. At best this can be valid for infinitely small systems, in which the corresponding light times are negligible. If we want to *separate* two systems, then already long before it disappears, their interaction will cease to be approximable through absolute Coulomb and other such laws. And there we are at our wit's end. The process of separation cannot be understood at all in the orthodox scheme. If one thereupon tells me: it can, you do know how beautifully Dirac and Heisenberg can describe the radiation interaction of systems quite according to your book, just by including eigenoscillations of the radiation field [Hohlraumeigenschwingungen] into the system – then I answer that I have not managed yet to investigate this more complicated system to see whether the contradiction obtains also there.

Please forgive the cigarette hole on p. 2, and warmest regards

[F]rom your sincerely devoted

Schrödinger

[Handwritten addendum on reverse of last page:]

The uniqueness proof is terribly simple. One just needs to ask the question: what condition must the complete orthogonal system $u_n(x_1)$ satisfy in order for

[4] ['Dein eines Beispiel': presumably a slip for 'Dein eigenes Beispiel' (*eds.*).]

the expansion coefficients $c_n\psi_n(x_2)$ to satisfy the condition that the $\psi_n(x_2)$ are normalised and mutually orthogonal? This condition written out explicitly says that for each k, the function of (x_1):

$$\int dx_2 \int dx_1' u_k(x_1') \Psi^*(x_1', x_2) \Psi(x_1, x_2)$$

must be orthogonal to *all* u_n, except possibly u_k itself. Thus it must be a multiple of u_k, and more specifically must equal $|c_k|^2 u_k$. If one writes it out, it means in words: the u_k and the $\frac{1}{|c_k|^2}$ are eigensolutions and eigenvalues of the homogeneous integral equation

$$u(x_1) = \lambda \int u(x_1') \mathcal{F}(x_1', x_1)$$

with the Hermitian symmetric kernel

$$\mathcal{F}(x_1', x_1) = \int dx_2 \Psi^*(x_1', x_2) \Psi(x_1, x_2).$$

Hence follows uniqueness, except in the case where multiple values of λ coincide.

Teller to Schrödinger, 11 June 1935*

79 Gower Street,
LONDON,
11th June, 1935.

Very revered Herr Professor!

Upon reconsidering our conversation of yesterday, it seems to me that I have brought superfluous complications into the discussion. Of course, I do not hope to settle the problem through this letter. I think, however, that one can pose the question somewhat more generally than as done by Einstein, and that the more general formulation improves clarity and simplicity.

To begin with I want to formulate briefly the essential points of what I have to say:

First: In order to arrive at paradoxes of the kind as in the Einstein paper, it is not necessary to consider *two* bodies. Rather, the same paradoxical situation always arises when one measures something for which quantum mechanically one has only probabilistic statements.

I want to try to make this generalisation plausible, but do not want to write about the difficult question of what one should do about this paradox of a more general nature.

Above all, however, I want to enquire into a point that I misapplied yesterday. Namely, I want to show that:

Second: In the measurement of a certain quantity, this [quantity] is *not* influenced in an uncontrollable way by the measurement procedure; only the canonically conjugate quantity is uncontrollably disturbed by the measurement.

I begin with the second statement and give as examples the position measurement and the momentum measurement.

In a position measurement, the object to be measured is illuminated by a light beam and the scattered light is observed or photographed. The scattering process causes an uncontrollable change of the momentum of the observed object. One cannot speak of a classically describable[1] disturbance of the position itself, however, since the experiment can be carried out in arbitrarily short time. One best sees that the transformation of a spread-out wavepacket into a localised one [ein konzentriertes] is not traceable to the uncontrollable disturbance during the measurement,

* [From AHQP 92, section 2-115, and Alpbach, 6.5, Korrespondenz, Vorordner (typewritten, 6pp., in German). Some corrections in pen (clearly Teller's), two marks in pencil or finer pen (presumably Schrödinger's). Transcription in Meyenn (2011, vol. 2, pp. 530–532) (*eds*.).]

[1] [beschreibenden/beschreibbaren (*eds*.).]

if one assumes the initial packet to be very highly spread out. Then indeed a very large change in position would have to occur during the measurement, and the more the initial packet was spread out, the more the influence of the measurement would have to be large as well as long-lasting.

In a momentum measurement, one illuminates the object again and enquires after the Doppler shift of the reflected light. The momentum of the object is in fact somewhat changed in the reflection, but this change can be kept arbitrarily small, if one only uses light of sufficient wavelength in the experiment. A certain change in momentum remains of course inevitable, but for this one can again show it is largely uncontrollable, for the wavelength and hence the momentum of the light used are indeed known. What is uncontrollable in this experiment is the instant when the in-principle known change of momentum (and hence of velocity) occurs. That makes the velocity *during* the experiment unknown, which induces an uncertainty in the position *after* the experiment. The uncontrollable disturbance has thus now influenced the position and not the momentum.

Now to the justification of the first statement. Consider a state with indeterminate[2] position and momentum (Δq and Δp). I can determine the position without disturbing the position. Determination of position means according to Einstein, however, that one gives the position physical reality. Again according to Einstein, this physical reality would have existed already before, since the measurement cannot bring with it an influence on the quantity to be measured (i.e. here position) and consequently not an influence either on the reality of the quantity to be measured. The same holds for momentum, and we have arrived at the Einsteinian paradox for a single body.[3]

In the absence of uncontrollable disturbance in the measurements with which we are concerned, the clarifying power of the uncertainty relation cannot be applied here. The uncertainty relation and the uncontrollable disturbances must be invoked in those cases in which one has to do with apparent contradictions between classical measurements. If, e.g., I measure the momentum of a particle, then carry out a position measurement and lastly measure again the momentum, the difference in the result of the first and second momentum measurement is attributable to the uncontrollable disturbance during the position measurement. This consideration therefore solves the paradox that arises from the different results of the first and second momentum measurement. In this way, however, one gets no answer to the question why one can obtain different momenta starting from the same quantum mechanical state.

[2] [unbestimmtem/unbestimmten (*eds.*).]

[3] [In the original this sentence is emphasised by a line in the margin (*eds.*).]

I finally would like to try and consider the Einsteinian example from the point of view of uncontrollable disturbances, from which, however, one should expect no more a solution to the difficulty than with the 'one-body-paradox'.

Let us first take the momentum measurement. The sum of the momenta is known, but the difference in momenta has become unknown through the knowledge of the difference in coordinates. One can e.g. imagine that one has determined the difference in coordinates just by joining them with a rigid rod. This rigid rod[4] has brought about an uncontrollable exchange in momentum between the two bodies. From the law of action and reaction it follows, however, that the disturbances in the momenta of both bodies were equal and opposite. Now if afterwards I make a momentum measurement on body I, I can thereby conclude how large the momentum disturbance on body I has been during the determination of the coordinate difference. Thus in turn by reversing the sign I get the momentum disturbance on body II during the measurement of the coordinate difference, and there vanishes the uncertainty in the knowledge of the momentum of body II.

It is similar for coordinate measurements. The location of the two bodies during the momentum measurement has become unknown, but the unknown amount is the same for both bodies. Hence it suffices to measure the coordinate of one of the bodies in order to come to know how strongly the coordinate of the other body was influenced.

For the sake of completeness it is still necessary to show how, say, the momentum measurement on I makes the determination of position on II unsharp. During the measurement of the total momentum, II has received the same disturbance in location as I. If one now measures the momentum of I independently, one disturbs the location of I for a second time, and no later measurement of the location of I can any longer determine how strongly the location of I (and thereby the location of II) was disturbed during the *first* measurement. The independent momentum measurement on I has thus *independently* made unsharp the location of I and of II.[5] All these considerations of course do not change the fact that the momentum measurement on I can not alter the momentum or in fact the 'reality' of the momentum of I. It is all the more anschaulich that in this momentum measurement also the momentum of II remains unchanged.

I must add that the best way out of these paradoxes seems to me not to speak of 'reality' in this connection. But I know that one cannot deal with this point in a letter. I would be glad if you would give me the chance to come back to it in conversation.

[4] [The transcription includes a 'dann' which is crossed out in the original (*eds.*).]

[5] [Addition of a mark in the original, presumably '¡<<!' (*eds.*).]

Please forgive me for this long letter about things well known to you. I mostly wanted to fight my own confusions.

With many greetings,

Your very devoted,

E. Teller

Schrödinger to Teller, 14 June 1935*

[Oxford,] 14th June 1935

Dear Herr Teller,

[T]hank you very much[1] for your long letter. Yes of course if for some reason you think you know that a Q-measurement only just disturbs the P-values but not the Q-values, then the paradox appears already for a single system. – I am happy to see, however, that upon detailed reflection you also in fact do not find a resolution of the difficulty.

Reducing the entire contradiction to a question of terminology looks unseemly to me. Whether one happens to use the word reality or not is quite irrelevant. The alleged crime is as follows:

Two dice lie before me on the table, each covered by a leather cup. Someone gives me a description of the two dice that contains no specification as to which sides are facing up. He asserts that this is the most complete description possible of the two dice. But I know that someone else has ways and means, without any material influencing of the dice, to correctly complete the description in this point. He can open his mouth and add to what the first has said: in addition, 5 lies face up on the left and 3 on the right. Thereby, my knowledge of the dice has been positively increased. The first description was thus not the most complete conceivable, the corresponding assertion is false. The usual Copenhagen excuse is that through the use of the ways and means that have procured me the new knowledge, my knowledge of the dice has been decreased in a different respect. But since in point of fact I still *possess* the first description, supplemented by the knowledge of the top sides 5 and 3, and the latter is correct, a decrease can only have occurred in that the first description has become false in some respects – through a process that does not materially affect the dice. This assumption stands on a level [steht auf den Standpunkt] with the savage, who believes that he can harm his enemy by piercing his image with a needle.

One[2] can turn the thing also another way. According to quantum mechanics, the preparation of a system, whereby it is brought into a certain given state, does not merely consist in material treatment of the system with tools of all kinds, but, rather, what happens afterwards depends on what one does with the tools – whether one burns them, melts them down, tramples them or preserves them in a museum – but

* [From AHQP 92, section 2-116, and Alpbach, 6.5, Korrespondenz, Vorordner (carbon copy, 2pp., in German). Transcription in Meyenn (2011, vol. 2, pp. 532–533) (*eds.*).]

1 [Danke/danke sehr (*eds.*).]

2 [Ich/Man (*eds.*).]

in particular whether and what kind of attention one pays to the signs of wear on the tools. It would be rather difficult to formulate accurately the condition of *what* one must do with the tools to ensure that the description one has eventually arrived at might not after all be subject to a subsequent correction, if someone takes one of the employed tools out of the museum or the compost heap and examines it thoroughly. And *if* a quantum theorist tried to formulate these conditions, he would find in fact that they hopelessly resemble the precautions of a criminal who carefully tries to wipe out all traces of his deed.

I cannot help it: the principle that one should limit oneself strictly to describing the results of possible experiments, condemning everything else as metaphysics, may be wonderful, uncommonly impartial and neutral; but in the application it has found here, it leads us rather a bit much in a circle.

Most cordial greetings from your most devoted

[E. Schrödinger]

[P.S.] (As a matter of course, I shall always be very happy to discuss things with you face to face. Probably I shall also write something about this soon.)

Pauli to Heisenberg, 15 June 1935*

Zürich, 15 VI 1935

Dear Heisenberg,[1]

I know of nothing new in physics. Weisskopf is on Delbrück's problem, and the attendant issues of subtraction physics get very ugly. – I do not want to report on this today, but rather on two *didactic* questions, on which you could possibly intervene publicly.

1) *Einstein* has once again spoken out publicly on quantum mechanics, specifically in the issue of the *Physical Review* of 15th of May (jointly with Podolsky and Rosen – not good company, by the way). As one knows, every time this happens it is a disaster. 'Weil, so schliesst er messerscharf – nicht sein kann, was nicht sein darf' (Morgenstern).[2]

At least I will concede to him that if a student in his early semesters were to raise such objections with me, I would consider him very intelligent and promising. – Since there is a certain risk of the public opinion getting confused by the publication – specifically in America – it would perhaps be appropriate to send a response to it to the *Physical Review*; for which I should like to enjoin *you*.

So perhaps it is after all worth it if I waste paper and ink in order to formulate the facts necessitated by quantum mechanics that give Einstein particular mental anguish.

He has now understood this much, that one cannot simultaneously measure two quantities corresponding to non-commuting operators, and cannot simultaneously assign them numerical values. But what he objects to in this context is the way in which in quantum mechanics one composes two systems into a total system. – One sees this in the following example, which he (essentially) also invokes.

We have a total system consisting of two mass points, 1 and 2. Let them be force-free and possibly be situated at a very great distance from each other. According to quantum mechanics, there is a possible state of the total system in which – at least at a certain instant t_0 – both the distance $x_2 - x_1$ between the particles and the

* [From CERN-ARCH-PLC heisenberg_0215 (handwritten, 6pp., in German) (*eds.*).]

[1] [Pauli uses the informal 'Du' to address Heisenberg (*eds.*).]

[2] [This is the famous last line of Christian Morgenstern's poem, 'Die unmögliche Tatsache' ('The Impossible Fact'), originally published in Morgenstern (1910). The line as it is usually translated is: 'For, he reasoned pointedly / That which must not, can not be'. A more faithful translation would be: 'For, he reasons trenchantly / That which must not, can not be' (*eds.*).]

centre of mass momentum $p_1 + p_2$ of the particles have definite numerical values (since $p_1 + p_2$ commutes with $x_2 - x_1$). Let these numerical values be P_0 and x_0.[3]

$$p_1 + p_2 = P_0, \quad x_2 - x_1 = x_0. \tag{1}$$

(Eigenfunction $\psi(x_1, x_2) = e^{\frac{i}{\hbar}P_0\frac{x_1+x_2}{2}}\delta(x_2 - x_1 - x_0)$ (for $t = t_0$))

Let the state be produced by some mechanism – e.g. previous interaction between 1 and 2, or otherwise. Einstein now emphasises the *correct* fact that measurements on particle 2 certainly cannot disturb particle 1, and furthermore the [following] claims of quantum mechanics[:] A) If, through a measurement on 2, I have found the value x_2^0 for the coordinate, then I can predict with certainty the outcome of the x_1-measurement on 1, namely the value $x_0 - x_2^0$. After the measurement on 2, 1 is in the state with eigenfunction $\delta[x_1 - (x_0 - x_2^0)]$.

B) If, on the other hand, I have ascertained through a measurement on 2 the value p_2^0 for the momentum p_2, then I can predict with certainty the outcome of the p_1-measurement on 1, namely the value $P_0 - p_2^0$. After the measurement on 2, 1 is in the state with eigenfunction $e^{\frac{i}{\hbar}(P_0 - p_2^0)x_1}$.[4]

Now comes the 'deep feeling' and [he] goes on to say: 'Since the measurements on 2 cannot disturb particle 1, there must be a[5] Something called "physical reality", namely the state of the particle 1 *in itself* – independently of which measurements one has made on 2. It is just absurd to assume that particle 1 can be transformed via measurements on 2, i.e. can be carried from one state to another. In truth, the quantum mechanical description of 1 is correct but *incomplete*. A complete description would have to assign to the state of particle 1 features that also already include all those properties of 1 that – *after* possible measurements on 2, which do not disturb 1 – can be predicted with certainty'.

A pedagogical response to the line of thought placed in '. . . ' must, I think, clarify the following concepts.

1) The distinction between the following statements[:]

a) Two systems 1 and 2 have no interaction with each other (= lack of interaction energy).

Definition. This is the case if after maximal observation of 1 the expectation values of all quantities of 1 have *the same time development* as though 2 were not present. (N.B. For sufficiently short times the concept of interaction does not play a role anyway.)

[3] [The transcription occasionally has $\mathfrak{p}_1$ and $\mathfrak{p}_2$ for p_1 and p_2 (*eds.*).]

[4] Einstein is familiar with the fact that even through measurements on 2 one cannot *simultaneously* predict for 1, say, values of p_1 and x_1, and grants this completely.

[5] [ein/eine (*eds.*).]

b) The total system is in a state where the subsystems 1 and 2 are *independent*. (Decomposition of the eigenfunction into a product.)

Definition. This is the case if, after carrying out a measurement on 2 of an arbitrary quantity F_2 with *known* result $F_2 = (F_2)_0$ (number), the expectation values of the quantities F_1 of 1 remain the same as without a measurement carried out on 2.

Quite independently of *Einstein*, it seems to me that in a systematic construction of quantum mechanics one should *start from* the composition and separation of systems more so than has been done to date (e.g. in *Dirac*). – This is indeed – as Einstein has *rightly* felt – a very fundamental point in quantum mechanics, which furthermore has a direct connection to your considerations about the *cut* and the possibility of its relocation to an arbitrary place. (*Maybe you could briefly present these once not in popular form – with formulas*; there is something similar[6] by Neumann in his book, but too epsilontic.) One must distinguish several layers of reality: An R that contains all information one can find out by measurements on 1 *and* 2, and an r (deducible from R) that contains only the information one can obtain by measurements on 1 alone. One must then show how disclosure of a measurement outcome on 2 has discontinuous change of r ($r \rightarrow r_A$ or $r \rightarrow r_B$ etc.) as its result (unless the particle systems were independent). And that one will necessarily run into contradictions if one attempts to describe these changes classically or semi-classically without reference to 2 – perhaps as 'hidden properties' of 1.

In any case, older gentlemen like *Laue* and *Einstein* are haunted by the idea that quantum mechanics should indeed be *correct* but *incomplete*. That one should be able to make it *complete through new*[7] *statements it does not contain, without changing the statements it does contain.* (I call a theory with such a property *incomplete*, in the logical sense. Example: the kinetic theory of gases.) Perhaps you could – on the occasion of the reply to Einstein – make it clear for once in an authoritative way that such a completion of quantum mechanics is not possible without modifying its contents.

N.B. Maybe I have only taken such pains on these matters that are trivialities for us, because some time ago I received an invitation to Princeton for the next winter semester. It will be a lot of fun for me to go. In any case, I want to make the Morgenstern motto popular there!

---

2) A second didactic question arises through Wick's paper by now published in the *Rendiconti Lincei* about the magnetic moment of the proton.[8] Though I must

6 [ähnliche/ähnliches (*eds.*).]
7 [The transcription omits the word 'neue' (*eds.*).]
8 [Wick (1935).]

admit that because of the Italian language I have perhaps not understood everything, it nevertheless appears to me that he has bungled all points.

1) That only a non-zero mean magnetic moment of the intermediate state results if one calculates relativistically with respect to the *heavy* particle (introduction of the Dirac matrices for the heavy particle), which purely logically goes well beyond Fermi's assumptions about β-decay.

2) That one first gets ∞ then if one does no cut-off.

3) That transitions to negative protons must be excluded in order to obtain a non-0 magnetic moment for the neutron.

4) That in the first-order relativistic wave equation (Dirac equation) for the heavy particles, additional terms of the type $\alpha\gamma^{\mu}\gamma^{\nu}F_{\mu\nu}\psi$ appear in an external field, as has already been presumed multiple times.

Perhaps you could after all briefly publish in the *Zeitschrift für Physik* your own calculations about this – going further than the remarks in the Zeeman Festschrift[9] – as a clarification. It is not good if a kind of secret science develops on the topic.

Now I fear I have bored you greatly with this letter. But about quantum electrodynamics I know nothing. The longer I think about it, the darker it becomes to me; and also the more unsatisfactory appears to me your idea of interpreting the introduced Δs as the effect of the measurement apparatus.

Many greetings

Your W. Pauli

[9] [Heisenberg (1935a).]

Einstein to Schrödinger, 17 June 1935*

Old Lyme Connecticut. 17. VI. 35

Dear Schrödinger![1]

Szilard is right in a sense. I think that Flexner would be glad to appoint you at the Institute; I also spoke with him about it once. But he evidently hesitates to set up competition for the University so to speak, and – so it seems – also avoids discussing it with the university people. So it goes, when one quite needlessly puts oneself into a position of moral dependence! By the way, he almost never discusses things with me. I also cannot confidently judge to what extent Veblen and Weyl have influence on him. (The former will presumably become his successor in a couple of years.) If you go to Graz, I should think that Flexner would make you an offer, because the offer from Princeton University would thereby so to speak 'expire'; but one cannot count on it. I can understand that you do not fancy being 'stored away' in Oxford, so to speak, although it is all for the best for one of us if he is not disturbed from outside. –

I did not go to Spain, because then I should also go to France and England, and I shudder at that many people and commitments. Here I sit in solitude on an enchanting sea bay, and have my little sailing boat and my peace; there just can be nothing better.

I do not believe in the higher-order electromagnetic terms. Sound is something so un-elementary, that the parallel with the electromagnetic processes in the vacuum appears to me almost as 'blasphemy'.[2]

I have found that the general-relativistic neutral massive particle and electric particle can be represented without additional terms as singularity-free fields.[3] There is thus[4] a serious possibility of representing atomic theory [die Atomistik] in relativistic–field-theoretic terms, even if mathematically it appears exceedingly difficult to advance to the many-body problems. From a point of view of principle, I absolutely do not believe in a statistical base for physics in the sense of quantum mechanics, no matter how fruitful this formalism may prove to be in detail. I do not believe that one can develop such a theory general-relativistically. But apart from this, I find the renunciation of a spatiotemporal comprehension of the Real idealistic–spiritualistic. This epistemology-drenched orgy will have to wear itself

* [From AHQP 92, section 2-106, and Alpbach, 6.2, Korrespondenz, EPR-Korrespondenz (handwritten, 2 pp., in German). Transcription in Meyenn (2011, vol. 2, pp. 535–536) (*eds.*).]

[1] [Einstein uses the familiar second person 'Du' (*eds.*).]

[2] [Einstein is here referring to an argument in a previous letter from Schrödinger. For details, see p. 80 (*eds.*).]

[3] [This is a reference to Einstein and Rosen (1935) (*eds.*).]

[4] [aber/also (*eds.*).]

out . . . You are surely smiling at me and thinking that already many a young whore has become an old bigot, and many a young revolutionary an old reactionary. Nothing doing!

With warm greetings from your

A. Einstein.

Einstein to Schrödinger, 19 June 1935*

Old Lyme Connecticut. 19. VI. 35.

Dear Schrödinger!

I was very pleased about your detailed letter dealing with the little paper. For linguistic reasons this was written by Podolsky after many discussions. But in fact what I actually wanted has not come out so well; instead, the main issue is so to speak buried by erudition.

The actual difficulty lies in the fact that physics is a kind of metaphysics: physics describes 'reality'. But we do not know what 'reality' is; we know it only through the physical description!

All of physics is description of reality; but this description can be 'complete' or 'incomplete'. The meaning of these expressions is at first also problematic. I wish to explain them using the following analogy.

Before me stand two boxes with hinged lids into which I can look[1] if they are opened; the latter is called 'making an observation'. In addition, there is a ball, which is always found in one box or the other when one makes an observation.

Now I describe a state thus: *the probability that the ball is in the first box is* $\frac{1}{2}$. – Is this a complete description?

No: A complete statement is: the ball *is* in the first box (or it is not). This then is how the characterisation of the state should look in a complete description.
Yes: Before I lift the lid of the box, the ball is not at all in *one* of the two boxes. This being in a definite box only comes about through my lifting of the lid. Only through this does the statistical character of the world of experience, or rather of its empirical regularities, come about. The state before lifting the lid is *completely* characterised by the number $\frac{1}{2}$, whose meaning, however, when carrying out observations manifests itself only as a statistical result. The statistics only come about because the observation introduces insufficiently known factors foreign to the described system.

We are faced with the analogous alternative when we want to interpret the relationship of quantum mechanics to reality. In the case of the ball system, of course, the second 'spiritualist' or 'Schrödingerian' interpretation is so to speak insipid,

* [From AHQP 92, section 2-107, and Alpbach, 6.2, Korrespondenz, EPR-Korrespondenz (handwritten, 4pp., in German). Transcription in Meyenn (2011, vol. 2, pp. 537–539) (*eds.*).]

[1] [hineinsehen/hineinschauen (*eds.*).]

and the man on the street [der Bürger] would take seriously only the first 'Bornian' one. The Talmudic philosopher, however, laughs at 'reality' as one would at a simpleton's bogeyman, and declares the two views to be different only in manner of speaking.

Now *my* way of thinking is like this: strictly, one cannot prevail upon the Talmudist without the aid of an additional principle: 'separation principle'. Namely: 'the second box, together with everything that pertains to its contents, is independent of what happens with respect to the first box['] (separate subsystems). If one holds fast to the separation principle, one thereby excludes the second ('Schrödingerian') view, and only the Bornian one remains, according to which, however, the above description of the state is an *incomplete* description of reality or of the real states.

The above comparison corresponds only very imperfectly to the quantum theoretical example of the paper. But it is suitable for making clear the point of view that matters to me. – One describes in quantum theory a real state of a system by a (normalised) function ψ of the coordinates (of configuration space). The time evolution is given unambiguously through the Schrödinger eq[uation]. One would now like to say the following: ψ corresponds bijectively to the real state of the real system. The statistical character of the measurement outcomes is to be imputed exclusively to the measurement apparatus, or the process of measurement. If this works, I talk of a complete description of reality through the theory. If, however, such an interpretation is not feasible, I call the theoretical description 'incomplete'. At first, the Talmudist is quite right in pronouncing this stipulation to be without content, but it will soon make good sense.

Now let us describe the total system composed of the subsystems A and B by its ψ-function ψ_{AB}. Let the description refer to a point in time when the interaction has practically ceased. The ψ of the total system can then be put together from the normalised eigen-ψ $\psi(x_1)$, $\chi(x_2)$ belonging to the eigenvalues of the 'observables' (or comm[uting][2] systems of observables) α and β, respectively. We can write

$$\psi_{AB} = \sum_{mn} c_{mn} \psi_m(x_1) \chi_n(x_2) \qquad (1)$$

If one now makes an α-measurement on A, this expression is reduced to

$$\psi_B = \sum_{n} c_{mn} \chi_n(x_2) \quad \ldots . \qquad (2)$$

[2] [The transcription amends 'vert.' to 'vertauschte'; it ought to be 'vertauschbare' (*eds.*).]

This is the ψ-function of subsystem B, in the case I have made an α-measurement on A.

Now instead of expanding in terms of the eigenfunctions of the observables α and β, I can also do so in terms of the eigenfunctions $\underline{\alpha}$ and β, where $\underline{\alpha}$ is a system of different commuting variables:

$$\psi_{AB} = \sum_{mn} \underline{c_{mn}\psi_m(x_1)}\chi_n(x_2) \quad \ldots \tag{1a}$$

From which after measurement of $\underline{\alpha}$ one obtains

$$\underline{\psi_B} = \sum_{n} \underline{c_{mn}}\chi_n(x_2) \quad \ldots\ldots \tag{2a}$$

What is essential now is only that ψ_B and $\underline{\psi_B}$ are at all different from each other. I claim that this difference is irreconcilable with the hypothesis that the ψ-description corresponds bijectively to physical reality (to the real state).

Indeed, after the collision the real state of $(A\,B)$ consists of the real state of A and the real state of B, two states that have nothing to do with each other. *The real state of B now cannot depend on what measurement I perform on* A. ('Separation principle' from above[.]) Then, however, corresponding to the same state of B there are two (in fact arbitrarily many) equally legitimate ψ_B, which contradicts the hypothesis of a bijective or complete description of the real states.

Note: I don't care a *fiddle* [ist mir *wurst*] whether the ψ_B and $\underline{\psi_B}$ can be considered eigenfunctions of observables β, $\underline{\beta}$.

Now I only note that I do not believe that we need content ourselves with an 'incomplete' description of real things,[3] rather that we should search for a complete description.

The way out hinted at in the last paragraph of your letter, that this ought to be related to the fact that one works with an infinite c, does not look promising to me. But in any case, the preceding argument is valid for non-relativistic quantum mechanics, which is the only consistent theory available.

Warm greetings from your

A. Einstein.

[3] [Zustände/Dinge (*eds.*).]

Born to Schrödinger, 28 June 1935*

246, Hills Road, Cambridge. 28. 6. 35.

Dear Erwin,[1]

I had already heard sensational things about the Einstein–Podolsky–Rosen paper, but I read it only yesterday – after your letter[2] –, and I am deeply disappointed. It is an age-old swindle, which I already thought about for my adiabatic paper (which contains the statistical interpretation) about 10 years ago.

2 systems I and II,	uncoupled	from $t = -\infty$ to $t = 0$,
	coupled	from $t = 0$ to $t = T$,
	uncoupled	from $t = T$ to $t = \infty$.

On system I the quantities	A,	eigenvalues a_k,	eigenfunct[ions] $u_k(x_1)$
	and B,	[eigenvalues] b_r,	[eigenfunctions] $v_r(x_1)$

A state of the total system *after* T is

$$\begin{aligned} \Psi(x_1, x_2) &= \textstyle\sum_n \psi_n(x_2)u_n(x_1) \\ \text{or} \quad &= \textstyle\sum_s \varphi_s(x_2)v_s(x_1) \end{aligned}$$

If A is measured with result a_k, then $\Psi = \psi_k(x_2)u_k(x_1)$
[if] B [is measured with result] b_r, [then] $\Psi = \varphi_r(x_2)v_r(x_1)$.

This means: a measurement on system I *alone* determines the state of system II; it is $\psi_k(x_2)$ or $\varphi_r(x_2)$, depending on whether the result of the measurement on I was $u_k(x_1)$ or $v_r(x_1)$ – although the two systems are not coupled at all: mysterious action at a distance.

But in fact one can measure only either A or B on I. After, say, A has been measured, one has to first set the system back to the state $\Psi_0(x_1, x_2)$, which obtained *before* the coupling and from which after the coupling has arisen $\Psi(x_1, x_2)$, and then one must measure, say, B.

Thus, indeed, it only makes sense to talk of 'predictions' of the values of quantities on II, if one sees the thing like this: one compares *two* experiments, both starting from the same state $\Psi_0(x_1, x_2)$ *before* the coupling; to fix $\Psi_0(x_1, x_2)$ one needs of course in each experiment measurements on *I and II*. When $\Psi_0(x_1, x_2)$ has evolved by way of the coupling to $\Psi(x_1, x_2)$, then in order to determine the latter it is enough to perform a measurement on sub-system I *alone* (II is then already codetermined, since of course $\Psi(x_1, x_2)$ has evolved from the fixed $\Psi_0(x_1, x_2)$). Depending on

* [From AHQP 92, section 2-102, and Alpbach, 6.2, Korrespondenz, EPR-Korrespondenz, and 6.5, Korrespondenz, Vorordner (handwritten, 5pp., in German). Transcription in Meyenn (2011, vol. 2, pp. 540–543) (*eds.*).]

[1] [Born addresses Schrödinger by his first name, but uses the formal 'Sie' (*eds.*).]

[2] [We have been unable to locate Schrödinger's letter to Born; thanks to Nancy Greenspan for assistance (*eds.*).]

whether one measures $A \to a_k$ or $B \to b_r$ on I, one can predict which value the complementary quantity P resp. Q of II has in that case, quite irrespectively of whether P, Q commute or not.

I cannot understand what is supposed to be obscure here. After all, it is just the same in the classical theory. Imagine two linear oscillators moving past each other; let the mean motions [die Bewegung der Centren] be uniform. Let them be coupled, say, electrically.

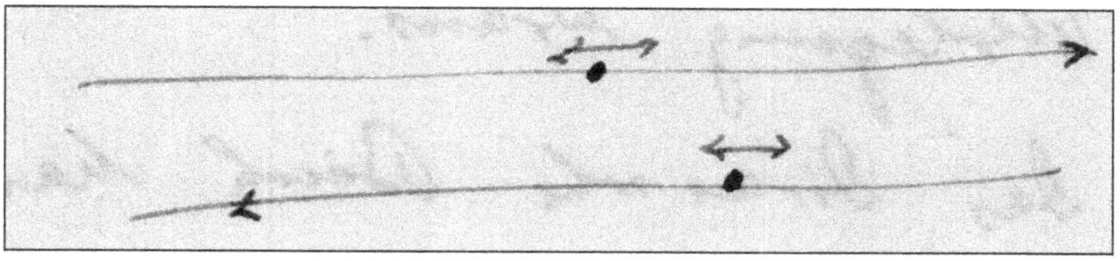

Fig. 26.1 [Born's sketch of coupled harmonic oscillators (*eds.*).]

Then also the final state depends on the totally arbitrary phase. If one only specifies the energies, then one can also only make statistical claims about the final state. This corresponds more or less to specifying the state Ψ_0 in quantum theory. I do not know whether one can pursue this example much further still, by analogy to quantum theory.

Your reasoning on the light quantum seems to me to go much deeper. I agree completely with your interpretation of the contraction of the wavepacket. Claims about *1* light quantum are indeed always statements about independent experiments with many light quanta (I mean – one each time). Just like playing roulette: the occurrence of red says nothing about the chances in the next throw. That is indeed also the gist of my above criticism of Einstein's reasoning.

I bought Dirac's book yesterday and will look at it.[3] For the time being, I am trying to get somewhat deeper into v[on] Neumann's book. He is indeed the sharpest of all. You should[4] feel free to publish your lectures, even if there is some overlap with Dirac. Moreover, you know how to write readably, and you have certainly thought through details that are elegantly bypassed in Dirac.

I was in Glasgow for 8 days at my publisher's to finish the popular book. It has turned out very well. Now I want to rest for 8 days.[5] Then Infeld is coming. I want

[3] [Presumably the second edition of Dirac's textbook (Dirac 1935a or 1935b) (*eds.*).]

[4] [sollen/sollten (*eds.*).]

[5] [noch 8 Tage ausruhen/mich 8 Tage ausruhn (*eds.*).]

to try then again in earnest to extract the square root in field theory using Dirac matrices. I believe it must work.[6]

Won't you come over sometime? It is rather quiet here now.

The whole household sends warm regards. Your

Max Born.

[6] [The 'popular book' was *The Restless Universe* (Born 1935a). As Greenspan (2005, p. 201) notes, Born had taken a break from work with Infeld on field theory to work on this and on his *Atomic Physics* (Born 1935b). For Schrödinger's interest in Born and Infeld's work, see pp. 80 and 205, note 13 (*eds.*).]

Schrödinger to Born, 29 June 1935*

[Oxford,] 29th June 1935.

Dear Max![1]

Thank you *very much* for your letter. As a matter of fact, lacking anything better to do, I am currently busying myself provoking various clever people about this thing. Indeed, I should like to write an essay that at least explains as clearly as possible what we have now, and I am very glad to get pointers as to how I should write it so that one can understand what people like Laue, Einstein and myself are on about. (Pardon me for the arrogant line-up, but these happen to be the most renitent obstructionists.)

Now look out, perhaps I can provoke you again. Indeed, at first go you do it pretty much like all the others (e.g. London, Teller, Szilard[2]). You set out the circumstances with clarity. And after you have said for two-and-a-half pages only things that I could have said exactly the same, all of a sudden you say: I cannot understand what is supposed to be obscure here. You do not even consider Einstein's central claim. He does not say that 'there is something obscure here'. He says that while the quantum mechanical description may well be perfectly good and correct, its claim of being, at least in certain cases (so-called 'pure case'), the conceivably most complete can be refuted. – A remark that you make later, by the way, sounds as if you would grant him this point. Namely you say: indeed, it is just the same in the classical theory when one specifies only the energies (not the phases) – and that this corresponds more or less to specifying [Ψ_0] in quantum mechanics – so one can only make statistical statements about the final state.

Indeed: *this corresponds more or less to specifying* [Ψ_0] *in quantum mechanics*. So there! We are happy with that. Because you are surely clear about it that in classical mechanics specifying only the energies is not a complete description.

But no, I do not want to ensnare you with words. I merely want you to see that – forgive the unkind phrase – our problem begins where your arguments leave off. The reference to the classical case, where the phases certainly have meaning and ignoring them makes the description incomplete, is surely not at all appropriate to explain that and why the quantum theoretical phases are meaningless and ignoring them does *not* affect the completeness of the description.

* [From AHQP 92, section 2-103, and Alpbach, 6.5, Korrespondenz, Vorordner (carbon copy, 3pp., in German). Transcription in Meyenn (2011, vol. 2, pp. 543–546) (*eds.*).]

1 [Schrödinger addresses Born by his first name but uses the formal 'Sie' (*eds.*).]

2 [London and Szilard were both in Oxford in this period (*eds.*).]

If you continue thinking about this matter, I also do not want you to ascribe to the opponent the foolish opinion that the simultaneous specification of canonical conjugates is *in fact* imperative for a complete description. That would be childish, it is definitely not like *that.* Note the very cautious form of the conclusion of that paper: if one accepts the present interpretation with the addendum that the description (in the pure case) is complete, then one arrives at the following contradiction: 1) conjugates are not determinable simultaneously[;] 2) conjugates must be determinable simultaneously. Thus, what one has accepted is inconsistent. Statements 1) and 2) are deduced only to bring out this contradiction. Not, say, to pronounce 1) to be false and to put 2) in its place as correct. That would be a claim that goes too far and, as I am firmly convinced, is false – for which the given argument does not provide the slightest support.

I have not made myself intelligible with my recent, as it seems very unclearly-communicated opinion about the light quantum and the reduction of wavepackets. For you have understood my opinion to be: 'Claims about one light quantum are indeed always statements about independent experiments with many light quanta (I mean – one each time)'. With that you saddle me precisely with the generally accepted interpretation. That is just what I do *not* mean. After the word 'experiments' I would have to continue with 'with an unknown number of light quanta varying from run to run whose average value *a priori* is One'. Here I take the *a priori* probability for [o]ne, two, three etc. light quanta to be:

[missing formula]

(These are the prob[abilitie]s that are valid, e.g., for a small subvolume of a large volume, if the subvolume is occupied on average by one particle and the particles are distributed randomly in the large volume.) The advantage of this conception compared to the usual one you have described is this. Consider, e.g., a light beam split in two. Then if you have found a light quantum in one sub-beam, this does not reduce *at all* the p[robability] for also finding one in the other sub-beam. (It corresponds to your having searched half of your subvolume. It does not change the p[robability] for the other half at all.)

I have the slight feeling that this conception corresponds to the so-called second quantisation. I would be happy if its formalism provided any support for justifying a necessary uncertainty of the total number.

Thank you very much for your invitation – my aunt (sister of my mother) has announced she will be here in the coming week, I do not know exactly when, and for the weekend Bertl Sachsel. I hope it is possible at some later time. Are you all remaining in Cambridge for the present?[3]

With warmest greetings to you

[F]rom your sincere

[Erwin Schrödinger]

[3] [The aunt Schrödinger mentions is presumably Rhoda Bauer (1864–1954), who used to spend much time in Leamington Spa, where her mother was from (cf. Moore 1989, p. 9, and www.hinterberger.org/wiki/index.php?title=Rhoda_Arzberger_geb_Bauer, consulted 30 August 2017). Bertl Sachsel (1901–1989) was an Austrian-born photographer active in Britain from the 1930s (cf. www.fadingimages.uk/photoGa.asp, consulted 30 August 2017). She also appears in the guest list of one of the Schrödingers' *Würstel*-evenings in Berlin in February 1932 (cf. Alpbach, 12.1, Heiteres, Würstelabend) (*eds.*).]

Berliner to Schrödinger, 1 July 1935*

Berlin W 9, 1 July 1935.
Linkstrasse 23–24
Publishing house of Julius Springer

To Herr Professor Dr. E. Schrödinger
Oxford

Dear Herr Professor Schrödinger!

I actually got the idea of the Innominato because you had not used the name [Einstein] yourself.[1] Of course there is not the least objection to mentioning the name, in whatever form. You will have seen in the last two years that I have not yet learnt to be afraid, nor have I any intention to be instructed on the subject. I am glad that you are going to take part in the discussion and am all the more glad about it as presumably all theoretical physicists of stature will intervene. We can expect something special from friend Pauli. He seems to be sharpening his already sharp tongue for this purpose particularly. Upon incidentally asking Weisskopf, during a different correspondence, what Pauli thought about the article,[2] I got a card from Pauli himself, in which he speaks against the article with his characteristic edge, judges Laue's position with respect to the whole of quantum mechanics with the greatest scepticism, and expects of me that I consult also Heisenberg on the matter. He takes a stand in particular against Laue's earlier paper in *Naturwissenschaften*.[3] One can apply to Pauli a nice quote from *Faust*: 'How lewd! Yet lewd as it might be, it was not lewd enough for me'. Change lewd to shrewd and it all fits Pauli's way of seeing things he does not quite fully agree with.[4]

Finally, let me repeat myself by saying that the non-mentioning of the Innominato is not a recommendation, just an idea that happened to come to me reading your letter. Whether you mention the author in a footnote or otherwise remains entirely at your discretion.

* [From Alpbach, 3.1, Manuskripte ES, Berliner (typewritten, with one correction and last half-sentence handwritten, 2pp., in German). Transcription in Meyenn (2011, vol. 2, pp. 548–549) (*eds.*).]

1 [The Innominato (the Unnamed) is a character in Manzoni's historical novel *I promessi sposi* (Manzoni 1840) (*eds.*).]

2 [As Meyenn points out, in 1935 Weisskopf published a three-part essay on problems of quantum field theory in *Naturwissenschaften* (Weisskopf 1935) (*eds.*).]

3 [Laue (1934).]

4 [Gretchen's lines in *Faust I*, Scene 17 (Goethe 1808), in which she recalls how quick she used to be in condemning other young women who had fallen prey to temptation. Berliner's original word play is between 'schwarz' (black) and 'scharf' (sharp) (*eds.*).]

And now another question or rather a favour: at Laue's suggestion I have requested for review a little book (ten signatures thick) by Douglas McKie and [N. H. de] V. Heathcote, with a preface by C. Andrade, *The discovery of specific and latent heats*, because Laue thinks this book is particularly interesting.[5] Would you review the little book and may I send it to you? It is certainly no effort for you, it is probably going to be quite a fun read and will prompt a couple of remarks from you that will interest all physicists.

With heartiest regards, your

A. Berliner

[P.S.] And Graz?

[5] [McKie and Heathcote (1935).]

Heisenberg to Pauli, 2 July 1935*

Leipzig, 2. 7. 35

Dear Pauli!

Excuse me for answering your letter so late. First, about the pedagogical question: I heard from Copenhagen that Bohr is going to publish a reply to Einstein in the *Physical Review*. This reply already exists in manuscript in Cop[enhagen]. The essential point of Bohr's reply is roughly the following: it can be shown that the Einstein thought experiment is identical in principle to the repeatedly discussed screen with two slits. The separation between the slits fixes $x_1 - x_2$ of the light quanta, the momentum measurement on the screen determines $p_{x_1} + p_{x_2}$. One then applies the usual arguments a[nd] shows that the simultaneous existence of interference and conservation laws leads to the Einsteinian paradoxes.

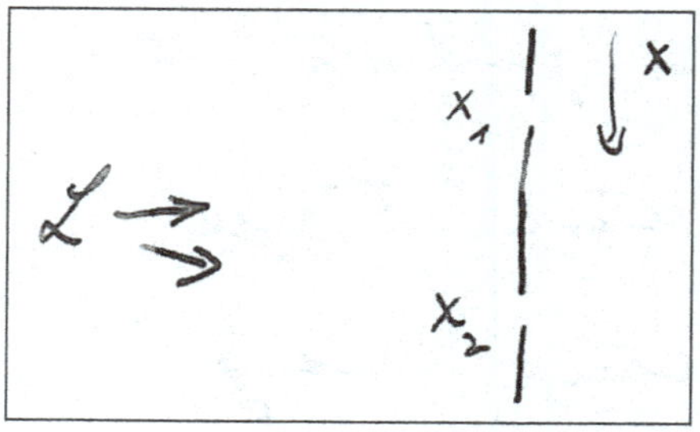

Fig. 26.2 [Heisenberg's sketch of a double-slit experiment (*eds.*).]

Since this reply is very different from the thoughts that one might for instance express in connection with the 'cut', I actually still want to write something about the cut in line with your suggestion. In particular I would like to treat the question of the incompleteness of quantum theory. By the way, a half-philosophical essay on this point by Miss Grete Hermann has appeared recently, which has grown out of discussions in Leipzig, and which I find reasonable on the whole (but for the philosophically-rationalist tendency). (Title: 'Die naturphilosophischen Grundlagen der Qu[anten] M[echanik]'; Verlag 'Öffentliches Leben' Inh[aber] Erich Irmer, Berlin 1935[.])[1]

During the holidays, I first want to finish a paper about nuclear physics, treating in more detail the question of whether α-particles are present in heavy nuclei. I have worked out a procedure here that is significantly more precise than Majorana's,

* [From CERN-ARCH-PLC heisenberg_0017-092r (handwritten, 3pp., in German). Also in AHQP 80, section 2. Transcription in Pauli (1985, pp. 407–409) (*eds.*).]

[1] [Hermann (1935c).]

a[nd] with which one can, I believe, fully clarify this question. When I have finished the paper, I want to send to you a carbon copy. After that I have time for the paper on the 'cut'. However, I should also write an essay for 'Fysisk Tidskrift' for Bohr's 50th birthday. –[2]

Wick's paper is doubtless very full of gaps.[3] However, I do not actually believe that the calculations should be published yet again. Perhaps a correction of Wick's note can be made within a paper by Weizsäcker, who has calculated the exchange forces and corresponding spin–spin interactions for a series of hypotheses about possible fermionic interactions. The publication of this paper will however still take some time.

At the end of July I want to travel to Switzerland. If you were in Zurich, say, in the period of 20th–24th July, I would be glad to visit you sometime during this time. By the way, Bethe will probably also be in Switzerland in those days.

Many greetings

Your W. Heisenberg

[2] [The papers referred to are, respectively Majorana (1933), Heisenberg (1935c) and Heisenberg (1935d) (*eds.*).]

[3] [Wick (1935).]

Schrödinger to Pauli, early July 1935*

[Oxford, early July 1935]

Dear Pauli![1]

Arnold Berliner has written to me that you reacted very temperamentally to the three-man paper in *Physical Review* (without writing to me exactly how, but I can well imagine that, too[2]).

Well. For me, this note was the motivation for thinking through again very accurately this case (which essentially we all knew long ago[3]) and other related matters, specifically from the point of view that 'really having a value', 'really being thus and thus' and such are not meaningful expressions. Einstein 'doubles down' even. He has a representation of reality on a map with little flags. To each thing of reality there must correspond a little flag on the map, and vice versa.

Well. I would be very glad to know whether you would agree with the following portrayal of the évenement, which – as hard as I try – I cannot make sense of [hinter die ich beim besten Willen nicht zurückkann].

It is well known that there are preparative measurement methods by which one can make a system maximally known or turn it into a 'pure case'. I claim: there are methods of preparation in which after definitively isolating the system to be prepared it still lies within the power of the experimenter to extend the method in such a way that a pure case of type A will definitely result, or else to extend it in such a way that a pure case of type not-A will definitely result.

So far you ought to agree with me. Yes? Would you also allow me[4] to call the pure case a state and to say: after the system is already isolated, it thus still lies within the power of the experimenter to bring about, or to prevent, its getting to a state of type A?

Now, while 'state' is a word that everybody uses, even the holy PAM,[5] this does not make it more meaningful. But one can quite easily see that, and in what sense, a system *itself* changes if its psi-function changes. Indeed, each psi-function conveys the *same amount* of knowledge. When it changes, it is never only a gain but always also a loss of knowledge. But one can only acquire, not lose, knowledge (except

* [From AHQP 92, section 2-113, and Alpbach, 6.2, Korrespondenz, EPR-Korrespondenz, and 6.5, Korrespondenz, Vorordner (carbon copy, with handwritten addendum, 2pp., in German). Transcriptions (identical except for the emendation 'Dublin' in the first, 'Oxford' in the second) appear in Pauli (1985, pp. 406–407), and in Meyenn (2011, vol. 2, pp. 549–551) (*eds.*).]

[1] [Schrödinger addresses Pauli by his surname but uses the familiar 'Du' throughout (*eds.*).]

[2] [dazu schon denken/schon dazu denken (*eds.*).]

[3] [im wesentlichen längst kennen/längst im wesentlichen kannten (*eds.*).]

[4] [Da Du mir auch noch erlaubst/Ob Du mir auch noch erlaubst (*eds.*).]

[5] [I.e. Dirac (*eds.*).]

by going soft in the head, which we shall ignore here). A loss of knowledge can only come about if the circumstances have changed. In this sense I say: different psi-functions definitely correspond to different circumstances – or states. I do not consider this an illegitimate dragging in of the reality dogma. But I would very much like to know what you think about this. And whether you really think that the Einstein case[6] – let us call it like this – gives rise to no further worry, but is totally clear and simple and self-evident. (To date everyone I spoke for the first time about it thought so, because they had learned well their Copenhagen credo in unum sanctum. Typically, three days later there came the message: what I said recently was of course entirely incorrect, much too complicated. Or it was (Szilard): I first need to think *what* it is that I must forbid you.[7] But I still have not had a clear explanation of why everything is so clear and simple. Somebody[8] even thought: it is just almost exactly like in the classical case; if there I know only the energies but not the phases of two oscillators,[9] of course I do know not what comes out........; and the whole objection is directed just against completeness!)

Warm regards, dear friend

[F]rom your old

[Schrödinger]

[Handwritten addendum:]

I am aware that 'pure case' (or 'definite[10] ψ-function') is in fact also an *ensemble*; further, that this ensemble comes about each time through *selection* made *on the basis* of the reading from the (already separated) measurement apparatus. And in the two cases *based on different* selection principles. Thus[11] everything seems understandable.

But I think, each quantum mechanical ensemble consists entirely of fully identically prepared [gelagert[12]] individual cases. If one grants me differences among the individual cases constituting the quantum mechanical ensemble, then one already admits with that alone the incompleteness of the quantum mechanical 'empirical[13] description'. Incomplete, because I am ordered to regard cases as identical, even though I know that they are not; because I am forbidden from enquiring further, even though I know that there is still something to enquire about.

E. S.

[6] [der Einsteinfall/dass der Einsteinfall (*eds.*).]

[7] [Szilard was in Oxford at the time, so these must have been discussions rather than correspondence (*eds.*).]

[8] [Schrödinger is obviously referring to the example in Born's letter of 28 June (see p. 301) (*eds.*).]

[9] [Wenn ich von zwei/Wenn ich da von zwei (*eds.*).]

[10] [bekannte/bestimmte (*eds.*).]

[11] [Damit/Somit (*eds.*).]

[12] [The choice of word in German is unusual, but the same word is used also in the Schrödinger cat paper (Schrödinger 1935f) (*eds.*).]

[13] [mechanischen/empirischen (*eds.*).]

Pauli to Schrödinger, 9 July 1935*

Zürich, 9. VII 35

Dear Schrödinger!

Thanks for your letter. In the evaluation of the 'Einstein case' we are basically in agreement – at least if I have always understood correctly what you say – although in certain respects there may be small differences in shading. In the meantime, Bohr has sent me the manuscript of a note that has now been sent off to *Physical Review* in response to Einstein. Perhaps I shall first sketch the essentials of the content, which, incidentally, includes nothing new to me. By the way, I think indeed that the 'Einstein case' contains nothing but very elementary a[nd] direct[1] consequences of the indeterminacy relation.

Now Bohr points to the following examples: 1.) Let a particle with given momentum in the z-direction pass through an opening L found in a screen that as a whole is free to move in the x-direction.[2]

Further, let the momentum p_x of the screen in the x-direction be known *before* the particle has passed through L. *After* the particle has passed through L, I now still have the *free choice* – in both cases *without* disturbing the particle mechanically – ***either*** to measure once again p_x of the screen: then I can predict with certainty the magnitude a[nd] direction of the particle's momentum *after* its passage through L – [***or***] after the passage of the particle through L, I can measure the *position* x of the screen S; then I can predict also the position of the particle at least an 'arbitrarily short' time after the position measurement of S, as this will then coincide with that of S.

One readily sees that after the *one* (p_x) ***or*** the *other* (x) measurement on S, the position x or respectively the momentum of the *particle* can no longer be predicted on the basis of *further* measurements on S, since in the first case the screen S has experienced an indeterminable displacement, in the second case an indeterminable change in momentum.

2.) Now one can consider a screen with *two* openings, L_1[3] and L_2, a[nd] send two particles *simultaneously* through L_1 and L_2, a[nd] measure the momentum of S in the x-direction before a[nd] after.

Then immediately after the passage I know *first* the *difference* of the position coordinates $x_2 - x_1 = l$ (at a certain moment in time!) of the particles, second the

* [From AHQP 92, section 2-114, and Alpbach, 6.2, Korrespondenz, EPR-Korrespondenz (handwritten, 6pp., in German). Transcribed in Pauli (1985, pp. 419–422) but not in Meyenn (2011). Double underlinings in the original are rendered in bold italics (*eds.*).]

[1] [elementar-direkte/elementare u. direkte (*eds.*).]

[2] [Given that knowledge of the particle's momentum in the x-direction is what is crucial in the argument, we assume Pauli here means the particle's momentum lies *wholly* in the z-direction. That is, a plane wave is approaching the screen perpendicularly, and has zero momentum parallel to it (*eds.*).]

[3] [L,/L_1 (*eds.*).]

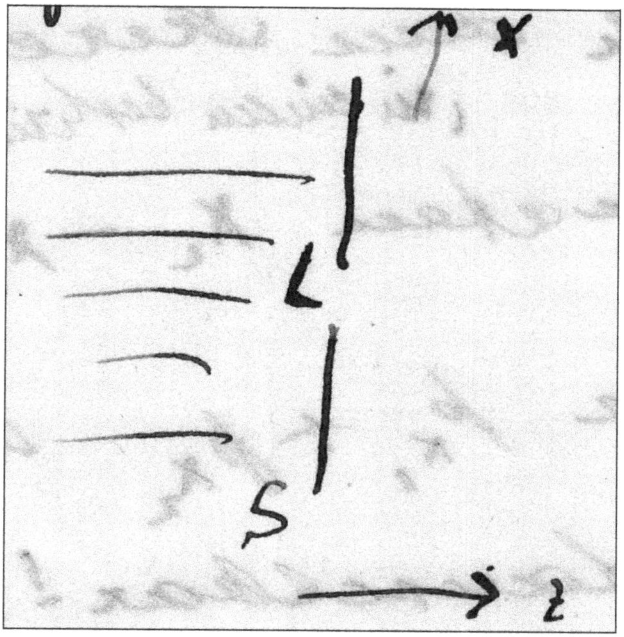

Fig. 26.3 [Pauli's sketch of Bohr's single-slit experiment (*eds.*).]

sum of the momenta $p_{x_1} + p_{x_2}$ of the particles. (In fact, $x_2 - x_1$ commutes with $p_{x_1} + p_{x_2}$!)

That is the Einstein example. Now I still have the *free choice*, ***either*** to *infer* x_2 through subsequent measurement of x_1 ***or*** to *infer* p_{x_2} through subsequent measurement of p_{x_1} – but without ever being able to infer x_2 and p_{x_2} simultaneously. *Thus far Bohr.*

Now, whether one should describe [the] pure case as a *state*? Already in his textbook on wave mechanics Kramers has substituted the term 'state' throughout with the term 'physical situation'.[4] A pure case of A is a global situation in which the results of particular measurements on A (a maximal set [im maximalen Umfang]) are predictable with certainty. I have nothing against calling this the 'state' – but even then it *is* the case that changing the state of A – i.e. that which is predictable of A – lies within the ***free choice*** of the experimenter even without directly disturbing A itself – i.e. even *after* isolating A.

This circumstance required by quantum mechanics appears to me *inescapable* in order to avoid contradictions, if the conservation laws as well as the superposition

[4] [Kramers (1933).]

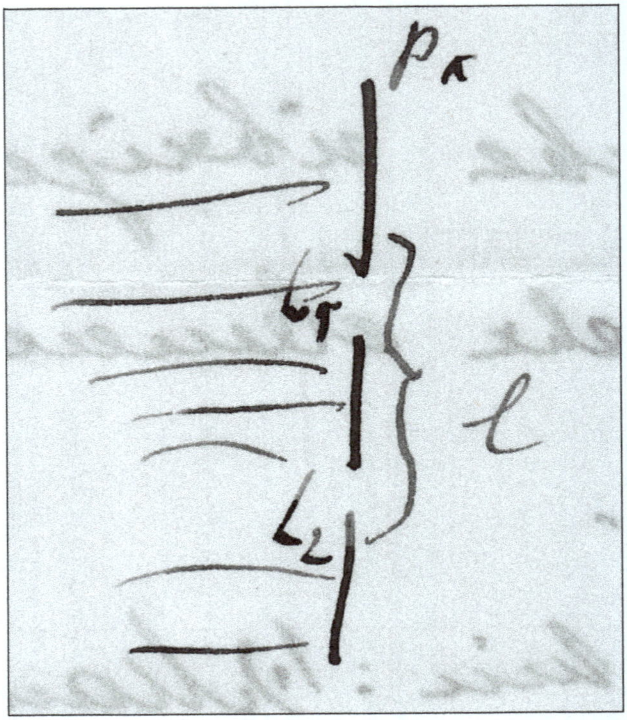

Fig. 26.4 [Pauli's sketch of Bohr's double-slit experiment (*eds.*).]

principle are valid. (Of course, this is precisely a ***difference*** as compared to the *classical* concept of state.) In my opinion ***there is indeed no problem here*** – and one knows the fact in question even without the Einstein example.[5]

Now I would also like to examine more closely the general question raised in your handwritten addendum – independently of the Einstein example. To what extent is it possible to make the additional assumption that presently still unknown features exist that distinguish among the individual cases of the quantum mech[anical] ensembles?

I would quite like to take this question from the level of the prohibition 'thou shalt not' to that of a statement that is logically decidable within qu[antum] mechanics: *What kind of additional assumptions are possible without changing the* ***statistical*** *consequences of* (= *statements of*) *quantum mechanics itself?*

[5] This is what I wanted to gloss rather dramatically when I quoted to Berliner as a characterisation of Einstein's paper the lines taken from a poem by *Morgenstern*: '[Weil, so schliesst er messerscharf – nicht sein *kann*, was nicht sein *darf*!' [see note 2, p. 291 (*eds.*)]].

This question is very important since there seems to be an incorrect analogy that is incurably stuck in the minds of the conservative older gentlemen (Einstein, Laue, etc.) with the kinetic theory of gases (classical statistical mechanics). The latter *can* indeed be completed – without disrupting the accuracy of its statistical statements – through additional deterministic axioms for the mechanism of the micro-processes. – The issue is *not* about forcing nonbelievers to believe in qu[antum] mechanics; it is also not an issue of generally denying that new degrees of freedom could still be discovered.[6] – But: ***if one subdivides the quantum mech[anical] ensembles*** *mentally into sub-ensembles and in the latter* ***simultaneously*** *assigns numerical values with greater accuracy than the uncertainty relation allows to quantities A, B belonging to non-commuting operators, then one comes into conflict with statistical claims of qu[antum] mechanics about the total ensemble.*

Once again, in other words: A difference in the individual cases constituting an ensemble would in principle be possible, but the following is *not* possible: If you have e.g. a pure case that corresponds to a given numerical value of the momentum, then you cannot subdivide it conceptually into individual cases corresponding to different value ranges of q. While the expectation value of any function of q alone or of p alone – or more generally of a sum $F_1(p) + F_2(q)$ – would not be changed by this subdivision, the *value distribution* of

$$F = F_1(p) + F_2(q)$$

would; or the expectation value of the squared error of F:

$$\overline{(F - \overline{F})^2}$$

has a different expectation value [*sic*] in the quantum mech[anical] total ensemble than in the mixture of the imagined sub-ensembles with subdivided q-values a[nd] fixed p. (E.g. the operator $F = p^2 + \omega^2 q^2$ familiar from the case of the oscillator – or the angular momentum $P_z = p_y q_x - p_x q_y$;[7] for a given $p = p_0$, according to the imagined subdivision one has a contin[uous] distribution of values[8] for F or P_z, but according to qu[antum] mech[anics] a discrete one.)

In general, the hypothetical distinguishing feature of the individual cases – even if it is only an imagined feature – must itself not contradict the uncertainty relation.

I also often like to mention the following example: When I study the diffraction pattern of *two* slits, it is not the *superimposing* of the diffraction pattern at each

[6] E.g. one can initially assume that a nucleus has no spin. A more precise analysis shows, however, that it indeed has one. Ensembles that one had initially taken for pure cases thereby turn out to be mixtures. (Introduction of new variables $\xi_1, \xi_2, \ldots$ into the ψ-function that commute with all previously known quantity-operators.)

[7] [$P_z = p_y x - p_x y / P_z = p_y q_x - p_x q_y$ (*eds.*).]

[8] [Verteilung/Werteverteilung (*eds.*).]

individual slit. If it is *indeterminate* through which slit the light quantum has gone, it is not the same as when I know that only *one* of the two holes was open, but for whatever reason I do not know *which* of the two holes was open. – One can indeed[9] *not* – as the conservative old gentlemen would like – declare the statistical statements of quantum mechanics (wave mechanics) to be correct, and *still* prop them up with a hidden causal mechanism!

In this sense the system of quantum mechanical laws seems to me logically closed (complete in the axiomatic sense) – in *contrast* to the kinet[ic] theory of gases – open instead are the problems of *relativistic* quantum theory (and here I do not have as much faith in the 'holy PAM').[10]

In non-relativistic wave mechanics, however, I think only pedagogical questions remain.

Well, I hope you are satisfied with this rather detailed answer.

Warm greetings from your

W. Pauli

[9] [aber/eben (*eds.*).]
[10] [I.e. Dirac (*eds.*).]

Schrödinger to Einstein, 13 July 1935*

[Oxford,] 13. July 1935.

Dear Einstein!

You have given me great pleasure with your two lovely letters of the 17th and 19th of June and the greatly detailed discussion of very personal things in the one and very impersonal things in the other. Thank you very much. But I still find the greatest pleasure in the *Physical Review* paper itself, because it acts so much like a pike in a pond of carp and stirs everyone up.

From your letter I can see that I completely agree with you in terms of the attitude toward the *existing* theory. Indeed, after you pointed them out already years ago in Berlin, we discussed the issues in the seminars often and heatedly. Again and again the others pointed out to me that there is no magic in a crass sense, say, in which the 'American' system gives $q = 6$ when I do *nothing* or the *one* thing with the 'European' system (you see, we emphasised the separation!), instead $q = 5$ when I do the *other*; again and again I said: it need not be *that* bad in order to cause trouble. In fact, by abusing the European system I can steer the American one *at will* into a 'q-sharp' state or into one that definitely does *not* belong to this class, e.g. into a 'p-sharp' one. That is *also* magic!

Now I am having fun and taking your note as an opportunity to prod a variety of clever people on this topic – London, Teller, Born, Pauli, Szilard, Weyl. I got the most out of Pauli yet, who admits at least that using the word 'state' for the psi-function is very fishy. – What I have seen so far in printed reactions is not very clever. I understand E. C. Kemble least, in *Phys. Rev.* 15th June[1] – indeed, he does not even mention the *one* case that gives us headaches. It is as if one says: in Chicago it is bitterly cold, and another answers: that is a fallacy, it is very hot in Florida. – Then in Nature on 22nd June there was an equally rather vacuous report, I think by Flint (it was only signed with initials).[2] I foamed with rage about N. Bohr's letter to *Nature* of 13th July.[3] He *just* makes one curious, reveals not a word of what he thinks, and refers to a forthcoming article in *Phys. Rev.*[4]

The great difficulty in my coming to any mutual understanding with the orthodox ones on the issue has led me to try to analyse the present interpretational situation for once *ab ovo* in an extended piece. I do not know yet whether or what I shall

* [From AHQP 92, section 2-108, and Alpbach, 6.5, Korrespondenz, Vorordner (carbon copy, 5pp., in German). Transcription in Meyenn (2011, vol. 2, pp. 551–555). For the quotations from von Neumann (1932) we follow the standard translation (von Neumann 1955) except where noted (*eds.*).]

[1] [Kemble (1935); see Chapter 17.]

[2] [Flint (1935); see Chapter 19.]

[3] [Bohr (1935a); see Chapter 21.]

[4] [Bohr (1935b); see Chapter 16.]

publish of this, but for me it is always the best way to make the issues really clear to myself. A few things in the current foundations have emerged from this as very peculiar.

1) It is funny indeed – here is a *new* theory that departs starkly from everything prior by the claim of indeterminacy, the claim of acausality and by many more specific claims. But on one point it has not budged from the spot: the *only* way it manages to interpret the sole reality in the world, the measurement results, is well and truly classical, as the measurement of a variable in a classical model. When the elastic theory of light was replaced by the electromagnetic theory, one stopped talking about the density and elasticity of the medium; one no longer understood the state at one point of the medium as a deformation, rather as something entirely different. The goal of a sensibly designed measurement procedure could no longer possibly be to measure *these* things, if the designer was *truly* basing himself on the new theory. In quantum mechanics it is different – supposedly one goes on happily measuring the same things as before, supposedly because our language is utterly incapable of capturing anything else. All that is novel is that 'mere probabilities' are predicted (and of course the way they are calculated). The events *for* which one predicts the probability are classical. Yet, one has the feeling that it is precisely the most important statements of the new theory that can really be squeezed into these Spanish boots only with difficulty. For instance: the probability of finding the value E for the energy of an oscillator is always zero, except when it is a value from the series E_1, E_2, E_3 etc. – and in fact, this is valid independently of the initial state! Officially this procedure is tailored 'to predict from the result of previous measurements the p[robability] of later ones'. In very important special cases the previous measurements do not matter at all and the fundamental discreteness of the energy levels must be expressed in this twisted way: the p[robability] of finding a different value is zero.

Even worse if one needs to say: the p[robability] of finding the angular momentum of a point-mass relative to the coordinate origin (sic!) to have a value other than $n(n+1)h/2$ (with $n = 1, 2, 3$ etc.) is zero. Here again it does not depend on the state at all, and also not on the choice of coordinate origin!

Well and truly, one has the feeling here that one is dealing with new features of the system, which no longer have much in common with the classical ones, and which by their nature may only have these values.

2) The psi-function at *one* instant, as is well-known, can take an arbitrary form. Then only its time evolution is determined by the differential equation. Nevertheless, one says that the psi-function is uniquely determined by (at most) n appropriately selected and ideally precise measurements (except for an inessential

phase factor and perhaps normalisation). A function of n variables is uniquely determined by n individual values. I know of course how the hocus pocus works mathematically. But I do not like such a theory; something is still wrong here.

3) 'Canonical variables have as eigenvalues all numbers from minus-infinity to plus-infinity', one reads in Dirac.[5] And now I say: any real measurement consists in the reading of a scale. Every scale has a finite number of gradations. If I add the designation 'off the scale' to these, then I can say that a finite number of possible measurement outcomes is 'stipulated' in advance for each measurement. The measurement merely serves to decide among these. Even in an ideal measurement I can of course imagine this number to be arbitrarily large, but not 'actually infinite'. Or let us say: at least I may *not* do this *if* for the purpose I have in mind it makes a difference whether I do the one or the other.

And it does make a difference if I ask myself whether the measured quantity can have a canonical conjugate together with which it satisfies $pq - qp = 1$. Indeed, for finite matrices p and q the *trace* of $pq - qp$ is always zero, and of course it does not budge from zero if I make the number of columns and rows as large as I wish.

J. von Neumann, who is far and away the clearest and cleanest of all quantum mechanicians, has indeed noticed these things. In his very clever, if also difficult book one finds pronouncements such as this: (p. 214) 'Now it is a fundamental fact with macroscopic measurements that everything which is broadly measurable at all [so überhaupt messbar], is also simultaneously measurable'. In this sentence really only the 'broadly' is irritating.[6] Because it looks as though 'macroscopic' in fact is not quite being equated with 'performable by a human'. I very much appreciate also his note at the end (n. 126 on p. 251).[7] There he very honestly criticises the idealising way in which he has tried to capture the 'inexact measurement' mathematically. Namely, he has done it as follows: he has divided the domain of the quantity to be measured into small intervals, and assumed that the measurement determines with absolute certainty which interval. Now he self-criticisingly adds '[a]ctually, the boundaries [...] are indistinct'. And in the end he claims no more than: '[n]evertheless, our method of description appears to be the most convenient one mathematically, at least for the present['].

[5] [Schrödinger here is translating from the second German edition of Dirac's book (Dirac 1935b, p. 94). For more on this point cf. Schrödinger's note on 'The observation of canonically conjugates'; see Chapter 20 (*eds.*).]

[6] [The standard translation (von Neumann 1955, p. 402) omits this irritating little word (*eds.*).]

[7] [Von Neumann (1955), footnote 126 on p. 222 (the German edition has endnotes rather than footnotes) (*eds.*).]

I think that here our Jonny has already applied the chisel to the spot where reform is wanted. It is actually pretty clear to me what has happened. We had in fact *lost* the classical model. We not only found no other one, but also encountered the greatest difficulties of principle standing in the way of any model at all. So we said: pshaw, we shall keep the classical one after all, declare that all its variables are in principle measurable, and add with a philosophically wise air that these *measurements* are all that is real, what lies beyond is metaphysics. Then it does not bother us at all that our claims *about the model* are monstrous. We have indeed forsaken it – and may therefore continue to use it all the more carefree. The error was this: *if* one wishes to adopt this highly philosophical standpoint, then one must really do it, i.e. one must declare truly performable measurements or their idealisations to be the 'sole reality'. – I do not know *what* a gamma-microscope really measures; but according to the above *definitely not* a variable that has a canonical conjugate (in the sense of $pq - qp = 1$).

I have already blathered at you too much, and therefore do not want to say any more today about my 'blasphemous' acoustical analogy. I fully understand your aversion, but I think that I am still correct. For my purposes, in lieu of sound, a system of formulae from mathematical population statistics could have served [herhalten] just as well. It is only a matter of realising by the example that the beloved method in electrodynamics for deriving the conservation laws is *so good* that it also perfectly captures the effect of the terms that one has struck out. And the formal analogy is so confoundedly close that the suspicion naturally arises that something is missing also in electrodynamics. Besides, somebody once said to me: If everything in the world proceeded linearly, nothing would act on anything.[8]

Excuse my poor typescrawling. It is still more legible than my handwriting.

Warm greetings to you from your sincerely devoted

[Schrödinger]

[8] [This is possibly a reference to the now famous letter from Lorentz to Schrödinger of 27 May 1927, in which Lorentz commented extensively on Schrödinger's wave mechanics. Among other things, Lorentz commented on Schrödinger's remarks trying to account for the observed radiation frequencies $\nu_i - \nu_k = \frac{1}{h}(E_i - E_k)$: 'if these fundamental equations are also linear then the superposition of two solutions, $\psi_1 = a_1 \cos(2\pi\nu_1 t + b_1)$ and $\psi_2 = a_2 \cos(2\pi\nu_2 t + b_2)$ will lead to beats; no instrument (resonator, grating) whose operation is completely determined by linear equations would respond to these beats as it would to vibrations of frequency $\nu_1 - \nu_2$' (Przibram 1967, p. 49) (*eds.*).]

Schrödinger to Berliner, 25 July 1935*

[Oxford,] 25th July 1935.

Dear Herr Doktor Berliner!

Excuse me for not answering your lovely letter of 1 July for such a long time. I wanted to tell you at once further details of the essay, which has changed rather a lot every time I put my hands on it. I can best characterise the change like this: the quote from the article in *Phys. Rev.* and its discussion come only after the third quarter, at the beginning of the fourth quarter.

I have tried to do the following. A clean exposition of how things are done now in quantum mechanics, without formulas. In particular the role of the classical model, what it is still used for and what it is no longer used for. Then the precise theory of the measurement process and of the 'reduction of the wavepacket', without formulas – by the way, without waves either. The psi-function is described and explained as a catalogue of expectations. For instance, a Bädecker [*sic*] is used as an intuitive image, which is skillfully[1] divided up into its components, *one* of which is then valid for the rest of the journey. The Einstein case then falls into place quite by itself, as a special measurement process with weird features. Critical remarks and explanations are interspersed from the very beginning, I hope without disturbing the exposition of how it is done [dessen, wies gemacht wird (*sic*)], but such that one always immediately sees what the criticism is directed at. I hope it is going to be understandable; I have put in a lot of effort.

Here comes a critical point, however: the scale has become monstrous. I have counted: 10230 words of text, without the §-titles, which are still missing. This word count gives 13.3 *pages* of *Naturwissenschaften*, so not much short of 14. Dividing across two issues would be obvious. (I would not like it *very* much across three, because if one substitutes logical thoughts for all the formula stuff, it is indeed very important to keep the whole connected.) But one asks oneself, since as you said several will take up the pen on this, whether one person can take up so much room, in particular one who is somewhat refractory to the *general* opinion. There will be replies or even articles and *Naturwissenschaften* might be choked for a while with quantum epistemology, to the annoyance of the readers who are not only interested in that.

* [From Alpbach, 3.1, Manuskripte ES, Berliner (carbon copy, 2pp., in German). Transcription in Meyenn (2011, vol. 2, pp. 555–557). Most of p. 557 of the transcription appears to have been included in error, and to be the last page of a subsequent letter to Berliner's successor as editor of *Naturwissenschaften*, Hans Matthée. The original of the present letter possibly included at the bottom of the second page only handwritten greetings and the signature (*eds.*).]

1 [sinngerecht/kunstgerecht (*eds.*).]

I must leave that to you, possibly after inspection of the manuscript (which I still need to type up, which will still take a while).

I also want to say that recently an Italian bookseller (Mandadori [*sic*]) has been interested in essays of mine, to collect them together with some by Heisenberg and by Fermi in a book.[2] The thing is still in the very first stages: he has not even written to me himself. That would not involve any difficulties, nor publication in a different language in the first place? It is not my ideal, it gives one so terribly much to do. But one cannot have more than one view at one point in time, since thank goodness one is not a quantum mechanical system oneself. And if then people indeed want something, one would rather give them the latest stuff.

I shall gladly sniff at and munch on the specific heats (Douglas McKie & [N. H. de] V. Heathcote[3]), even though Laue actually ought to be punished for thinking it is interesting. Or maybe in fact it is?[4]

[Greetings missing]

[E. Schrödinger]

[2] [A publication project with Mondadori does not appear to have proceeded further (*eds.*).]

[3] [McKie and Heathcote (1935).]

[4] [As noted in Meyenn (2011, vol. 2, p. 557), this book review was eventually taken over by A. Eucken (*eds.*).]

Schrödinger to Laue, 25 July 1935*

Oxford, 25th July 1935.

Dear Laue![1]

Thank you for your lovely card. The reason why I have not answered for so long, apart from my usual laziness in writing, is that London has not yet passed me the offprint, counting on my having the *Naturwissenschaften* of course. There I had indeed already seen your note, but at that moment not thought about it properly – and in the meantime the issue [of the journal] has made itself scarce.[2]

Over the past weeks I have written a long essay on quantum mechanics for Berliner, if it is not too long for him. The initial impulse came from the Einstein–Podolski–Rosen [*sic*] note, but now it has become a presentation, meant to be generally understandable, of the whole currently common procedure [(w]ay the classical model is used, measurement theory, reduction of the wave packet) with critical commentary. In the *Physical Review* case I have by the way also worked out something that I think is interesting. It [the *Physical Review* case] consists, does it not, in the fact that one knows

$$\begin{aligned} x_1 - x_2 &= x' \\ p_1 + p_2 &= p' \end{aligned} \tag{a}$$

x_1, x_2, p_1, p_2 are quantum variables, x', p' are ordinary numbers. Then one can show: *any arbitrary* measurement on 1 can be prognosticated from an *appropriate* measurement 2. Let the measurement on 1 be denoted by the 'well-ordered' Hermitian-symmetric polynomial $F(x_1, p_1)$. I claim that then the following operator on 2 gives the same measurement value:

$$F(x_2 + x', p' - p_2).$$

This is *not* quite trivial, because the equations (a) are not in fact operator *identities*. But one can show that the operator

$$O = F(x_2 + x', p' - p_2) - F(x_1, p_1),$$

* [From Meyenn (2011, vol. 2, pp. 558–560). The source there is given as the ZBP-Wien, but the letter is not in the digitised Nachlaß and could not be found in the collections (*eds*.).]

[1] [With Laue, Schrödinger uses the formal 'Sie' (*eds*.).]

[2] [As Meyenn notes, Laue's postcard and Schrödinger's comment here refer to a short communication by Laue in the 6 June issue of *Naturwissenschaften* (Laue 1935a). Cf. also the more extended Laue (1935b) (*eds*).]

applied to the ψ-function of the total system, always gives zero.[3] Introducing the commutative quantum variables x, p

$$\begin{aligned} x_1 - x_2 &= x \\ p_1 + p_2 &= p \end{aligned} \tag{b}$$

I write it:

$$O = F(x_1 + x' - x, p_1 + p' - p) - F(x_1, p_1).$$

I want to show that it can be simplified in such a way that each polynomial has either a factor $(x' - x)$ or $(p' - p)$ on the *right*: for I know *this* of the ψ-function of the total system, that such a factor annihilates it. Thus in the simplification I shall just delete terms satisfying this condition. And now I take an arbitrary polynomial of F, e.g.

$$\begin{aligned} \ldots (x_1 + x' - x)(p_1 + p' - p)(p_1 + p' - p) \\ \cdot (x_1 + x' - x)(x_1 + x' - x)(p_1 + \cancel{p' - p}). \end{aligned} \tag{2}$$

In the last factor I have already stricken out the annihilating factor $p' - p$. *Its partner* p_1 *now commutes with all the others*, I move it through to the left into *first* place. Then in the penultimate factor I strike the $x' - x$ through. x_1 commutes with all the brackets, is moved to the left into *second* place. And so on. In the end we are left with the polynomial in the *reverse* order of factors, *cleansed* of the $(p' - p)$ and $(x' - x)$:

$$p_1 x_1^2 p_1^2 x_1 \ldots.$$

This must appear also in $F(x_1, p_1)$ because of the Hermitian symmetry. So everything indeed cancels out.

I have struggled a lot with this proof. In fact, for non-Hermitian [polynomials] the theorem does not hold. And also not *then*, when $\sqrt{-1}$ appears explicitly, e.g. for the *Hermitian* $\sqrt{-1}(p_1 x_1 - x_1 p_1)$ the sign is wrong. One can make the proof somewhat simpler if one introduces δ-functions for the ψ-function of $x - x'$ and $p - p'$, but following Dirac this is wrong, the wavefunction *itself* is not a δ-function.

What interests me here is this. System No. 1 'clearly already intends' to give a very definite answer in an x_1-measurement, should this be performed first (for I can extract the answer without interfering with it); similarly it has prepared an answer for p_1; but now similarly also for each $F(x_1, p_1)$, for instance for

$$p_1^2 + a^2 x_1^2 \qquad (a = \text{arbitrary positive number}).$$

[3] [The left-hand side should presumably read O rather than 0 (which is the reading of the transcription), since as Schrödinger has just remarked, the two terms are not equal as operators. The transcription has $F(x_1 + x', p' - p_2)$ instead of $F(x_2 + x', p' - p_2)$ (*eds.*).]

Here the prepared answer must be a number from the sequence

$$\frac{ah}{2\pi}, \frac{3ah}{2\pi}, \frac{5ah}{2\pi}, \frac{7ah}{2\pi}, \ldots.$$

From that I can see that the prepared answers do *not* stand in the algebraic relationship one would perhaps expect. For, letting the number x'_1 be prepared for x_1 and the number p'_1 for p_1, then of course

$$\frac{2\pi}{h} \frac{p'^2_1 + a^2 x'^2_1}{a} = \text{odd number}$$

cannot be the case for every positive value of a. My devilish system knows all its answers; whichever I ask it first, it will give. But I have not the foggiest idea by what mnemonic technique it masters all these answers. Nevertheless: the 'agreement in x_1 and p_1' (as I can compactly say) guarantees already the agreement with respect to all other variables.

Warm regards to you all from your sincere

Schrödinger

[P.S.] Have a good rest and much fun in the mountains! I hear you are going with Thirring. Please give him my best regards.

Berliner to Schrödinger, 29 July 1935*

Berlin W 9, 1 July 1935.
Linkstrasse 23–24
Publishing house of Julius Springer

To Herr Professor Dr. E. Schrödinger
Oxford

Dear Herr Professor Schrödinger!

I am answering immediately your interesting and lovely letter of 20 July [*sic*] to tell you how I am looking forward to your essay and that I reserve for you as much space as you see necessary, and in two immediately successive issues. I am also not worried if someone or other claims a lot of space. I cannot wish for anything better than the stars of first magnitude in the physics firmament finding their way into the *Naturwissenschaften*, like in previous years at the time of the dispute over the theory of relativity or at the time of Bohr's theory. After all, our journal is there just for such things, and I am all the happier to put the *Naturwissenschaften* at the disposal of pure science without any reservation, given that other journals – you do know what I mean.

Should the journal *Current Science*, published in India, come to your notice, and should you to your surprise find me named there as associate editor, I ask you not to count that too heavily against me. I wanted to decline, but have been urged by everyone else whom I unfortunately asked for their advice (which also includes Dr Ferdinand Springer) not to refuse this 'honour'. This for your amusement – hopefully also the following sheet [Blatt].[1]

With best regards, your [Stamp: *Die Naturwissenschaften*]

A. Berliner

* [From Alpbach, 3.1, Manuskripte ES, Berliner (typewritten, 2pp., in German). Transcription in Meyenn (2011, vol. 2, pp. 560–561) (*eds.*).]

[1] [In his reply Schrödinger comments also on this enclosure, but it is not clear what it consisted of (*eds.*).]

Einstein to Schrödinger, 8 August 1935*

Old Lyme, Conn. 8. August 1935.
White House

Professor Erwin Schrödinger
24 Northmoor Road
Oxford
England

Dear Schrödinger:

You are indeed the only person with whom I truly enjoy having a discussion. Almost every Joe fails to look at the theory starting from the facts, but only looks at the facts starting from the theory; they cannot get out of the conceptual net once they have adopted it, but can only quaintly flop about inside it. You, however, look at it as you wish – from the outside and from the inside. Nevertheless, in our conception of the way ahead we are perfect opposites.

My solution to the paradox given in our paper is as follows: the ψ-function does not describe a state of *one* system, rather (statistically) an ensemble of systems. Compared to a ψ_1, a linear combination $c_1\psi_1 + c_2\psi_2$ means an *extension* of the *collection* of systems. Conversely, in our example of the system comprised of two parts A, B, the change that the ψ-function undergoes when I make an observation on A means the detaching of one subcollection from the whole ensemble. The detaching simply takes place according to a different criterion, depending on the *choice* of the quantities I measure on A. The result is then an ensemble for B that likewise depends on this choice.

Of course, this interpretation of quantum mechanics brings out with particular clarity the fact that by limiting ourselves to statistical statements we have acquired the ability of incompletely representing real states and processes. –

You instead view something totally different as the reason for the intrinsic difficulties. You see in ψ the representation of reality, and want to modify or completely abolish the connection with the concepts of ordinary mechanics. Indeed, only thus could the theory truly stand on its own legs.

This point of view is certainly coherent; but I do not think it is able to remove the ills we perceive. I want to justify this using a crude macroscopic example.

Let the system be a substance in a chemically unstable equilibrium – say, a pile of gunpowder that can ignite itself by internal forces, where the average lifetime is

* [From AHQP 92, section 2-109, and Alpbach, 6.5, Korrespondenz, Vorordner (typewritten, 3pp., in German). Transcription in Meyenn (2011, vol. 2, pp. 561–562) (*eds.*).]

of the order of one year. In principle this is very easy to represent quantum mechanically. Initially the ψ-function characterises a sufficiently definite macroscopic state. Your equation, however, ensures that this is no longer the case after a year has passed. The ψ-function then describes rather a kind of mixture of a not-yet and an already-exploded system. Through no interpretational artistry can this ψ-function be made into an adequate description of a real state of affairs; in truth there just is no middle thing between exploded and unexploded. Thus your equation certainly cannot provide the description of the actual process as you in fact imagine. (On the other hand, in the statistical sense it [er (*sic*)] can certainly reproduce correctly the changes in a collection of systems.) With this example I want to suggest that your attempt at an interpretation fails with respect to what we know with certainty from coarse macroscopic experience.

Have a look at the little paper that I have recently published with Mr Rosen in the *Physical Review* on a conceivable relativistic analysis of matter. This could lead to something if the mathematical difficulties can be overcome.[1]

I would very much like to have you at Princeton, and I will strive to dispose of the 'diplomatic' hindrances. In the meantime, warm greetings from

Your

A. Einstein.

[1] [This is a reference to Einstein and Rosen (1935) (*eds.*).]

Schrödinger to Berliner, 11 August 1935*

[Oxford,] 11 August 1935.

Dear Herr Doktor Berliner!

I did not thank you immediately for your extremely lovely letter of 29 July because after you promised it such a lovely welcome, the ms. bugged me twice as much. I have just now packed it up and will dispatch it tomorrow together with this letter. Born and London have read it. In fact I shall be able to talk with Born extensively only in a couple of days. But as things stand I can indeed send it off and it can be printed immediately – *if you wish.* You know that *your* discretion to judge what really belongs in the *Naturwissenschaften* is sacrosanct to me.

[I wonder] whether this letter reaches you in Berlin itself or whether you are now in fact resting a tad. The semester here finished two months ago already. But after the beautiful trip to Spain I very much felt the need once again to earn my right to live by toil. Therefore I was very happy that such was found – for often one sits around with the best intentions and nothing comes of it.

If you are on holiday or are going soon, I wish you all that is best and restful for it – and if not then I wish it for the purpose you wish it for.

The mathematical excerpt you sent me interested me very much. In this connection I would speak of a sudden interruption in the secretory activity of the mucous membranes of the mouth. At one point, however, I could not quite follow. *One* metamathematical theorem is formulated thus: from the proof of the absurdity of a claim there does not follow the correctness of the claim. If this is not a typo, then this theorem is indeed strong support for the subsequent [sentence], in which it is stated: we have finally got to philosophy.

Very warm greetings again from your faithful

[E. Schrödinger]

* [From Alpbach, 3.1, Manuskripte ES, Berliner (carbon copy, 2pp., in German). Transcription in Meyenn (2011, vol. 2, p. 563) (*eds.*).]

Berliner to Schrödinger, 14 August 1935*

Dr A. Berliner
Berlin W 10, Herkulesufer 11
14 August 1935.

To Herr Professor Dr. E. Schrödinger
Oxford

Dear Herr Professor Schrödinger!

As with every letter coming from you, I was truly pleased to receive your lovely letter with the friendly trust you put in my judgement. This time, however, with a large percentage of pathos, because it is the last letter you will have sent me as the editor of the *Naturwissenschaften*. Since midday yesterday I no longer am. You know the reasons without me having to elaborate. I cannot tell yet whether possibly I shall be able to continue doing something for the journal, perhaps as an intermediary. Already a long time ago I had mentioned Professor Matthée to Dr Ferdinand Springer as a possible successor to myself, the editor of the *Zeitschrift für den physikalischen und chemischen Unterricht* [*Journal for Physical and Chemical Education*] published with Julius Springer. Matthée is a man full of insight and perhaps even too modest. He has told me time and again he does not feel capable of continuing to manage the journal in the way I managed it all these years. However, that will now have to be left to the future, and really here the eating is the proof of the pudding.[1] I assume Matthée will accept becoming my successor only under the condition that I can support him in some way. But I consider it not improbable that this will be prevented through some plotting or other. I assume that the manuscript, which according to your letter of 11 August you have already sent off, will come out of Professor Matthée's hands into mine,[2] and then your preferred order of business is vouchsafed. Vederemo![3]

You can rather imagine that the departure from the journal is difficult for me. You surely no longer remember that our relations started already in the year 1917, in fact with an essay on atomic and molecular heat,[4] and even less that we were

* [From Alpbach, 3.1, Manuskripte ES, Berliner (typewritten with manuscript corrections and additions, 2pp., in German). Transcription in Meyenn (2011, vol. 2, pp. 564–565) (*eds.*).]

[1] [The words 'the eating the proof of the pudding' are in English in the original (*eds.*).]

[2] [Added in the margin in Berliner's handwriting: 'has just happened' (*eds.*).]

[3] [Rarer variant of Italian 'vedremo' [we shall see] (*eds.*).]

[4] [Schrödinger (1917).]

recommended to you by our mutual friend in Princeton.[5] And I hope our friendly relations will not change.

I have just informed Max Born of everything and have taken the opportunity to ask him whether he might not attract the attention of an English publisher to the excellent review in *Nature* of the fifth edition of my book. Maybe it might prompt the publisher to print an English translation. This would be important for me now because it is going to be all over for my book here in Germany. Should you think of anything useful I could still contribute, let me know. I am only old in years, otherwise I have such a need for activity as if I was only half as old. Do you think it conceivable to set up a journal like the *Naturwissenschaften* in Switzerland?

With best regards,

Your A. Berliner

[Handwritten addendum in the margin:]

Can you write to Planck to prompt him to do something on the part of the K[aiser] W[ilhelm] G[esellschaft], whose organ the journal [is], to preserve the *NW* [*Naturwissenschaften*] in its current status, for otherwise it is presumably doomed.

[5] [The phrase 'our mutual friend' is in English in the original (*eds.*).]

Schrödinger to Einstein, 19 August 1935*

19. August 1935.

Dear Einstein!

Many thanks for your lovely letter of 8 August. I do not believe it will work to refer the ψ-function to an ensemble and thereby solve the antinomy or the paradox. Actually I am not at all happy with the parlance of 'it will not work', because it is used with partiality by the people with blinders who regard certain formal stuff [rechnerische Dinge] as inevitably fixed, since they cannot imagine things without them. But here it is [indeed] the case:

The antinomy in your example, where one can infer *p or x*, as desired, from a measurement on the *other* system, is in fact much worse still. One can, if one follows the present interpretation (and one must of course do this, if one wants to show that it is amiss) not only ascertain p or x according to one's choice, but instead also e.g. $p^2 + x^2$, more generally any $F(x, p)$, through *one* measurement on the other system; indeed, the current Q.M. presupposes that for every such operator there is a measurement procedure, and I have proven that in your case corresponding measurements on the two systems agree no matter what one chooses for F. (This is not *quite* obvious; the proof has even made me sweat a bit.) According to the principles of Q.M., however, a measurement of $p^2 + x^2$ must always yield an odd multiple of h. (Not only when the thing is an oscillator, but always.) And not just the measurement of $p^2 + x^2$ but also that of $\frac{p^2}{b} + bx^2$. I say that current Q.M. dictates that the measured value for this operator must always be an odd multiple of h. And, by the way, this is true for any arbitrary value of the constant b (positive and finite). Thus [we have]: a series of operators; the system apparently 'knows' entirely accurately what it will answer when it is first directly asked 'about the value of one of these operators'; it must know this because I can, of course, predict the answer through an appropriate question which I put to the *other* system, with which ours cannot communicate. And this answer must in every case, i.e. for every b, be an odd multiple of h. That this requirement cannot be satisfied even for an ensemble without modifying the correspondence to the concepts of ordinary mechanics is clear. This is not to say, of course, that the latter must necessarily happen; one of the postulates of Q.M. that I have used above can be rejected – and then maybe it works with an ensemble. But the postulates used are such that if one is rejected,

* [From AHQP 92, section 2-110, Alpbach, 6.5, Korrespondenz, Vorordner (carbon copy, 5pp., in German), and EA 22-051-3 (original typescript with added footnote). Transcription in Meyenn (2011, vol. 2, pp. 565–568) (*eds.*).]

then Q.M. is thoroughly upset. Referring to an ensemble instead of to an individual system is then the least [upsetting].

I am long past the phase when I thought one can consider the ψ-function somehow directly as a description of reality. In an extended essay that I have just written, I give an example that is very similar to your exploding powder keg. I just really wanted to ground the whole issue in an indeterminacy[1] that, in the current understanding, is really 'Heisenbergian' and not 'Boltzmannian'. A Geiger counter is locked in a steel chamber, set up with a tiny amount of uranium – so little, that in the next hour *one* atomic decay is to be expected with equal probability as none. An amplifying relay ensures that the first atomic decay smashes a small flask with prussic acid. This, and – horribly – a cat, are also found in the chamber. After an hour, then, *sit venia verbo*, the living and the dead cat are smeared in equal parts in the ψ-function of the total system.

I had already read your essay with Rosen about the 'bridges'.[2] My memory from before is that the spatial metric in the vicinity of a mass point (if for purposes of visualisability one limits oneself to two spatial dimensions) can be described as something like half a rotation hyperboloid, where the smallest circle at the 'neck' represents the mass point with its finite dimension, in which 'time stands still'. Since the hyperboloid has no singularity there, I toyed with the idea that the extension should be 'allowed' too, but in a somewhat crazy way, which I felt calculationally impotent to pursue. I wondered whether one positive and one negative particle might not be connected 'underground' by a kind of 'handle'.[3]

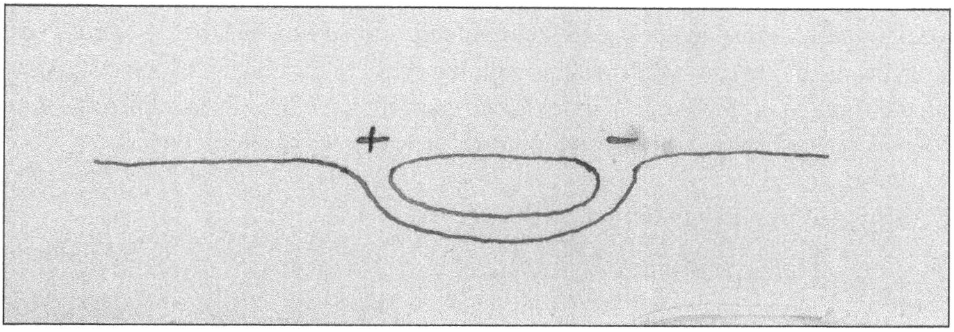

Fig. 26.5 [Schrödinger's drawing of the cross-section (*eds.*).]

[1] [eine Unbestimmtheit ins Spiel zu bringen/das Ganze auf eine Unbestimmtheit zu gründen (*eds.*).]

[2] [Einstein and Rosen (1935); the word 'bridges' is in English in the original (*eds.*).]

[3] [Here the original in the Albert Einstein Archives (EA 22-051-3) contains a handwritten footnote with a drawing (Fig. 26.5 here), labelled 'Im Querschnitt:' ['Cross-section:'] (*eds.*).]

Not that I want to return to this specific fantasy. But I have the feeling that one might not succeed with *two* Riemannian surfaces in which exactly the same happens, i.e. which are faithful copies of one another. But I understand much too little of this. – I find the thing with the sign very strange: that one needs to insert the negative Maxwell tensor into the gravitational equations. I do not at all see what that means. Of course it is not unimportant, otherwise one could not in one case get a singularity-free solution, and in the other case not. I ask myself in vain what the physical difference is between the two signs.

And now something completely different: Arnold Berliner has been removed out of the blue from the editorial desk at *Naturwissenschaften*. It left a particularly crass impression on me; I had exchanged several letters with him concerning the article alluded to above, and when I sent him the ms. about a week ago, a couple of days later I received from him a touching letter: he expected in fact that he would be assigned to handle the ms. himself, since he was still able to assist as a consultant, but as of 24 hours earlier he was no longer editor. My first thought was: pull out. However, since in the same letter he touchingly suggests that he not only means still to do all that he can for his work, but also asks us to do the same, I did not do it. It is one of the most emotionally shocking cases. Berliner asked if one might not interest an English publisher in translating his textbook, which has just appeared in the 5th edition, and which had a brilliant review in *Nature*. His aim is to find some activity by working through the translation. –

Many thanks for all the kindness you express to me. Yes, I should be terribly glad! But perhaps the path truly runs through Graz. It seems to me at the moment that I shall accept there, naturally without regard to the speculation just mentioned, just in order to return to having an appointment instead of a benefice. It would be a bit awkward in the case of Princeton with the moving expenses. If I get Austria to pay me some of it, of course I cannot leave again right away without repaying them.

However, one should not muse too much. It will work out somehow.

Warm greetings from your devoted

Schrödinger

Einstein to Schrödinger, 4 September 1935*

Old Lyme, Conn. White House 4. 9. 35

Dear Schrödinger:

If I claimed to have understood your last letter, then that would be a fat lie. You talk of an arbitrary function $F(x, p)$. Does that refer to the first subsystem, or the second, or to both? Of course I do not want both subsystems to be considered as identical (same function F). But then the claim that in the mentioned special cases the measurement results (the measured one and the prophesied one) are equal is meaningless.

Leaving aside everything I do not understand, I do not see why it is contradictory to assume that the ψ-function relates to a statistical ensemble. In any case, one can thus avoid the paradox mentioned in our paper. Taking the measurement on system A into consideration then indeed implies going over to a sub-ensemble, which of course can proceed according to different criteria.

By the way, your cat example shows that we fully agree in assessing the character of the present theory. A ψ-function which includes both the living and the dead cat cannot indeed be considered a description of a real state. Instead, this same example suggests that it makes sense to assign the ψ-function to a statistical ensemble, which includes both systems with a living cat and ones with a dead cat.

As regards my relativistic conception of corpuscles, the denominator-less form of the covariant equations is crucial, since only in this case does the vanishing of the determinant (g_{uv}) imply no singularity. Thus one needs a true generalisation of the theory, however obvious this may seem *a posteriori*. I have now found a method that appears to make the treatment of the many-body problem feasible. What speaks in favour of the theory is that it provides from the outset a kind of explanation for the atomistic nature of ponderable and electrical mass. A theory that involves a field-like binding of a positive particle of electricity with a *specific* negative one seems even worse to me than the institution of marriage in human society.

I also found the sign change of the electrical terms rather curious. One should consider, however, that there is no energy-momentum conservation at all in the sense of the present theory. I think what is concerning is only that I have no mathematical equivalent at all for this important fundamental law; then again, the quantum phenomena exclude a localisation of the energy in the field as in Maxwell's theory, so that in any case here we must get ready for a big adventure.

I am immensely sorry for Berliner. I have been able to arrange for Berliner's textbook to be translated here. Flexner has let me know that the physicist Löb, the son of the deceased great physiologist, has taken charge of this translation. I will

* [From AHQP 92, section 2-111, and Alpbach, 6.5, Korrespondenz, Vorordner (typewritten, 3pp., in German). Transcription in Meyenn (2011, vol. 2, pp. 569–570) (*eds.*).]

find out how the matter stands. Also, I shall endeavour to have Berliner invited to America; but this will be rather difficult, because people here do not easily do things just out of the goodness of their heart. As regards your possible call to our institute, I want to write to Flexner now. I believe that given the present situation Flexner could indeed venture to issue a call after checking with the university people.

I am actually sorry that you still publish in Germany. In my opinion, the righteous should terminate relations for as long as the local scientists tolerate this shameful regime. The risk of a barbarisation of the foreign countries that are still healthy is not at all slight; any person of consequence can contribute something to the collapse [of the regime], by indirectly influencing the German intellectuals, through a consistent negative attitude.

Warm greetings from your

A. Einstein.

Bohr to Heisenberg, 10 September 1935*

10 Sept[ember] 35.

Dear Heisenberg,

I am very sorry that I have not answered your kind letter with the manuscript of your article about the completeness of quantum mechanics, but as my wife wrote, I have had jaundice for several weeks and have begun to get up only in the last days and then to get to grips with the work again. I have just sent off corrections of my response to Einstein in 'Physical Review', where, as I wrote in summer, I have tried to put forward the principled point of view a bit more clearly with some small changes. In order to give you an idea about what it is about I shall send you in a few days a copy of the revised manuscript that has just been typed.

I have also in these days read your article with interest. There are however some points in the logical structure that I do not completely understand and that I shall write to you about, when I have thought about it a bit more, at the same time as I send you the manuscript.

Today I just wanted to give a little sign of life from me and say that it is nice to get up again, even though I still feel rather shaky, and I shall in fact need to get away for a few weeks to recuperate before I am really able to work again. However, I shall be back by October, and I am very much looking forward to your visit and to having the opportunity to discuss with you in more detail all the various matters.

With many kind regards from all of us

Yours

[Niels Bohr]

* [From AHQP-BSC 20, section 2 (carbon copy, 1p., in Danish). Thanks to Terri Seager for help with the translation (*eds.*).]

Heisenberg to Margrethe Bohr, 13 September 1935*

Leipzig 13. 9. 35.

Very revered dear Mrs Bohr!

Excuse me for sending only such a late thanks for your letter. I was very glad to hear from you and was naturally even gladder to learn from the last letter that in the meantime the illness has ended. I was very sorry about your having had so many difficulties this summer; I did not know at all that jaundice can be contagious. Hopefully now your husband will recover quite properly in the coming weeks. For the moment please say a heartfelt thank you to your husband for the last letter; there is no rush at all with the manuscript, in any case it is not being sent off until there is agreement over the content.

I am looking forward very much to the visit in Copenhagen; but I only wish to come if I am sure that you are all fully well again and I am not disturbing.

With many cordial greetings,

Your gratefully devoted

Werner Heisenberg.

* [From BSCSupp-HEI-350913t (handwritten, 2pp., in German). With many thanks to Felicity Pors for providing a copy of this letter (*eds*.).]

Bohr to Heisenberg, 15 September 1935*

15. September 35.

Dear Heisenberg.

As I promised a few days ago, I send you today the improved manuscript for my paper in 'Physical Review', which corresponds to the corrections inserted in the proofs. As you will see, these are of a formal nature only; without entering more closely into the mathematical formulation of quantum mechanics, I have aimed to show that the complementarity feature in the description is unavoidable if the description is to cover the characteristic elementary quantum phenomena and, at the same time, make correspondence-like connection to the classical mechanics.

It is in this connection that I have experienced difficulties in trying to understand more closely the argumentation in your article. Because I am not quite sure that I fully understand the importance you attach to the freedom of shifting the cut between object and measuring apparatus. Any well-defined quantum-mechanical problem must be concerned with a certain classically described experimental setting; and if one changes the kind or use of the measuring instruments, and thus of the [experimental] setting, the phenomenon will usually change completely. I therefore believe that for a given experimental setting the cut is determined by the very nature of the problem (of course, except for a shift within those domains of the experimental setting that are described unambiguously by the classical theory).

Moreover, I do not fully understand how the mentioned narrow possibility of shifting the cut in any given experimental setting can be used in connection with the particular problem of measurement which you proceed to deal with. I believe in fact that the assumption that between the systems *A* and *B* there is only an interaction for a definite value [q'_A] of the coordinate [q_A] belonging to *A*, must already imply that the experimental setting provides the means to determine a definite value *q*. If, for example, [q_A] is an ordinary space coordinate, then I believe that it is required that the position of *B* relative to those parts of the experimental setting that determine the coordinate system is uniquely determined; if so it does not matter if *B* is directly connected with the solid frame, or if it stands in interactions with other parts of the experimental setting,[1] in such a way that its position relative to the frame can be uniquely determined. Therefore, any use of *B* for measuring the complementary quantity [p_A] requires a substantially different experimental setting, and then the possibility of defining a value [q'_A] in a unique way is lost. No one, who like Laue

* [From AHQP-BSC 20, section 2 (carbon copy, 2pp., in Danish). Thanks to Helge Kragh and Terri Seager for help with the translation (*eds*.).]

[1] [Bohr leaves a space here for a symbol to be handwritten in (*eds*.).]

and Schrödinger is familiar with the theory, will deny that this result, which is an immediate consequence of the very possibilities of definition of the classical concepts used, is also a consequence of quantum mechanics; for this reason I fear that they will not be more convinced of the rationality of the theory because of your argumentation. Their criticism, as far as I know it, is simply that they are dissatisfied with an acausal element of description; and I fail to see that one can answer this in any other way than [by pointing out] that according to the very nature of our experience all elements of the description are classically defined and that nevertheless the classical theory leaves no room for the quantum mechanical laws.

You understand, I am sure, that I have no intention of being quarrelsome and that I am quite prepared [to admit] that I have either misunderstood your argumentation or that I, in a pedagogical sense, have a much too one-sided attitude to the problems of atomic physics. I have never doubted the completeness of quantum mechanics ever since your foundational work more than a decade ago. Quite apart from the mathematical advances, which are included in the transformation theory etc., for me any endeavour to clarify our attitude in this domain has been exclusively epistemological, not to say of a psychological kind. This is also the background for the form of my reply to Einstein, which all the same I hope is not too 'un-psychological' in a pedagogical sense.

I shall be glad to know what you think of the content of this letter, which requires – 'in the sense of Pauli' – to be judged mildly and overbearingly; I look forward in particular to when we can soon really discuss these questions when you arrive here in October. I hope that by then I will have rested enough that I can take up in an open-minded way a whole new discussion of all the psychological and pedagogical aspects behind the unreasonable criticism of quantum mechanics.

With many kind regards,

Yours

[Niels Bohr]

Heisenberg to Bohr, 19 September 1935*

Leipzig 19. 9. 35

Dear Bohr!

Many thanks for your letter, I was very glad to see – as one can well read from the style of the letter – that you are again feeling much better!

Regarding first the physical questions in your letter, of course the shifting of the cut that is treated in my manuscript is indeed the one inside the region in which classical physics can be applied. (I have also said this explicitly at the end of §1.) Further, we agree that a system B which only reacts for a given value q'_A is not suitable for measuring the coordinate complementary to q_A; it is also not necessary that system B is the same in both cases. (If one wants B to still allow for a measurement of q_A *or* p_A, then, as we have indeed often discussed in the past, one can let B e.g. be composed of a source of light and a lens in a rigid frame. (Let A be an electron in a fixed object plane.) Then one can place the system C, namely the photographic plate, either in the image plane or in the focal plane and measure position in the one case, momentum in the other. However, I did not want to enter at all into the specific measurement arrangements, since they have in fact been discussed on various occasions already.) The easiest way to explain why I find the movability of the 'cut' so particularly important is this: You correctly write that 'all elements of the description are classically defined and that nevertheless the classical theory leaves no room for the quantum mechanical laws'. This statement appears plainly contradictory to physicists who are accustomed to thinking within the formalism, as I know e.g. from discussions with Mr von Laue. Therefore it seemed important to me to emphasise the feature of the formalism that ensures that no contradiction arises here; and that seems to me to be indeed the movability of the cut. If this were to fail, then there would be just two categories of physical systems, classical and quantum-theoretical, and one could *never* apply classical concepts to the latter. That is more or less how von Laue thinks of the matter. I also think it would be difficult then to argue against the hope of a later causal completion. – In any case, I think if one wants to understand within the formalism why it does not just lead to new concepts that are entirely different from the classical ones, one can learn this most directly from the movability of the cut.

But I would rather not engage with the psychological questions of how to convince a von Laue or an Einstein; because maybe understanding quantum theory is in-principle complementary to understanding the psychology of those who do not

* [From AHQP-BSC 20, section 2 (handwritten, 4pp., in German). On the reel this letter is misdated as 29 September 1935 (*eds.*).]

want to understand quantum theory. Rather, I am looking forward to an extensive discussion of this in Copenhagen. (Of course I shall also not send off the manuscript before that, since I have the feeling that you are still unhappy with this more formal way of treating quantum theory; but we shall indeed see in Copenhagen!)

Besides, I would also like to say that the purpose of the manuscript was to bring out a certain important property of the formalism more than before, while in fact I have no doubt that your response to Einstein is an entirely clear and perhaps psychologically better account of the issue; as indeed I have already written earlier to you.

And now I shall wish you a very good recovery, so that the psychological discussions in Copenhagen can be carried out with sufficient vigour.

Many greetings to you all!

Your Werner Heisenberg.

Schrödinger to Einstein, 4 October 1935*

4. October 1935.

Dear Einstein!

I was very annoyed at having lately expressed myself so unclearly. So, to begin with, the thing with the statistical interpretation of the [ψ]-function. I have made it so needlessly complicated. I want to give the simplest example. The system is *one* mass point in a plane. Present quantum mechanics asserts that the angular momentum of this mass point with respect to the coordinate origin is always an integer multiple of h. But not just with respect to the coordinate origin – the angular momentum should always be an integer multiple of h also with respect to any other fixed point. One cannot satisfy this requirement with a statistical ensemble. Each element of the ensemble can have an integer momentum only with respect to some privileged reference points. For an arbitrarily chosen reference point, the momentum will turn out to be a non-integer. This is *terribly* trivial. But still it shows that by mere reinterpretation into statistical ensembles – without quite radical modifications – one can do nothing.

Concerning the other story with the $F(x, p)$, I am sending the carbon copy of a little note.[1] What I was talking about is in Section 3, on p. 7. But, pray, do not labour over it yourself, it is not written very well. If one of your young people reads it through, he can explain it orally to you in 2 minutes. Perhaps you will find it very trivial then. But I messed around with it a lot for two days, and did not at all feel certain whether the thing is correct. I was very keen to use only the operator equations[2] (10) and not to write down expressions that are mathematically meaningless. Not out of some cleanliness obsession against [δ]-functions.[3] But this [Ψ] *is* no [δ]-function. Only the square of its absolute modulus is one.

I had a good laugh about your comparison of the handle-shaped tubes with the institution of marriage. No, no, I do not want to support the former. An indissoluble marriage between individual particles would be truly even more unbearable than the human one. On the other hand, the concept of individual particles is surely somehow really wrong (Fermi statistics, Pauli principle). I am afraid you will reject this. But how else should one understand the hocus-pocus that yields, say, the correct spectrum for the Helium atom? Why are there only two spectra and not four?

* [From AHQP 92, section 2-112, and Alpbach, 6.5, Korrespondenz, Vorordner (carbon copy, 3pp., in German). Transcription in Meyenn (2011, vol. 2, pp. 570–572) (*eds.*).]

[1] [This is Schrödinger (1935e). The discussion and equation Schrödinger refers to are on pp. 559–560 (see pp. 168–170) (*eds.*).]

[2] [Singular in the transcription (*eds.*).]

[3] ['ψ-functions' in the transcription (*eds.*).]

I was so happy that the issue with Dr Berliner worked out so quickly, and that he and Laue are now with yourselves. Concerning your objection against publication in Germany, I honestly thought that it was not applicable to the *Naturwissenschaften* as long as Berliner was still the editor there and courageously stood clear of any 'conforming'. My article was already sent off when I learned of the evildoing. And shortly thereafter I received a letter from Berliner that clearly indicated he did *not* wish for the retraction.

I have in fact demonstrated through my actions that I share your opinion with respect to the goal to be reached. I have also repeatedly contended with people like Heisenberg, Planck, Debye e.g. when they went round and round trying to keep German science somewhat on its feet by filling the gaps as well as possible, etc. In this sense I also share your criticism of German scientists. However, one cannot really require those who are physically exposed to the brutal barbarity to openly rebel against it. I am a coward myself and therefore do not reproach cowardice in anyone. These people work, it seems, with – humane methods. Indeed, one cannot properly call them bestial, because a poor good beast does not come up with such ideas.

At present, the Institute for Theoretical Physics in Berlin stands empty, as you will have already heard. I find that an excellent joke. I am instantly happy when I think about the empty rooms.

Warmest greetings to you from your sincerely devoted

[Schrödinger]

Schrödinger to Bohr, 13 October 1935*

13 October 1935.

Dear and revered Mister Bohr!

A few days ago I read the carbon copy of your *Phys. Rev.* article (the reply to the Einstein paradox).[1] I confess that reading it gave me a lot to think about. But I think that in the end I just cannot rest content with it.

I want to discuss the situation in the second experiment. One had measured very precisely the momentum of the double-slit diaphragm, then one let the two electrons pass through, then measured again very precisely the diaphragm's momentum. In this case one is dealing with *three* bodies. Firstly the two particles. Thirdly the rough, rigid, massive coordinate system (perhaps an iron platform) from which *all* measurements are made. I will call this the initial situation S_I. Now I want to discuss the *possible* situations in which *one* particle and a rough rigid coordinate frame can stand to each other; call these situations S'. In classical mechanics, an S' is described completely by giving the coordinates and momentum components of the particle. In classical mechanics, it is impossible to obtain two different S' from the same initial situation S_I by manipulations that have no appreciable physical effects on either the particle or the coordinate frame. In quantum mechanics, an S' is already completely described by giving the coordinates alone or the momentum components alone (leaving aside other more complicated situations). And in quantum mechanics it *is* possible to get to different S' for our coordinate frame and, say, our particle No. 1 from one and the same S_I, through manipulations that, while appreciably affecting particle No. 2, have only arbitrarily small effects on the coordinate frame.

This possibility seems paradoxical to me and I find one cannot rest content with it. – In the argument, the last point could give cause for concern: perhaps making a very precise coordinate or momentum measurement of particle No. 2 *from the coordinate frame* (standing on it, so to speak) does not after all have an arbitrarily small effect on the coordinate frame, no matter how heavy and massive and solid the coordinate frame might be. But this objection in fact does not apply. For had we (starting from situation S_I) first established the situation[2] S' of particle No. 1 through a *direct* measurement, we would not believe that this S' can be modified through further measurements that we make from the same iron platform on other light particles,

* [From AHQP 92, section 2-101, and Alpbach, 6.5, Korrespondenz, Vorordner (carbon copy, 5 pp., in German). Transcription in Meyenn (2011, vol. 2, pp. 573–576) (*eds.*).]

[1] [Bohr (1935b); see Chapter 16.]

[2] ['S'' missing in transcription (*eds.*).]

e.g. also on particle No. 2. The physical influence of such a measurement on the platform is therefore considered to be negligible. –

Actually I did not even want to talk about this point, at least not expecting you to give me a reply – but about something else. You keep expressing in the most definite terms the conviction that measurements must be described through classical concepts. E.g. on p. 61 of the 1931 booklet published by Springer: It is essential to a physical observation that all experience ultimately must be expressed with the help of classical concepts, neglecting the quantum of action'.[3] And *ibid.*, p. 74: 'the use of classical concepts required by the essence of the measurement'. And also here you speak again of 'the indispensable use of classical concepts in the interpretation of all measurements'.[4] But now, soon after that you say: 'The removal of any incompleteness in the present methods of atomic physics ..... might indeed only be effected by a still more radical departure from the method of description of classical physics, involving the consideration of the atomic constitution of all measuring instruments, which it has hitherto been possible to disregard in quantum mechanics'.[5]

From this one could understand that what formerly was described as essential to all physical observation, as 'indispensable necessity',[6] is after all only an expedient tool of enquiry, so far luckily still admissible, which one will presumably at some time be compelled to give up. If this were your opinion, I would happily agree. But the subsequent tight and clear comparison with relativity theory leads me to doubt whether I correctly capture your position with what I just said. For in relativity theory, considered as a conceptual construction in itself, unrelated to quantum mechanics, one would surely never be able to give up the sharp distinction between space and time *in a measurement*. But it might be that through the necessary mutual modification of the two theories *both* would be forced to shake off their classical shells – and that *this* is your opinion.

Be that as it may, there must be very definite, clear reasons that prompt you to repeatedly assert that one *must* interpret observations classically, that it is utterly part of their essence. When you say this, you always say it so definitely and clearly in the indicative, without any qualifying 'surely' or 'might' or 'we must expect

[3] [Schrödinger is referring to Bohr's *Atomtheorie und Naturbeschreibung* [*Atomic Theory and the Description of Nature*] (Bohr 1931). In his own copy (Alpbach, 2.2, Separata anderer: Bohr), Schrödinger had emphasised this passage in the margin, and underlined and put a question mark under 'must'. In the English edition (Bohr 1934), this and the next passage quoted are on pp. 94–95 and 114, respectively, but the translation here is ours (*eds.*).]

[4] [Quoted in English in the original; compare the penultimate paragraph of Bohr's reply (1935b, p. 701; see p. 252): 'the indispensable use of classical concepts in the interpretation of all proper measurements' (*eds.*).]

[5] [Quoted in English. This passage is *not* present in the published version of Bohr's reply (*eds.*).]

[6] [The phrase 'indispensable necessity' is again quoted in English (*eds.*).]

that', as if it were the most definite thing in the world. It must be part of your own most firm beliefs – and I cannot understand what it is based on.

It cannot just be this (which you told me very forcefully already in person in 1926), that our traditional language and inherited concepts are totally inappropriate for expressing the things that we face here now.[7] For this has surely been the case again and again over the course of development of our science (and of mathematics), from its first inceptions up to the state at the end of the nineteenth century. If the break with the past now appears greater than ever before, then surely we must expect that a certain temporal perspective contributes to forming that impression, according to which *that* development we experience ourselves stands out as more substantial and significant compared to the earlier ones, that we only know from history and whose stages we usually learn backwards. In that context it is often difficult for us to transport ourselves to an *earlier* mindset. And although the difficulty of *re*tracing such a historical step in our thought really most eloquently speaks to just *how* large it must have seemed to the pioneers during the first advance – sometimes we just cannot resist the feeling: 'Incredible, how limited people were back then!' That is where the underestimation due to one's time perspective is most visible.

So, I mean: it cannot possibly be that the fact that our thinking and its forms of expression are not yet adapted to the new theory, that this justifies the conviction that experiments must always be described classically, by neglecting the essential traits of the new theory. It may be a childish example, but just to say briefly what I mean: after the elastic theory of light was supplanted by the electromagnetic one, the idea was not that experimental findings still had to refer to elasticity and density of the ether, to displacements, deformation states, velocities and angular velocities of ether particles.

Forgive me for being so verbose. The idea is whether you might not clarify this point fully in the more detailed paper that you announce in the *Phys. Rev.* note: Why do I emphasise time and again that it is essential to a measurement that it can only be interpreted classically? And above all: is this temporary modesty, or are we somehow able to know that we will never proceed beyond it?

I would like so much to see you and talk to you again. But the times now are so little suited to pleasure trips. Soon the philistine desire to be someplace permanently once again will stir in me, i.e. to know with significant probability what one will do in the next five or ten years. It is awfully philistine and jars with the times;

[7] [For the 1926 discussions between Bohr and Schrödinger, see e.g. Bacciagaluppi and Valentini (2009, section 4.6.2) and references therein (*eds*.).]

probably only a reflection of the local environment, which rolls along freely quite undisturbed by the world's affairs.[8]

With warmest greetings and good wishes from our home to yours

Your very devoted

[E. Schrödinger]

[8] [For how Schrödinger felt in Oxford, see Moore (1989, chapter 8) (*eds.*).]

Bohr to Schrödinger, 26 October 1935*

University Institute for Theoretical Physics
Blegdamsvej 15, Copenhagen Ø[sterbro]
26 October 1935.

Dear Schrödinger.

I thank you and your wife warmly for the friendly wishes for my birthday, and thank yourself[1] for the friendly and interesting letter that I received a week ago. Regarding my answer to Einstein's article, I believe it has already come out, and as you will see, I have tried in several places to formulate the thoughts somewhat more clearly, and I hope that your first objection regarding the measurement setup is thereby also answered. Further, I have omitted reference to the possible importance of the atomic structure of all measuring instruments for the elucidation of the yet unsolved difficulties in the electron theory, because I am just finishing a paper together with Rosenfeld on the measurement problems of the electron theory, which illuminates this question somewhat further.[2] But these considerations have no close relation to the Einstein paradoxes and the question of the limit of the causal description of quantum phenomena. Here I must confess that I cannot share your misgivings. My emphasis on the unavoidableness of the classical description of experiments in the end just reduces to the apparently self-evident fact that the description of every measurement setup must essentially contain the arrangement of the apparatus in space and its functioning in time, if we are to be able to say anything at all about the phenomena.[3] The argument is in fact primarily that the measuring devices, if they are to serve as such, cannot be included in the proper domain of application of quantum mechanics. It amused me very much that you noticed what even I did not know myself, that precisely at this and only this point, I do not say 'might'; I will also follow your advice and treat this fundamental point as clearly and in detail as possible in the paper on the application of space-time concepts in the atomic theory announced in my article. As soon as this paper is finished, I will send you a carbon copy.[4]

* [From AHQP 92, section 2-102, and Alpbach, 6.5, Korrespondenz, Vorordner (typewritten, with corrections in pen and underlining in blue pencil, 2pp., in German). Transcription in Meyenn (2011, vol. 2, pp. 576–577) (*eds.*).]

[1] [vielmals/selbst (*eds.*).]

[2] [This is presumably the paper reproduced in Bohr (1996, pp. 195–209). See the discussion in Section 5.2 above, esp. note 9 (*eds.*).]

[3] [Starting with 'the apparently. . . ', this passage is underlined in blue pencil, presumably by Schrödinger, since Bohr's corrections are in black pen (*eds.*).]

[4] [No such paper appears to be extant, but there are fairly extensive notes on this topic in AHQP-BMSS 14, section 1 (handwritten, 58pp., in Danish). See again section 5.2, note 9 (*eds.*).]

With the friendliest greetings and best wishes to you and your wife from my wife and

Your

Niels Bohr

Envoi

Bohr to Einstein, 25 October 1935*

25 October 1935

Dear Einstein,[1]

I need not say how happy you made me by the kind words you wrote to a Danish newspaper for my 50th birthday.[2] All the kindness I have received on this occasion has touched me deeply, and not least was I glad to feel through your words the friendship that you have extended me throughout so many years and which has always been such an encouragement to me. I often think of our lively discussions about the foundations of physics in particular together with Ehrenfest, whom we all miss so much. You will surely have seen by now that I have used the article on physical reality by yourself and your coworkers to take up this discussion again, and I am very eager to hear sometime what you think of it. Maybe Pauli, who is in Princeton now, will write me a few words about it.[3] I hope very much that it will not be too long until we see each other again and send you and your wife the kindest regards and warmest wishes from my wife and

[Y]our

[Niels Bohr]

* [From BSCSupp-EIN-351025f (carbon copy, 1p., in German) (*eds.*).]

1 [Bohr addresses Einstein with the formal 'Sie' (*eds.*).]

2 ['Den geniale Skaber' ['The Brilliant Creator'], *Politiken*, 7 October 1935 (EA 89-911, in Danish) (*eds.*).]

3 [No such letters from Pauli appear to be extant. See also note 6 on p. 121 above (*eds.*).]

References

Aharonov, Y. and Vaidman, L. (2007). The Two-State Vector Formalism of Quantum Mechanics: An Updated Review. arXiv: https://arxiv.org/abs/quant-ph/0105101v2.

Bacciagaluppi, G. (1999). Nelsonian Mechanics Revisited. *Foundations of Physics Letters* **12** (1), 1–16.

Bacciagaluppi, G. (2008). The Statistical Interpretation According to Born and Heisenberg. In: *HQ-1: Conference on the History of Quantum Physics,* Vol. 2. Ed. by C. Joas, C. Lehner and J. Renn. Berlin: MPIWG, 269–288.

Bacciagaluppi, G. (2013). Insolubility Theorems and EPR Argument. *European Journal for Philosophy of Science* **3**, 87–100.

Bacciagaluppi, G. (2014). Insolubility from No-Signalling. *International Journal of Theoretical Physics* **53**, 3465–3474.

Bacciagaluppi, G. (2015). Did Bohr Understand EPR? In: *One Hundred Years of the Bohr Atom*. Ed. by F. Aaserud and H. Kragh. Copenhagen: Royal Danish Academy of Sciences and Letters, 375–394.

Bacciagaluppi, G. (2016a). An Everett Perspective on Bohr and EPR. In: *Niels Bohr in the 21st Century*. Ed. by J. Faye and H. J. Folse. Chicago: University of Chicago Press, 289–302.

Bacciagaluppi, G. (2016b). Einstein, Bohm, and Leggett–Garg. In: *Contextuality from Quantum Physics to Psychology*. Ed. by E. Dzhafarov, S. Jordan and R. Zhang. Singapore: World Scientific, 63–76.

Bacciagaluppi, G. (2017). Bohr's Slit and Hermann's Microscope. In: *Grete Hermann: Between Physics and Philosophy*. Ed. by E. Crull and G. Bacciagaluppi. Berlin: Springer, 135–147.

Bacciagaluppi, G. (2022). The Statistical Interpretation: Born, Heisenberg and von Neumann, 1926–27. In: *Oxford Handbook of the History of Quantum Interpretations*. Ed. by O. Freire *et al.* Oxford: Oxford University Press, 567–585.

Bacciagaluppi, G. and Crull, E. (2009). Heisenberg (and Schrödinger, and Pauli) on Hidden Variables. *Studies in History and Philosophy of Modern Physics* **40** (4), 374–382.

Bacciagaluppi, G., Crull, E. and Maroney, O. J. E. (2017). Jordan's Derivation of Blackbody Fluctuations. *Studies in History and Philosophy of Modern Physics* **60**, 23–34.

Bacciagaluppi, G. and Valentini, A. (2009). *Quantum Theory at the Crossroads: Reconsidering the 1927 Solvay Conference*. Cambridge: Cambridge University Press.

Bassi, A. and Ghirardi, GC. (2000). A General Argument against the Universal Validity of the Superposition Principle. *Physics Letters A* **275**, 373–381.

Bell, J. S. (1966). On the Problem of Hidden Variables in Quantum Mechanics. *Reviews of Modern Physics* **38**, 447–452.

Beller, M. (1997). Against the Stream: Schrödinger's Interpretation of Quantum Physics. *Studies in History and Philosophy of Modern Physics* **28** (3), 421–432.

Belousek, D. W. (1996). Einstein's 1927 Unpublished Hidden-Variable Theory: Its Background, Context and Significance. *Studies in History and Philosophy of Modern Physics* **27** (4), 437–461.

Bitbol, M. (1996). *Schrödinger's Philosophy of Quantum Mechanics*. Dordrecht: Kluwer.

Bohm, D. (1951). *Quantum Theory*. Hoboken: Prentice-Hall.

Bohm, D. (1952). A Suggested Interpretation of the Quantum Theory in Terms of 'Hidden' Variables. I and II. *Physical Review* **85** (2), 166–179, 180–192.

Bohr, N. (1928a). Das Quantenpostulat und die neuere Entwicklung der Atomistik. *Die Naturwissenschaften* **16**, 245–257.

Bohr, N. (1928b). The Quantum Postulate and the Recent Development of Atomic Theory. *Nature* **121**, 580–590.

Bohr, N. (1931). *Atomtheorie und Naturbeschreibung*. Berlin: Springer. Translated as Bohr (1934).

Bohr, N. (1932). Chemistry and the Quantum Theory of Atomic Constitution (Faraday Lecture). *Journal of the Chemical Society* **1932**, 349–384.

Bohr, N. (1933a). Lys og Liv. *Naturens Verden* **17**, 49–59. Translated as Bohr (1933b).

Bohr, N. (1933b). Light and Life. *Nature* **131**, 421–423, 457–459.

Bohr, N. (1934). *Atomic Theory and the Description of Nature*. Cambridge: Cambridge University Press.

Bohr, N. (1935a). Quantum Mechanics and Physical Reality. *Nature* **136**, 65.

Bohr, N. (1935b). Can Quantum-Mechanical Description of Physical Reality be Considered Complete? *Physical Review* **48**, 696–702.

Bohr, N. (1949). Discussion with Einstein on Epistemological Problems in Atomic Physics. In: *Albert Einstein: Philosopher-Scientist*. Ed. by P. A. Schilpp. Evanston, IL: The Library of Living Philosophers, 199–241.

Bohr, N. (1958). Quantum Physics and Philosophy. Causality and Complementarity. In: *Philosophy in Mid-Century: A Survey – La philosophie au milieu du vingtième siècle: Chroniques*. Ed. by R. Klibansky. Firenze: La Nuova Italia, 308–314.

Bohr, N. (1985). *Niels Bohr's Collected Works, Vol. 6, Foundations of Quantum Physics I (1926–1932)*. Ed. by J. Kalckar. Amsterdam: North-Holland.

Bohr, N. (1996). *Niels Bohr's Collected Works, Vol. 7, Foundations of Quantum Physics II (1933–1958)*. Ed. by J. Kalckar. Amsterdam: North-Holland.

Bohr, N. and Rosenfeld, L. (1933). Zur Frage der Messbarkeit der elektromagnetischen Feldgrössen. *Kongelige Danske Videnskabernes Selskab, Mathematisk-fysiske Meddelelser* **12** (8), 3–65.

Born, M. (1926a). Zur Quantenmechanik der Stoßvorgänge. *Zeitschrift für Physik* **37**, 863–867.

Born, M. (1926b). Quantenmechanik der Stoßvorgänge. *Zeitschrift für Physik* **38**, 803–827.

Born, M. (1926c). Das Adiabatenprinzip in der Quantenmechanik. *Zeitschrift für Physik* **40**, 167–192.

Born, M. (1929). Über den Sinn der physikalischen Theorien. *Die Naturwissenschaften* **17** (7), 109–118.

Born, M. (1935a). *The Restless Universe*. Glasgow: Blackie & Son.

Born, M. (1935b). *Atomic Physics*. Glasgow: Blackie & Son.

Born, M. (1949). *Natural Philosophy of Cause and Chance*. Waynflete Lectures. Oxford: Clarendon Press.

Born, M. and Heisenberg, W. (1928). La mécanique des quanta. In: *Électrons et Photons: Rapports et Discussions du Cinquième Conseil de Physique Solvay*. Ed. by H. A. Lorentz and translated by J.-É. Verschaffelt. Paris: Gauthier-Villars, 143–181. Page references are to the English translation of the original German typescript in Bacciagaluppi and Valentini (2009), 372–401.

Born, M., Heisenberg, W. and Jordan, P. (1926). Zur Quantenmechanik II. *Zeitschrift für Physik* **35** (8–9), 557–615.

Born, M. and Infeld, L. (1934a). Foundations of a New Field Theory. *Proceedings of the Royal Society of London A***144**, 425–451.

Born, M. and Infeld, L. (1934b). On the Quantization of the New Field Equations I. *Proceedings of the Royal Society of London A***147**, 522–546.

Born, M. and Infeld, L. (1935). On the Quantization of the New Field Theory II. *Proceedings of the Royal Society of London A***150**, 141–166.

Born, M. and Schrödinger, E. (1935). The Absolute Field Constant in the New Field Theory. *Nature* **135**, 342.

Breit, G. (1929). The Effect of Retardation on the Interaction of Two Electrons. *Physical Review* **34** (4), 553–573.

Breit, G. (1932). Dirac's Equation and the Spin–Spin Interactions of Two Electrons. *Physical Review* **39** (4), 616–624.

Brillouin, M. (1933). Préface. In: *E. Schrödinger, Mémoires sur la mécanique ondulatoire*. Paris: Librairie Félix Alcan, v–xi.

Brown, H. (1978). Problems of Measurement in Quantum Theory. PhD Thesis. London: Chelsea College.

Brown, H. (1981). O debate Einstein-Bohr sobre a mecânica quântica. *Cadernos de história e filosofia da ciência* **2**, 51–89.

Brown, H. (1986). The Insolubility Proof of the Quantum Measurement Problem. *Foundations of Physics* **16**, 857–870.

Bub, J. (2010). Von Neumann's 'Hidden Variables' Proof: A Re-appraisal. *Foundations of Physics* **40** (9–10), 1333–1340.

Bub, J. (2011). Is von Neumann's 'No Hidden Variables' Proof Silly? In: *Deep Beauty: Understanding the Quantum World through Mathematical Innovation*. Ed. by H. Halvorson. Cambridge: Cambridge University Press, 393–408.

Busch, P., Grabowski, M. and Lahti, P. (1997). *Operational Quantum Physics*. Berlin: Springer.

Busch, P., Lahti, P. and Mittelstaedt, P. (1991). *The Quantum Theory of Measurement*. Berlin: Springer.

Busch, P., Lahti, P., Pellonpää, J.-P. and Ylinen, K. (2016). *Quantum Measurement*. Cham: Springer.

Camilleri, K. (2009). *Heisenberg and the Interpretation of Quantum Mechanics*. Cambridge: Cambridge University Press.

Campbell, N. R. (1921). Atomic Structure. *Nature* **107**, 170.

Campbell, N. R. (1926). Time and Chance. *The London, Edinburgh, and Dublin Philosophical Magazine and Journal of Science* **1** (5), 1106–1117.

Casimir, H. B. G. (2010). *Haphazard Reality: Half a Century of Science*. 2nd edition. New York: Harper & Row.

Conway, J. and Kochen, S. (2006). The Free Will Theorem. *Foundations of Physics* **36** (10), 1441–1473.

Courant, R. and Hilbert, D. (1930). *Methoden der mathematischen Physik*. 2nd edition. Berlin: Springer.

Crull, E. (2017). Hermann and the Relative Context of Observation. In: *Grete Hermann: Between Physics and Philosophy*. Ed. by E. Crull and G. Bacciagaluppi. Berlin: Springer, 149–169.

Crull, E. (2022). Grete Hermann's Interpretation of Quantum Mechanics. In: *Oxford Handbook of the History of Quantum Interpretations*. Ed. by O. Freire *et al.* Oxford: Oxford University Press, 567–585.

Crull, E. and Bacciagaluppi, G., eds. (2017a). *Grete Hermann: Between Physics and Philosophy*. Berlin: Springer.

Crull, E. and Bacciagaluppi, G. (2017b). Grete Hermann's Lost Manuscript on Quantum Mechanics. In: *Grete Hermann: Between Physics and Philosophy*. Ed. by E. Crull and G. Bacciagaluppi. Berlin: Springer, 119–134.

de Regt, H. W. (1997). E. Schrödinger, Anschaulichkeit, and Quantum Theory. *Studies in History and Philosophy of Modern Physics* **28** (4), 461–481.

Derakhshani, M. (2017). Stochastic Mechanics without ad hoc Quantization: Theory and Applications to Semiclassical Gravity. PhD Thesis. Utrecht: Universiteit Utrecht.

Dickson, M. (2002a). The EPR Experiment: A Prelude to Bohr's Reply to EPR. In: *History of Philosophy of Science*. Ed. by M. Heidelberger and F. Stadler. Dordrecht: Kluwer, 263–276.

Dickson, M. (2002b). Bohr on Bell: A Proposed Reading of Bohr and Its Implications for Bell's Theorem. In: *Non-locality and Modality*. Ed. by T. Placek and J. Butterfield. Dordrecht: Kluwer, 19–35.

Dickson, M. (2004). Quantum Reference Frames in the Context of EPR. *Philosophy of Science* **71** (5), 655–668.

Dirac, P. A. M. (1928). The Quantum Theory of the Electron. *Proceedings of the Royal Society of London A***117**, 610–624.

Dirac, P. A. M. (1930). *The Principles of Quantum Mechanics*. 1st English edition. Oxford: Oxford University Press.

Dirac, P. A. M. (1932). Relativistic Quantum Mechanics. *Proceedings of the Royal Society of London A***136**, 453–464.

Dirac, P. A. M. (1934). Discussion of the Infinite Distribution of Electrons in the Theory of the Positron. *Proceedings of the Cambridge Philosophical Society, Mathematical and Physical Sciences* **30** (2), 150–163.

Dirac, P. A. M. (1935a). *The Principles of Quantum Mechanics*. 2nd English edition. Oxford: Clarendon Press.

Dirac, P. A. M. (1935b). *Die Prinzipien der Quantenmechanik*. 2nd German edition. Leipzig: S. Hirzel.

Dirac, P. A. M. (1936). Does Conservation of Energy Hold in Atomic Processes? *Nature* **137**, 298–299.

Dirac, P. A. M., Fock, V. A. and Podolsky, B. (1932). On Quantum Electrodynamics. *Physikalische Zeitschrift der Sowjetunion* **2**, 468–479.

Eddington, A. (1928). *The Nature of the Physical World*. The Gifford Lectures. Cambridge: Cambridge University Press.

Einstein, A. (1921). Über ein den Elementarprozeß der Lichtemission betreffendes Experiment. *Sitzungsberichte der Preussischen Akademie der Wissenschaften* **1921** (July–December), 882–883.

Einstein, A. (1932). Über die Unbestimmtheitsrelationen. *Zeitschrift für angewandte Chemie* **45** (1), 23. Editorial report (anon.) of Berlin colloquium in 1931.

Einstein, A. (1936). Physik und Realität. *Journal of the Franklin Institute* **221**, 313–347. An English translation by J. Picard appears in the same volume, pp. 349–382.

Einstein, A. (1953). Elementare Überlegungen zur Interpretation der Grundlagen der Quantenmechanik. In: *Scientific Papers Presented to Max Born on his Retirement from the Tait Chair of Natural Philosophy in the University of Edinburgh*. Edinburgh: Oliver and Boyd, 33–40. Translated by D. Karanth as 'Elementary Considerations on the Interpretation of the Foundations of Quantum Mechanics', arXiv: https://arxiv.org/abs/1107.3701v1.

Einstein, A. (2018a). *The Collected Papers of Albert Einstein, Vol. 15, The Berlin Years: Writings and Correspondence, June 1925–May 1927*. Ed. by D. Kormos Buchwald *et al.* Princeton: Princeton University Press.

Einstein, A. (2018b). *The Collected Papers of Albert Einstein, Vol. 15, The Berlin Years: Writings and Correspondence, June 1925–May 1927 (English Translation Supplement)*. Ed. by D. Kormos Buchwald *et al.* Princeton: Princeton University Press.

Einstein, A., Podolsky, B. and Rosen, N. (1935). Can Quantum-Mechanical Description of Physical Reality be Considered Complete? *Physical Review* **47**, 777–780.

Einstein, A. and Rosen, N. (1935). The Particle Problem in the General Theory of Relativity. *Physical Review* **48**, 73–77.

Einstein, A., Tolman, R. C. and Podolsky, B. (1931). Knowledge of Past and Future in Quantum Mechanics. *Physical Review* **37** (6), 780–781.

Elby, A. and Bub, J. (1994). Triorthogonal Uniqueness Theorem and Its Relevance to the Interpretation of Quantum Mechanics. *Physical Review A* **49** (5), 4213–4216.

Everett, H. (2012). *The Everett Interpretation of Quantum Mechanics: Collected Works 1955–1980 with Commentary*. Ed. by J. A. Barrett and P. Byrne. Princeton: Princeton University Press.

Fermi, E. (1932). Quantum Theory of Radiation. *Reviews of Modern Physics* **4** (1), 87–132.

Fine, A. (1981). Einstein's Critique of Quantum Theory: The Roots and Significance of EPR. In: *After Einstein: Proceedings of the Einstein Centennial Celebration at Memphis State University, 14–16 March 1979*. Ed. by P. Barker and C. G. Shugart. Memphis: Memphis State University Press, 147–158. Reprinted in Fine (1986), 26–39.

Fine, A. (1986). *The Shaky Game: Einstein, Realism and the Quantum Theory*. Chicago: University of Chicago Press.

Fine, A. (1994). Schrödinger's Version of EPR, and Its Problems. In: *Patrick Suppes: Scientific Philosopher,* Vol. 2. Ed. by P. Humphreys. Dordrecht: Kluwer, 29–38.

Fine, A. (2007). Bohr's Response to EPR: Criticism and Defense. *The Jerusalem Philosophical Quarterly* **56**, 31–56.

Fine, A. (2017). The Einstein–Podolsky–Rosen Argument in Quantum Theory. *The Stanford Encyclopedia of Philosophy.* https://plato.stanford.edu/archives/win2017/entries/qt-epr/. First published 10 May 2014; substantive revisions 31 October 2017.

Flint, H. T. (1935). Quantum Mechanics as a Physical Theory. *Nature* **135**, 1025–1026.

Forman, P. (1973). Scientific Internationalism and the Weimar Physicists: The Ideology and Its Manipulation in Germany after World War I. *Isis* **64** (2), 151–180.

Frappier, M. (2017). 'In the No-Man's Land Between Physics and Logic': On the Dialectical Role of the Microscope Experiment. In: *Grete Hermann: Between Physics and Philosophy*. Ed. by E. Crull and G. Bacciagaluppi. Berlin: Springer, 85–105.

Furry, W. H. (1936a). Note on the Quantum-Mechanical Theory of Measurement. *Physical Review* **49**, 393–399.

Furry, W. H. (1936b). Remarks on Measurement in Quantum Theory. *Physical Review* **49**, 476.

Gamow, G. (1928). Zur Quantentheorie des Atomkernes. *Zeitschrift für Physik* **51** (3–4), 204–212.

Gamow, G. (1966). *Thirty Years that Shook Physics: The Story of Quantum Theory*. New York: Doubleday & Co. Inc.

Goethe, J. W. von (1808). *Faust. Eine Tragödie*. Tübingen: Cotta.

Greenspan, N. (2005). *The End of the Certain World: The Life and Science of Max Born*. New York: Basic Books.

Hansen-Schaberg, I. (2017). A Biographical Sketch of Prof. Dr Grete Henry-Hermann (1901–1984). In: *Grete Hermann: Between Physics and Philosophy*. Ed. by E. Crull and G. Bacciagaluppi. Berlin: Springer, 3–16.

Heisenberg, W. (1926). Quantenmechanik. *Die Naturwissenschaften* **14**, 989–994.

Heisenberg, W. (1927a). Über den anschaulichen Inhalt der quantentheoretischen Kinematik und Mechanik. *Zeitschrift für Physik* **43** (3–4), 172–198.

Heisenberg, W. (1927b). Schwankungserscheinungen und Quantenmechanik. *Zeitschrift für Physik* **40** (7), 501–506.

Heisenberg, W. (1930a). *Die physikalischen Prinzipien der Quantentheorie*. Leipzig: S. Hirzel. Translated as Heisenberg (1930b).

Heisenberg, W. (1930b). *The Physical Principles of the Quantum Theory*. Translated by C. Eckart and F. C. Hoyt. Chicago: University of Chicago Press. Reprinted New York: Dover, 1949. Page references are to the Dover edition.

Heisenberg, W. (1930c). Versammlung der Gesellschaft Deutscher Naturforscher und Ärzte. *Zeitschrift für angewandte Chemie* **43** (38), 854. Editorial report (anon.) of various colloquia, including one by Heisenberg discussing his indeterminacy relation.

Heisenberg, W. (1934). Wandlungen in den Grundlagen der exakten Naturwissenschaft in jüngster Zeit. *Die Naturwissenschaften* **22** (40), 669–675.

Heisenberg, W. (1935a). Bemerkungen zur Theorie des Atomkerns. In: *Pieter Zeeman, 1865 – 25 mei – 1935: Verhandelingen op 25 mei 1935 aangeboden aan Prof. Dr. P. Zeeman*. Den Haag: Martinus Nijhoff, 108–116.

Heisenberg, W. (1935b). Prinzipielle Fragen der modernen Physik. In: *Wandlungen in den Grundlagen der Naturwissenschaft: Acht Vorträge von W. Heisenberg*. Zürich: S. Hirzel, 35–46.

Heisenberg, W. (1935c). Die Struktur der leichten Atomkerne. *Zeitschrift für Physik* **96**, 473–484.

Heisenberg, W. (1935d). Udviklingen af Kwantenteoriens principielle Grundlag efter 1925. *Fysisk Tidsskrift* **33**, 96–101.

Heisenberg, W. (1969). *Der Teil und das Ganze. Gespräche im Umkreis der Atomphysik*. München: Piper. Translated as Heisenberg (1971).

Heisenberg, W. (1971). *Physics and Beyond: Encounters and Conversations*. New York: Harper & Row.

Heisenberg, W. and Pauli, W. (1929). Zur Quantendynamik der Wellenfelder. *Zeitschrift für Physik* **56** (1–2), 1–61.

Heisenberg, W. and Pauli, W. (1930). Zur Quantendynamik der Wellenfelder II. *Zeitschrift für Physik* **59** (3–4), 168–190.

Held, C. (1998). *Die Bohr-Einstein Debatte: Quantenmechanik und physikalische Wirklichkeit*. Paderborn: Schöningh.

Hermann, G. (1935a). Die naturphilosophischen Grundlagen der Quantenmechanik. *Abhandlungen der Fries'schen Schule* **6** (2), 75–152. Reprinted as Hermann (1935c), and translated in Crull and Bacciagaluppi (2017), 239–278.

Hermann, G. (1935b). Die naturphilosophischen Grundlagen der Quantenmechanik. *Die Naturwissenschaften* **23** (42), 718–721. Translated in Lumma (1999).

Hermann, G. (1935c). *Die naturphilosophischen Grundlagen der Quantenmechanik*. Berlin: Verlag Öffentliches Leben.

Herrmann, K., ed. (2019). *Grete Henry-Hermann: Philosophie – Mathematik – Quantenmechanik*. Berlin: Springer.

Heywood, P. and Redhead, M. (1983). Nonlocality and the Kochen–Specker Paradox. *Foundations of Physics* **13** (5), 481–499.

Holland, P. R. (2005). What's Wrong with Einstein's 1927 Hidden-Variable Interpretation of Quantum Mechanics? *Foundations of Physics* **35**, 177–196.

Howard, D. (1985). Einstein on Locality and Separability. *Studies in History and Philosophy of Science* **16**, 171–201.

Howard, D. (1990). 'Nicht sein kann was nicht sein darf', or the Prehistory of EPR, 1909–1935: Einstein's Early Worries about the Quantum Mechanics of Composite Systems. In: *Sixty-Two Years of Uncertainty: Historical, Philosophical and Physical Inquiries into the Foundations of Quantum Mechanics*. Ed. by A. I. Miller. New York: Plenum Press, 61–111.

Howard, D. (1994). What Makes a Classical Concept Classical? Toward a Reconstruction of Niels Bohr's Philosophy of Physics. In: *Niels Bohr and Contemporary Philosophy*. Ed. by J. Faye and H. J. Folse. Dordrecht: Kluwer, 201–230.

Howard, D. (2007). Revisiting the Einstein–Bohr Dialogue. *The Jerusalem Philosophical Quarterly* **56**, 57–90.

Jähnert, M. and Lehner, C. (2012). Early Debates about Interpretations of Quantum Mechanics. Unpublished manuscript.

Jammer, M. (1974). *The Philosophy of Quantum Mechanics: The Interpretations of Quantum Mechanics in Historical Perspective*. Hoboken: John Wiley.

Joas, C. (2011). Schrödinger's Methodology of Physics: Application and Interpretation. Talk delivered at New Directions in the Foundations of Physics, 29 April–1 May 2011 in Washington, DC.

Joas, C. and Lehner, C. (2009). The Classical Roots of Wave Mechanics: Schrödinger's Transformations of the Optical-Mechanical Analogy. *Studies in History and Philosophy of Modern Physics* **40** (4), 338–351.

Jordan, P. (1927). Über quantenmechanische Darstellung von Quantensprüngen. *Zeitschrift für Physik* **40**, 661–666.

Jordan, P. and Pauli, W. (1928). Zur Quantenelektrodynamik ladungsfreier Felder. *Zeitschrift für Physik* **47** (3–4), 151–173.

Kalckar, J. (1996). Part I – Introduction. In: *Niels Bohr's Collected Works, Vol. 7, Foundations of Quantum Physics II (1933–1958)*. Ed. by J. Kalckar. Amsterdam: North-Holland, 3–51.

Kemble, E. C. (1935). The Correlation of Wave Functions with the States of Physical Systems. *Physical Review* **47**, 973–974.

Klein, M. J. (1970). The First Phase of the Bohr–Einstein Dialogue. *Historical Studies in the Physical Sciences* **2**, 1–39.

Kramers, H. A. (1933). *Theorien des Aufbaues der Materie, I.: Die Grundlagen der Quantentheorie*. Hand- und Jahrbuch der chemischen Physik, Vol. I. Leipzig: Akademische Verlagsgesellschaft.

Landau, L. and Peierls, R. (1931). Erweiterung des Unbestimmtheitsprinzips für die relativistische Quantentheorie. *Zeitschrift für Physik* **69** (1–2), 56–69.

Laue, M. von (1931). Die dynamische Theorie der Röntgenstrahlinterferenzen in neuer Form. *Ergebnisse der Exakten Naturwissenschaften* **10**, 133–158.

Laue, M. von (1932). Zu den Erörterungen über Kausalität. *Die Naturwissenschaften* **20** (51), 915–916.

Laue, M. von (1933). Materie und Raumerfüllung. *Scientia* **54**, 402–412.

Laue, M. von (1934). Über Heisenbergs Ungenauigkeitsbeziehungen und ihre erkenntnistheoretische Bedeutung. *Die Naturwissenschaften* **22** (26), 439–441.

Laue, M. von (1935a). Der optische Reziprozitätssatz in Anwendung auf die Röntgenstrahlinterferenzen. *Die Naturwissenschaften* **23** (23), 373.

Laue, M. von (1935b). Die Fluoreszenzröntgenstrahlung von Einkristallen (mit einem Anhang über Elektronenbeugung). *Annalen der Physik, Fünfte Folge* **23** (8), 705–746.

Lehner, C. (2011). The Prehistory of Entanglement, 1926–35. Talk delivered at New Directions in the Foundations of Physics, 29 April–1 May 2011 in Washington, DC.

Lumma, D. (1999). The Foundations of Quantum Mechanics in the Philosophy of Nature. *The Harvard Review of Philosophy* **VII**, 35–44. Translation of Hermann (1935b), with introduction.

Majorana, E. (1933). Über die Kerntheorie. *Zeitschrift für Physik* **82**, 137–145.

Malament, D. B. (1996). In Defense of Dogma: Why There Cannot Be a Relativistic Quantum Mechanics of (Localizable) Particles. In: *Perspectives on Quantum Reality*. Ed. by R. Clifton. Dordrecht: Springer, 1–10.

Manzoni, A. (1840). *I Promessi Sposi*. 2nd revised edition. Milan: Guglielmini e Radaelli.

Margenau, H. (1932). Probability and Causality in Quantum Physics. *The Monist* **42** (2), 161–188.

Margenau, H. (1936). Quantum-Mechanical Description. *Physical Review* **49**, 240–242.

Margenau, H. (1937). Critical Points in Modern Physical Theory. *Philosophy of Science* **4** (3), 337–370.

McKie, D. and Heathcote, N. H. de V. (1935). *The Discovery of Specific and Latent Heats*. London: Edward Arnold & Co.

Meitner, L. (1927). Über den Aufbau des Atominnern. *Die Naturwissenschaften* **15** (16), 369–378.

Meitner, L. (1934). Atomkern und periodisches System der Elemente. *Die Naturwissenschaften* **22** (44), 733–739.

Meyenn, K. von, ed. (2011). *Eine Entdeckung von ganz ausserordentlicher Tragweite: Schrödingers Briefwechsel zur Wellenmechanik und zum Katzenparadoxon*. Vols. 1–2. Berlin: Springer.

Mises, R. von (1931). *Wahrscheinlichkeitsrechnung und ihre Anwendung in der Statistik und theoretischen Physik*. Leipzig: Deuticke.

Møller, C. (1931). Über den Stoß zweier Teilchen unter Berücksichtigung der Retardation der Kräfte. *Zeitschrift für Physik* **70** (11–12), 786–795.

Moore, W. (1989). *Schrödinger: Life and Thought*. Cambridge: Cambridge University Press.

Morgenstern, C. (1910). *Palmström*. Berlin: Bruno Cassirer.

Muller, F. A. (1997). The Equivalence Myth of Quantum Mechanics: Parts I and II. *Studies in History and Philosophy of Modern Physics* **28**, 35–61, 219–247.

Muller, F. A. (1999a). The Equivalence Myth of Quantum Mechanics (Addendum). *Studies in History and Philosophy of Modern Physics* **30** (4), 543–545.

Muller, F. A. (1999b). The Locality Scandal of Quantum Mechanics. In: *Language, Quantum, Music*. Ed. by M. L. Dalla Chiara, R. Giuntini and F. Laudisa. Dordrecht: Springer, 241–248.

Myrvold, W. C. (2003). On Some Early Objections to Bohm's Theory. *International Studies in the Philosophy of Science* **17** (1), 7–24.

Nelson, E. (1966). Derivation of the Schrödinger Equation from Newtonian Mechanics. *Physical Review* **150**, 1079–1085.

Nelson, E. (1985). *Quantum Fluctuations*. Princeton: Princeton University Press.

Neumann, J. von (1927). Wahrscheinlichkeitstheoretischer Aufbau der Quantenmechanik. *Nachrichten von der Gesellschaft der Wissenschaften zu Göttingen, Mathematisch-Physikalische Klasse* **1927**, 245–272.

Neumann, J. von (1932). *Mathematische Grundlagen der Quantenmechanik.* Berlin: Springer. Translated by R. T. Beyer as von Neumann (1955).

Neumann, J. von (1955). *Mathematical Foundations of Quantum Mechanics.* Princeton: Princeton University Press.

Norsen, T. (2015). Are There Really Two Different Bell's Theorems? arXiv: https://arxiv.org/abs/1503.05017.

Pais, A. (1982). *'Subtle is the Lord...': The Science and the Life of Albert Einstein.* Oxford: Oxford University Press.

Pauli, W. (1933). Die allgemeinen Prinzipien der Wellenmechanik. In: *Handbuch der Physik.* Ed. by H. Geiger and K. Scheel. 2nd ed. Vol. XXIV(1). Berlin: Springer, 83–272.

Pauli, W. (1979). *Wolfgang Pauli: Scientific Correspondence with Bohr, Einstein, Heisenberg a.o., Vol. 1, 1919–1929.* Ed. by K. von Meyenn. Berlin: Springer.

Pauli, W. (1985). *Wolfgang Pauli: Scientific Correspondence with Bohr, Einstein, Heisenberg a.o., Vol. II, 1930–1939.* Ed. by K. von Meyenn. Berlin: Springer.

Pauli, W. (1990). *Die allgemeinen Prinzipien der Wellenmechanik.* Berlin: Springer. Reprint of the second edition of Pauli (1933) with introduction and historical notes by N. Straumann.

Peierls, R. (1934). The Vacuum in Dirac's Theory of the Positive Electron. *Proceedings of the Royal Society of London A***146**, 420–441.

Planck, M. (1900a). Über irreversible Strahlungsvorgänge. *Annalen der Physik, Vierte Folge* **1** (1), 69–122.

Planck, M. (1900b). Zur Theorie des Gesetztes der Energieverteilung im Normalspektrum. *Verhandlungen der Deutschen Physikalischen Gesellschaft, Zweite Folge* **2** (17), 237–245.

Planck, M. (1906). *Vorlesungen über die Theorie der Wärmestrahlung.* Leipzig: J. A. Barth.

Popper, K. (1934). Zur Kritik der Ungenauigkeitsrelationen. *Die Naturwissenschaften* **22** (48), 807–808.

Popper, K. (1935). *Logik der Forschung.* Vienna: Springer.

Popper, K. (1959). *The Logic of Scientific Discovery.* London: Hutchinson & Co.

Przibram, K., ed. (1967). *Letters on Wave Mechanics.* New York: Philosophical Library. Originally published as *Briefe zur Wellenmechanik.* Wien: Springer, 1963.

Rosen, N. (1931). The Normal State of the Hydrogen Molecule. *Physical Review* **38**, 2099–2114.

Rosen, N. (1985). Quantum Mechanics and Reality. In: *50 Years of the Einstein–Podolsky–Rosen Gedankenexperiment.* Ed. by P. Lahti and P. Mittelstaedt. Symposium on the Foundations of Modern Physics, Joensuu, Finland, 16–20 June 1985. Singapore: World Scientific, 17–33.

Rosenfeld, L. (1967). Niels Bohr in the Thirties. In: *Niels Bohr: His Life and Work as Seen by His Friends and Colleagues.* Ed. by S. Rozenthal. Amsterdam: North-Holland, 114–137.

Ruark, A. E. (1935). Is the Quantum-Mechanical Description of Physical Reality Complete? *Physical Review* **48**, 466–467.

Scheibe, E. (1973). *The Logical Analysis of Quantum Mechanics.* Oxford: Pergamon Press. Translated by J. B. Sykes.

Schilpp, P. A., ed. (1949). *Albert Einstein: Philosopher-Scientist.* Evanston, IL: The Library of Living Philosophers.

Schrödinger, E. (1917). Die Ergebnisse der neueren Forschung über Atom- und Molekularwärmen. *Die Naturwissenschaften* **5**, 537–543, 561–567.

Schrödinger, E. (1926a). Zur Einsteinschen Gastheorie. *Physikalische Zeitschrift* **27**, 95–101.

Schrödinger, E. (1926b). Über das Verhältnis der Heisenberg-Born-Jordanschen Quantenmechanik zu der meinen. *Annalen der Physik, Vierte Folge* **79**, 734–756.

Schrödinger, E. (1926c). Quantisierung als Eigenwertproblem (dritte Mitteilung: Störungstheorie, mit Anwendung auf den Starkeffekt der Balmerlinien). *Annalen der Physik, Vierte Folge* **80**, 437–490.

Schrödinger, E. (1926d). Quantisierung als Eigenwertproblem (vierte Mitteilung). *Annalen der Physik, Vierte Folge* **81**, 109–139.

Schrödinger, E. (1927). Energieaustausch nach der Wellenmechanik. *Annalen der Physik, Vierte Folge* **83** (15), 956–968.

Schrödinger, E. (1928a). La mécanique des ondes. In: *Électrons et Photons: Rapports et Discussions du Cinquième Conseil de Physique Solvay*. Ed. by H. A. Lorentz and translated by J.-É. Verschaffelt. Paris: Gauthier-Villars, 185–213. English translation of the original German typescript and of the published discussion in Bacciagaluppi and Valentini (2009), 406–431.

Schrödinger, E. (1928b). *Abhandlungen zur Wellenmechanik*. 2nd enlarged edition. Leipzig: J. A. Barth. Translated by J. F. Shearer and W. M. Deans as Schrödinger (1928c).

Schrödinger, E. (1928c). *Collected Papers on Wave Mechanics*. Glasgow: Blackie & Son.

Schrödinger, E. (1928d). *Four Lectures on Wave Mechanics*. Glasgow: Blackie & Son.

Schrödinger, E. (1928e). Der erkenntnistheoretische Wert physikalischer Modellvorstellungen. In: *Jahresbericht des Physikalischen Vereins zu Frankfurt am Main 1928/29*. Frankfurt: Physikalischer Verein, 44–51. Translated by W. H. Johnston as 'Conceptual Models in Physics and their Philosophical Value' in Schrödinger (1935d), 119–132. Page references are to this translation.

Schrödinger, E. (1929a). Was ist ein Naturgesetz? *Die Naturwissenschaften* **17** (1), 9–11. Translated by J. Murphy as 'What is a Law of Nature?' in Schrödinger (1935d), 107–118. Page references are to this translation.

Schrödinger, E. (1929b). Neue Wege in der Physik. *Elektrotechnische Zeitschrift* **50**, 15–16.

Schrödinger, E. (1929c). Die Erfassung der Quantengesetze durch kontinuierliche Funktionen. *Die Naturwissenschaften* **17** (26), 486–489.

Schrödinger, E. (1929d). Das Gesetz der Zufälle. *Die Koralle* **5**, 417–418. Translated by J. Murphy as 'The Law of Chance' in Schrödinger (1935d), 33–42. Page references are to this translation.

Schrödinger, E. (1929e). Das Rätsel des Lichtes. *Die Koralle* **5**, 294–298.

Schrödinger, E. (1929f). Antrittsrede des Hrn. SCHRÖDINGER. *Sitzungsberichte der Preussischen Akademie der Wissenschaften, Physikalisch-Mathematische Klasse*, C–CII.

Schrödinger, E. (1930). Was ist eigentlich Elektrizität? *Die Koralle* **6**, 110–112.

Schrödinger, E. (1931a). Über die Umkehrung der Naturgesetze. *Sitzungsberichte der Preussischen Akademie der Wissenschaften, Physikalisch-Mathematische Klasse* **8–9**, 144–153.

Schrödinger, E. (1931b). Spezielle Relativitätstheorie und Quantenmechanik. *Sitzungsberichte der Preussischen Akademie der Wissenschaften, Physikalisch-Mathematische Klasse* **12**, 238–247.

Schrödinger, E. (1932a). Sur la theórie relativiste de l'électron et l'interprétation de la mécanique quantique. *Annales de l'Institut Henri Poincaré* **2**, 269–310.

Schrödinger, E. (1932b). Anmerkungen zum Kausalproblem. *Erkenntnis* **3**, 65–70.

Schrödinger, E. (1932c). Über Indeterminismus in der Physik. In: *Über Indeterminismus in der Physik. Ist die Naturwissenschaft milieubedingt? Zwei Vorträge zur Kritik der naturwissenschaftlichen Erkenntnis*. Leipzig: J. A. Barth, 1–24. Translated by W. H. Johnston as 'Indeterminism in Physics' in Schrödinger (1935d), 43–65. Page references are to this translation.

Schrödinger, E. (1932d). Ist die Naturwissenschaft milieubedingt? In: *Über Indeterminismus in der Physik. Ist die Naturwissenschaft milieubedingt? Zwei Vorträge zur Kritik der naturwissenschaftlichen Erkenntnis*. Leipzig: J. A. Barth, 25–62. Freely translated by J. Murphy as Schrödinger (1935b) and Schrödinger (1935c).

Schrödinger, E. (1932e). *Über Indeterminismus in der Physik – Ist die Naturwissenschaft milieubedingt? Zwei Vorträge zur Kritik der naturwissenschaftlichen Erkenntnis*. Leipzig: J. A. Barth.

Schrödinger, E. (1933). *Mémoires sur la mécanique ondulatoire*. Paris: Librairie Félix Alcan. Translation by A. Proca of Schrödinger (1928b), with a Preface by M. Brillouin (see Brillouin 1933), and a new introduction and notes by the author.

Schrödinger, E. (1934). Über die Unanwendbarkeit der Geometrie im Kleinen. *Die Naturwissenschaften* **22** (31), 518–520.

Schrödinger, E. (1935a). Der Grundgedanke der Wellenmechanik. In: *Les Prix Nobel en 1933*. Stockholm: Norstedt & Söner, 1–13. Translated by J. Murphy as 'The Fundamental Idea of Wave Mechanics' in Schrödinger (1935d), 133–154. Page references are to this translation.

Schrödinger, E. (1935b). Is Science a Fashion of the Times? In: *Science and the Human Temperament*. London: George Allen & Unwin, 66–85. Free translation by J. Murphy of sections I–II of Schrödinger (1932d).

Schrödinger, E. (1935c). Physical Science and the Temper of the Age. In: *Science and the Human Temperament*. London: George Allen & Unwin, 86–106. Free translation by J. Murphy of section III of Schrödinger (1932d).

Schrödinger, E. (1935d). *Science and the Human Temperament*. London: George Allen & Unwin.

Schrödinger, E. (1935e). Discussion of Probability Relations between Separated Systems. *Proceedings of the Cambridge Philosophical Society, Mathematical and Physical Sciences* **31** (4), 555–563.

Schrödinger, E. (1935f). Die gegenwärtige Situation in der Quantenmechanik, I–III. *Die Naturwissenschaften* **23** (48–50), 807–812, 823–828, 844–849. Translated in Trimmer (1980). Page references are to this translation and to its reprint in this volume.

Schrödinger, E. (1935g). Contributions to Born's New Theory of the Electromagnetic Field. *Proceedings of the Royal Society of London A***150**, 465–477.

Schrödinger, E. (1935h). *La nueva mecánica ondulatoria*. Cursos de la universidad internacional de verano en Santander, vol. 1. Madrid: Signo, 1–73.

Schrödinger, E. (1935i). ¿Son lineales las verdaderas ecuaciones del campo electromagnético? *Anales de la sociedad española de física y química* **33**, 511–517.

Schrödinger, E. (1936). Probability Relations between Separated Systems. *Proceedings of the Cambridge Philosophical Society, Mathematical and Physical Sciences* **32**, 446–452.

Schrödinger, E. (1957). *Science, Theory and Man*. New York: Dover. Augmented edition of Schrödinger (1935d).

Schrödinger, E. (1962). Die Wandlung des physikalischen Weltbegriffs. In: *Was ist ein Naturgesetz? Beiträge zum naturwissenschaftlichen Weltbild.* München and Wien: Oldenbourg, 18–26.

Schrödinger, E. (1984). *Gesammelte Abhandlungen.* Wien/Braunschweig: Österreichische Akademie der Wissenschaften/Vieweg.

Seevinck, M. (2017). Challenging the Gospel: Grete Hermann on von Neumann's No-Hidden-Variables Proof. In: *Grete Hermann: Between Physics and Philosophy.* Ed. by E. Crull and G. Bacciagaluppi. Berlin: Springer, 107–117.

Slater, J. C. (1929). Physical Meaning of Wave Mechanics. *Journal of the Franklin Institute* **207** (4), 449–455.

Soler, L. (2017). The Convergence of Transcendental Philosophy and Quantum Physics: Grete Henry-Hermann's 1935 Pioneering Proposal. In: *Grete Hermann: Between Physics and Philosophy.* Ed. by E. Crull and G. Bacciagaluppi. Berlin: Springer, 55–69.

Stairs, A. (1978). Quantum Mechanics, Logic and Reality. PhD Thesis. London, Ontario: University of Western Ontario.

Stöltzner, M. (1999). Vienna Indeterminism: Mach, Boltzmann, Exner. *Synthese* **119** (1–2), 85–111.

Tauschinsky, A. and van Dongen, J. (2008). Over lichtemissie: Albert Einstein en de vroege geschiedenis van de Nederlandse Natuurkundige Vereniging. *Nederlands Tijdschrift voor Natuurkunde* **74**, 90–93.

Temple, G. (1935). The Fundamental Paradox of the Quantum Theory. *Nature*, 957.

Trimmer, J. D. (1980). The Present Situation in Quantum Mechanics: A Translation of Schrödinger's 'Cat Paradox'. *Proceedings of the American Philosophical Society* **124** (5), 323–338.

Uffink, J. (2007). Compendium to the Foundations of Classical Statistical Physics. In: *Handbook of the Philosophy of Physics.* Ed. by J. Butterfield and J. Earman. Amsterdam: Elsevier, 924–1074.

Uffink, J. (2017). The Prehistory of Entanglement: Schrödinger and the Development of the Einstein–Podolsky–Rosen Experiment. Talk delivered at the 25th International Congress of History of Science and Technology, 23–29 July 2017 in Rio de Janeiro, Brazil.

Uffink, J. (2018). Schrödinger and the EPR Argument. Talk delivered at the Foundations of Physics Seminar, Utrecht University, 30 November 2018.

Uffink, J. (2020). Schrödinger's Reaction to the EPR Paper. In: *Quantum, Probability, Logic: Itamar Pitowsky's Work and Influence.* Ed. by M. Hemmo and O. Shenker. Berlin: Springer, 545–566.

van Dongen, J. (2007a). Emil Rupp, Albert Einstein, and the Canal Ray Experiments on Wave–Particle Duality: Scientific Fraud and Theoretical Bias. *Historical Studies in the Physical and Biological Sciences* **37** (supp.), 73–120.

van Dongen, J. (2007b). The Interpretation of the Einstein–Rupp Experiments and Their Influence on the History of Quantum Mechanics. *Historical Studies in the Physical and Biological Sciences* **37** (supp.), 121–131.

van Dongen, J. (2015). Communicating the Heisenberg Uncertainty Relations: Niels Bohr, Complementarity and the Einstein–Rupp Experiments. In: *One Hundred Years of the Bohr Atom.* Ed. by F. Aaserud and H. Kragh. Copenhagen: Det Kongelige Danske Videskabernes Selskab, 310–343.

van Fraassen, B. C. (1974). The Einstein–Podolsky–Rosen Paradox. *Synthese* **29** (1–4), 291–309.

Wallstrom, T. C. (1994). Inequivalence between the Schrödinger Equation and the Madelung Hydrodynamic Equations. *Physical Review A* **49**, 1613–1617.

Weismann, A. (1892). *Das Keimplasma: Eine Theorie der Vererbung*. Jena: Fischer.

Weisskopf, V. (1935). Probleme der neueren Quantentheorie des Elektrons. *Die Naturwissenschaften* **23** (37–39), 631–637, 647–653, 669–674.

Weizsäcker, C. F. von (1931). Ortsbestimmung eines Elektrons durch ein Mikroskop. *Zeitschrift für Physik* **70** (1–2), 114–130.

Wentzel, G. (1926). Zwei Bemerkungen über die Zerstreuung korpuskularer Strahlen als Beugungserscheinung. *Zeitschrift für Physik* **40**, 590.

Wheeler, J. A. and Zurek, W. H., eds. (1983). *Quantum Theory and Measurement*. Princeton: Princeton University Press.

Whitaker, M. A. B. (2004). The EPR Paper and Bohr's Response: A Re-assessment. *Foundations of Physics* **34**, 1305–1340.

Wick, G. C. (1935). Teoria dei raggi β e momento magnetico del protone. *Rendiconti della Reale accademia nazionale dei Lincei* **21**, 170–173.

Wolfe, H. C. (1936). Quantum Mechanics and Physical Reality. *Physical Review* **49**, 274.

Index

Printed by Integrated Books International,
United States of America